ARCHAEOLOGY OF EAST ASIA

A Zooarchaeological Study of the Haimenkou Site, Yunnan Province, China

JUAN WANG

BAR INTERNATIONAL SERIES 2902 VOLUME 1

BAR
PUBLISHING

Published in 2018 by
BAR Publishing, Oxford

BAR International Series 2902

Archaeology of East Asia, Vol. 1
A Zooarchaeological Study of the Haimenkou Site, Yunnan Province, China

ISBN 978 1 4073 1613 0

COVER IMAGE *Male Gaur (after Richard Lydekker (1888–1890)*
The New Natural History. Volume 2. *p. 176).*

Printed in England

BAR
PUBLISHING

BAR titles are available from:

BAR Publishing
122 Banbury Rd, Oxford, OX2 7BP, UK
EMAIL info@barpublishing.com
PHONE +44 (0)1865 310431
FAX +44 (0)1865 316916
www.barpublishing.com

ARCHAEOLOGY OF EAST ASIA

Series Editor: Anke Hein, University of Oxford, UK

This series provides a platform for data-rich studies on a variety of topics and materials from all over East Asia as well as conference proceedings reflecting the newest research insights and trends. The series takes a regional approach to East Asia, encouraging projects that cross national borders, even into adjoining regions, and/or cover areas usually overlooked in mainstream research. This includes all parts of China, Japan, Korea, Mongolia, the Russian Far East, the Tibetan Plateau as a whole and the northern reaches of Southeast Asia. Especially encouraged are submissions proposing and conducting new approaches and methods in all aspects of archaeology including scientific techniques, spatial analysis, various digital methods, but also theory and model-based or traditional chronology-focused studies.

If you are interested in publishing in the *Archaeology of East Asia* series, please contact editor@barpublishing.com.

Dedicated to Jianhua Wu

Acknowledgements

I am grateful to many people for help, both direct and indirect, in writing this book. My deepest gratitude goes first and foremost to my supervisors, Professor Li Liu, Associate Professor Richard Cosgrove, and Professor Tim Murray, for their guidance and constant supervision. I feel lucky to have met Professor Li Liu in China. She first gave me the precious opportunity to begin my PhD candidature at La Trobe University in Australia. It was also Professor Liu who provided me with another precious opportunity to attend Stanford University in the USA as a visiting scholar to finish writing my book. She has walked me through all the stages of the writing of this book. Without her consistent and illuminating instruction, this book would not have reached its present form. I would also like to thank Associate Professor Richard Cosgrove. Although he was very busy teaching and excavating, he still gave me a lot of valuable advice in terms of zooarchaeological analysis, as well as in written English. Deepest gratitude is also due to Professor Tim Murray, who generously agreed to be my principal supervisor after Professor Li Liu left La Trobe University. Without his strong support I would not have been able to continue my research in the USA. In addition, I also wish to express my gratitude to La Trobe University and Stanford University for providing me with scholarships, and fieldwork and travel grants.

I would also like to convey my thanks to the people who helped me carry out field work in China. First of all, I would like to thank Professor Rui Min of the Yunnan Provincial Institute of Cultural Relics and Archaeology. He was the team leader of the 2008 season excavation at the Haimenkou site, and provided me with the faunal materials for this zooarchaeological study as well as valuable information regarding the site. Also, I am grateful to Professor Xingcan Chen of the Institute of Archaeology, Chinese Academy of Social Science. He was the one who helped to build connections between Yunnan archaeologists and my professors, and convinced Professor Rui Min to let me do this project. I would like to express my sincere gratitude to Professor Xiaolin Ma of the Henan Museum, and Mr. Yanfeng Hou of the Henan Provincial Institute of Cultural Relics and Archaeology. Without their generous offer of modern comparative collections, the identification of animal bones would not have been completed. I am greatly indebted to Dr. Shuo Wang of the Capital Normal University, China. He helped me with the identification of bird bones recovered in this study. I am also thankful to Dr. Song Li of the Reference Collection Department of Kunming Institute of Zoology. He allowed me to use the animal skeleton collection in the laboratory to identify wild animal remains recovered from Haimenkou. In addition, I sincerely thank a number of zooarchaeologists including Dr. Richard H. Meadow and Dr. Ajita K. Patel of the Peabody Museum, Harvard University, Professor Quanjia Chen and Professor Zhuowei Tang of Jilin University, Professor Jiarong Yuan of Hunan Provincial Institute of Cultural Relics and Archaeology, Professor Yunping Huang of Peking University. Thanks to these zooarchaeologists for answering my questions and providing valuable advice via email when I was doing the field work alone in Yunnan Province, China.

Special thanks to my best graduate friend Sheahan Bestel and her sister Lorien Bestel, who read and edited my drafts again and again. It is difficult to write a book in a foreign language, but thanks to their efforts my English writing is more fluent and clear. I am also grateful to other colleagues and friends who helped check my drafts, including Mr. Adam Joseph Nazaroff and Ms. Sadie Weber of Stanford University, and Mr. Rob Gargett of the University of California Observatories.

My thanks is also due to Dr. Susan Lawrence, Dr. Nicola Stern, Mr. Ming Wei, and Ms. Jenna Thurlow, staff members in the Department of Archaeology at La Trobe University, for their cheerful encouragement and kind support during my studies inside and outside Australia. I also owe a great deal to Ms. Laura Rossi and Ms. Lancy Eang, staff members of the Archaeology Centre, Stanford University, for helping me work out my problems during my study at Stanford. I am thankful to Professor Lynn Meskell, director of the Archaeology Centre, Stanford University, for financial assistance and support during my stay at Stanford. I am also grateful to Connie Chin, Manager in the Department of East Asian Languages and Cultures at Stanford, for administrative help and support.

Finally, I am forever indebted to my family for their loving consideration, understanding, and support through these years. I also owe my sincere gratitude to my friends in China for their daily companionship and company via internet.

Table of Contents

List of Figures

List of Tables

List of Abbreviations

ALNCJU:	Archaeology Laboratory of North China, Jilin University
ATYPM:	Archaeological Team of Yunnan Provincial Museum
CCAJCYP:	Codification Committee of Annals of Jianchuan County, Yunnan Province
CCMC:	Culture Centre of Maowen County
CCXC:	Culture Centre of Xihe County
CMICRA:	Chengdu Municipal Institute of Cultural Relics and Archaeology
CPAMJP:	Committee for the Preservation of Ancient Monuments of Jiangxi Province
CPAMSP:	Committee for the Preservation of Ancient Monuments of Sichuan Province
CPAMTAR:	Committee for the Preservation of Ancient Monuments of Tibet Autonomous Region
CRMATAP:	Cultural Relics Management, Aba Tibetan Autonomous Prefecture
CRMATQAP:	Cultural Relics Management, Aba Tibetan and Qiang Autonomous Prefecture
CRMDP:	Cultural Relics Management, Dali Prefecture
CRMJC:	Cultural Relics Management, Jianchuan County
CSBMC:	Culture and Sport Bureau, Maerkang County
DAPKU:	Department of Archaeology, Peking University
DHSU:	Department of History, Sichuan University
ECBRLPYP:	Editorial Committee of Breeds Records for Livestock and Poultry in Yunnan Province
ECFSAS:	Editorial Committee of Fauna Sinica, Academia Sinica
ESBBC:	Editorial Section of the 'Bovine Breeds in China'
ESPBC:	Editorial Section of the 'Pig Breeds in China'
FPM:	Fujian Provincial Museum
GATIACAS:	Gansu Archaeological Team, Institute of Archaeology, Chinese Academy of Science (before 1977)
GPAT:	Gansu Provincial Archaeological Team
GPICRA:	Gansu Provincial Institute of Cultural Relics and Archaeology
GSYRRAT:	Gansu Squad, Yellow River Reservoir Archaeological Team
HCM:	Huangyuan County Museum
HLQPASS:	History Laboratory, Qinghai Provincial Academy of Social Science
HPICRA:	Henan Provincial Institute of Cultural Relics and Archaeology
IACASS:	Institute of Archaeology, Chinese Academy of Social Science
IVPPCAS:	Institute of Vertebrate Palaeontology and Palaeoanthropology, Chinese Academy of Sciences
KIZCAS:	Kunming Institution of Zoology, Chinese Academy of Science
LIACAS:	Laboratory of Institute of Archaeology, Chinese Academy of Science (before 1977)
LIACASS:	Laboratory of Institute of Archaeology, Chinese Academy of Social Science (after 1977)
POYPM:	Preparatory Office of Yunnan Provincial Museum
QPAT:	Qinghai Provincial Archaeological Team

SJAE:	Sino-Japanese Archaeological Expedition
SPICRA:	Sichuan Provincial Institute of Cultural Relics and Archaeology
YPAT:	Yunnan Provincial Archaeological Team
YPICRA:	Yunnan Provincial Institute of Cultural Relics and Archaeology
YPM:	Yunnan Provincial Museum

Abstract

This book is based on an analysis of the faunal assemblage recovered from the Haimenkou site during the 2008 field season in Yunnan Province, China. It aims to investigate the human-animal relationships at Haimenkou through a time span from the late Neolithic Period to the middle Bronze Age (*ca.* 5000–2400 B.P.). The animal exploitation patterns, local animal domestication processes, human subsistence strategies, and communication networks linking Haimenkou and other regions in prehistoric China are studied.

Located on the Yunnan-Guizhou Plateau in southwestern China, Yunnan Province adjoins several provinces. It also borders countries such as Myanmar in the west and Laos and Vietnam to the south. Due to its pivotal position connecting these surrounding regions, Yunnan was important in antiquity as a path for human migration and cultural communication. Haimenkou is situated south of the exit of Jianhu Lake, which is about one kilometre northwest of Haimenkou village, Jianchuan County, northwestern Yunnan Province. It covers an area of about 50 thousand square metres (5 ha.). Haimenkou was first discovered on March 25, 1957, and then excavations were conducted over three seasons in 1957, 1978, and 2008. In season 2008, the stratigraphy of Haimenkou was divided into ten layers, four phases, including late Neolithic Period, Early Bronze Age, Early to Middle Bronze Age, and Historic period.

The 2008 season excavation at Haimenkou revealed a large number of archaeological features, including house foundations, wooden poles, pits, postholes, and a human bone pit (not a grave). The largest pile-dwelling site discovered in China was also uncovered, and was represented by four thousand wooden columns and beams with mortise and tenon joint structures. More than three thousand artefacts were recovered, including pottery, bronze and iron objects, tools of stone, wood, bone and antler.

Around 2,800 specimens (bone fragments) were recovered during the 2008 excavation season. A total of 1,791 specimens, weighing approximately 54 kg, were identified to a taxonomic level. Of these identified specimens, more than 50 percent were from domestic animals. Detailed faunal analyses mainly focused on range and relative importance of identified taxa, skeletal part representation, bone modification, and kill-off patterns for domestic animals.

The results of this study suggest that domestic pigs and dogs arrived at Haimenkou with their keepers during the late Neolithic Period (*ca.* 5000–3800 B.P.). Domestic sheep or goats were introduced to Haimenkou from northwestern China in the early Bronze Age (*ca.* 3700–3200 B.P.). After inhabiting Haimenkou, people developed a mixed subsistence economy, consisting of rice, millet, and wheat multi-cropping, plant food gathering, animal husbandry, hunting, and fishing. They hunted numerous animal taxa, mainly including the gaur and different cervids. Both animal husbandry and hunting techniques improved considerably through time, especially between the late Neolithic Period and the early Bronze Age. In addition, cultural communication and human migration occurring between Haimenkou and other regions in prehistoric China, such as northwestern China (Qinghai and Gansu Provinces), Sichuan Province, Tibet, and the middle Lancang River Valley in southwestern Yunnan, were reflected in the assemblage of animal remains recovered from Haimenkou.

1

Introduction

1.1 Aims of Faunal Analysis at the Haimenkou Site

This book is based on the analysis of the faunal assemblage recovered from the Haimenkou site in Yunnan Province, China. A vast amount of faunal remains were unearthed from clear stratigraphic contexts at this site. They represent a time span from the late Neolithic Period (*ca.* 5000 B.P.) to the middle Bronze Age (*ca.* 2400 B.P.).

Haimenkou is located in Jianchuan County, Yunnan Province, southwestern China. The site is the most extensive water-logged settlement characterised by pile-dwelling constructions discovered to date in China (YPICRA, et al. 2009a:599). During three seasons of excavation (1957, 1978, and 2008) many artefacts and ecofacts were revealed, including a large number of animal remains.

The Haimenkou site is of special significance, particularly in the field of zooarchaeology. Firstly, as it was occupied for several thousand years, Haimenkou provides a great opportunity to explore the changes in human-animal relationships and human subsistence economy across this long period from the aspect of the faunal assemblage, such as the exploitation patterns for different animals and the processes of animal domestication.

Secondly, due to its pivotal position, Yunnan itself was an important region for cultural communication and human migrations in prehistory. The introduction and exchange of animals is an important topic for study. In the third excavation report for the Haimenkou site, it is noted that material remains are an indication of cultural interaction and population movement from other regions in China, such as northwestern China (YPICRA, et al. 2009b). This assumption will be evaluated by the close observation of animal remains from Haimenkou, in terms of the range of taxa, the distribution of faunal remains in different layers, and the phenomena reflecting animal domestication.

Thirdly, the study of the faunal assemblage from Haimenkou may provide important data for zooarchaeological study in Yunnan Province. So far, relatively limited research on faunal assemblages has been undertaken at Haimenkou and surrounding areas in Yunnan Province. From the 1920s onward, more than 300 Neolithic sites were discovered in Yunnan, and dozens of them were excavated (Yang, F., et al. 2010:1). These excavations undertaken in Yunnan have provided important zooarchaeological information and large quantities of faunal remains (Zhang, X. 1987:370). Based on these finds, a picture of animal hunting, domestication, and utilisation in prehistoric Yunnan was established. However, most of these faunal assemblages were simply mentioned as a list of animal species revealed in surveys or excavations in archaeological reports. This 'laundry list' style reporting is evident in the brief report of the Baiyangcun site (YPM 1981) and an earlier report of the Haimenkou site (YPM 1995). In other cases, even a complete list of animal species was absent, as in the Yinsuodao excavation report which stated that, 'a relatively large number of animal bones were found here, including fish, mice, and pig mandibles and teeth' (YPICRA, et al. 2009c:25). Only a few sites have provided specialised zooarchaeological reports, such as the Dadunzi site in Yuanmou County (Zhang, X. 1985). Although the artistic representations of many animals uncovered from several Bronze Age sites in Yunnan, such as Shizhaishan (Zhang, Zengqi 1998) and Lijiashan (YPICRA 2007), have been reported in detail and provided useful information for the investigation of subsistence patterns and human-animal relationships, further study based on animal bones themselves is needed for a fuller understanding of the complex human-animal relationships in ancient times. In sum, while other archaeological research, on pottery, stone implements, and bronzes has been undertaken in Yunnan Province, there is still a lack of research from the perspective of zooarchaeology.

Based on the study of the faunal assemblage recovered from the site, the aim of this book is to fill in this blank by exploring the human-animal relationships at Haimenkou. This includes a study of the animal exploitation patterns and local animal domestication processes, along with human subsistence strategies and communications between Haimenkou and other regions in prehistoric China. The research questions may be summarised as follows:

What is the mixed spectrum of animal taxa living at prehistoric Haimenkou? Among these animal taxa which were domesticated and which were wild? Where did the domestic animals come from? Were they locally domesticated or introduced from other locations in China?

How were these domestic animals exploited? What was the structure of the subsistence economy of the Haimenkou people? What changes in the Haimenkou people's subsistence economy occurred during the long occupation of Haimenkou? What role did Haimenkou play in cultural communication between other regions in prehistoric China?

The questions proposed above will be answered in subsequent chapters. In order to provide important background on previous zooarchaeological research in Yunnan, a brief introduction to zooarchaeological history in China and in

Yunnan itself will be presented in the following section.

1.2 Brief Introduction to the History of Zooarchaeology in China and Yunnan Province

1.2.1 A Brief Zooarchaeological History of China

The earliest zooarchaeological work conducted in China dates back to the 1930s (Teilhard de Chardin and Young 1936). This research may be considered pioneering work in the history of zooarchaeology in China. Nevertheless, research during this period was limited and was largely suspended during World War II. The main zooarchaeological history of China is typically considered to have started after the establishment of the People's Republic of China in 1949.

Yuan (1995a:84) suggested that the zooarchaeological history of China may be divided into two periods, with the 1950s to 1970s being the formative period, and the 1980s to recent times being the developing period. Based on this division, the Chinese zooarchaeological history is described below.

1.2.1.1 Formative Period (1950s–1970s)

From the 1950s onward, the focus of archaeological studies in China was mainly on the reconstruction of ancient historical temporal sequences and their spatial framework (Liu, L. and Chen 2001). As a result, Chinese archaeologists devoted considerable attention to the study of stone implements and pottery recovered in China, as well as comparisons among varying artefact assemblages from different regions and periods. The study of animal bones played a subordinate role in these contexts. In the formative period of zooarchaeology, faunal assemblages were relatively neglected and poorly collected. Scholars who engaged in zooarchaeological studies were most often palaeontologists with little archaeological background. During this period, a noteworthy zooarchaeological study allowed Y. Li and Han (1959) to introduce the Banpo site with its cultural and environmental backgrounds in their zooarchaeological report. They (Li, Y. and Han 1959) also described the faunal assemblage and identified the animal specimens to the most precise taxonomic level possible, as well as classifying them into three categories: domestic and probably domestic animals, game animals, and animals that may have burrowed or intruded into the site later. The methods employed in this report included morphological observation, age profile construction, and measurements of available specimens. At the end of the article, they discussed the site environment and animal domestication, and noted correlations between the faunal assemblage and ethnic groups. The format and methods employed in this article made a significant impact on subsequent zooarchaeological research in China and formed a methodological basis that is still widely followed in modern Chinese zooarchaeological reports. Nevertheless, the weakness in the Banpo research was the lack of specific stratigraphic information on the locations of the animal remains. This problem was already present in the Anyang report of the 1930s (Teilhard de Chardin and Young 1936), but was a tradition of palaeontologists in the era. The Banpo researchers were not aware of the importance of data comparisons among different periods in the Banpo research, or in their subsequent research on the faunal assemblage from the Zengpiyan site (Li, Y. and Han 1978).

1.2.1.2 Developing Period (1980s to Present)

Although the excellent example of Y. Li and Han (1959) was present in the formative period, some modern zooarchaeological reports persisted in only listing the taxa of identified faunal specimens. These are sometimes added in the form of appendixes to archaeological excavation reports and are lacking in any real interpretation (e.g., Li, T. 2001; Zhu 1987). These 'laundry lists' of faunal remains hinder zooarchaeological study. To address this concern, Qi (1983) proposed eight important issues for zooarchaeological investigation, including: the recovery and reconstruction of the environment of ancient settlements; the study of hunting techniques and prey selection; the procedures of food processing; the study of long-term or seasonal site occupation; the investigation of ancient social structure and settlement patterns; the study of ritual practice and taboo; the study of ancient trade; the investigation of sources of raw material of artefacts; and research into animal domestication. In addition, Qi (1988) was the first scholar who applied MNI (minimum number of individuals) in Chinese zooarchaeology. In the zooarchaeological report of Jiangzhai, a famous Neolithic site in Shaanxi Province, she (Qi 1988) completed a plan of the locations of the unearthed animal remains, and speculated about the human population distribution at this site based on the faunal assemblage.

A resource also worth mentioning in this developing period was the publication of the more-than-150-page monograph, 'Zhejiang Yuyao Hemudu Xinshiqi Shidai Yizhi Dongwuqun [The Fauna from the Neolithic Site at Hemudu, Zhejiang]' (Wei, et al. 1989). This introduced an extensive style of research writing to the Chinese zooarchaeological field. Approximately 61 taxa of vertebrate and invertebrate remains were recovered from the third and fourth layers of Hemudu during the excavation seasons of 1973–1974 and 1977–1978. However, only identifiable specimens were described in this monograph. Species, environmental exploitation, animal domestication, bone tool manufacture, and folklore reflected by the animal remains were discussed in this book based on the faunal assemblage.

In the 1980s the first article focusing on the study of fish remains was published as an appendix to an field report, 'Jiao Xian Sanlihe [The Sanlihe Site in Jiao County]', of a Neolithic site of Sanlihe in Jiao County, Shandong Province (Cheng 1988). In the article, Cheng (1988) identified fish remains, compared them with modern fish reference specimens, and made the interesting inference that the people who lived at this settlement would scale fish before eating them.

During the 1990s, besides normal laboratory work and report writing, Chinese zooarchaeologists began to pay more attention to the history and development of zooarchaeology both worldwide and of China. It was a time for Chinese zooarchaeologists to reflect upon developing the discipline in a more systematic way. Yuan (1995a) and other researchers summed up the history of Chinese zooarchaeology, while Qi and Yuan (1997) and others reviewed the history of zooarchaeological development in the Western world. Moreover, Yuan (1995b) and others discussed the theories, aims, and methods of zooarchaeology. At the same time, much translation work, from Japanese and English to Chinese, was underway (e.g., An and Long 1993; Li, T. 1992; Qi and Yuan 1997; Yuan 1992; 1993; Yuan and Jiao 1993; Yuan and Qin 1994). At the end of the 1990s, Yuan (1999) classified different characteristics of meat-acquiring patterns from several main Chinese Neolithic regions. He (Yuan 1999) classified them as dependence on wild animals, elementary animal domestication[1], and advanced animal domestication. Based on these three stages, he proposed a model of meat acquisition that demonstrated the basic developmental process of meat acquisition in Neolithic China. He believed the process could be summed up in a theory of passive development, because the sequential changes of meat acquisition patterns happened and developed due to the restrictions from surrounding natural and environmental sources. Yuan's article indicated the direction of zooarchaeological study in China for the next decade focussing on animal domestication, although this model is less commonly used nowadays.

Since the 21st century, Chinese zooarchaeological studies have concentrated on animal domestication and related issues. Yuan (2001) proposed criteria for the identification of domestic animals from the Neolithic sites in China and discussed when and where the earliest Chinese domestic animals may have appeared. These arguments were maintained in his subsequent publications concerning animal domestication (e.g., Yuan 2002; Yuan and Flad 2002). According to Yuan (2001; 2002), in China animal domestication was focused on pigs. Pigs have a domestication history which may be traced back to over 8,000 years ago in China (Luo and Zhang 2008). They have also been one of the most important food resources in ancient China (Chang 1997; Yuan 2001). Other zooarchaeologists have also shown a keen interest in this area of pig domestication, such as X. Ma (2007) and Luo and Zhang (2008). In the light of newly discovered material from the seventh season of excavation at the Jiahu site in Henan Province, Luo and Zhang (2008) reanalysed the pig remains from all seven seasons of this excavation. They (Luo and Zhang 2008) arrived at an important conclusion that domestic pigs had been bred by the Jiahu people from the first phase of the site. This suggested that the first appearance of domestic pigs in North China may be traced back to about 8,500 years ago (Luo and Zhang 2008). This result was later supported by other researchers (Cucchi, et al. 2011).

In addition to the research on domestic pigs, animals important in the zooarchaeological field of China include water buffaloes, cattle, sheep, horses, dogs, and other domesticates. For example, L. Liu and colleagues (2006) discussed the origins of domestic water buffaloes (*Bubalus mephistopheles*) in China based on measurements of buffalo specimens and a review of literature. They suggested that water buffalo remains recovered from the Neolithic and Bronze Age sites in China represented wild taxa. This conclusion is contrary to the previous belief that buffaloes were domesticated and played an important role in farmlands during the Neolithic period and the Bronze Age in China (Liu, L., et al. 2004; Liu, L., et al. 2006). In other examples, ancient DNA was employed in the study of animal origins and domestication, including in research on the remains of water buffalos (e.g., Yang, Dongya, et al. 2008), sheep (e.g., Cai, et al. 2007), horses (e.g., Cai, et al. 2009), and dogs (e.g., Okumura, et al. 1999).

1.2.2 Zooarchaeological Studies in Yunnan for Periods from the Early Holocene Epoch to the Bronze Age

As mentioned previously, more than 300 Neolithic sites have been discovered in Yunnan Province (Yang, F., et al. 2010:1). Despite this, fewer than 40 sites have been excavated and their details published (Wang, N. 1992; Yang, F., et al. 2010). Although a large number of faunal remains have been uncovered and collected from the archaeological excavations in Yunnan Province since the founding of the People's Republic of China, zooarchaeological studies based on faunal remains are scarce and neglected. Usually the animal species are listed or only mentioned in the description of a site, such as in the brief excavation report of the Baiyangcun site in Binchuan County, Yunnan (YPM 1981). Articles focusing on investigation and discussion of faunal remains are rare, but include zooarchaeological reports on the Huangjiaying site in Tonghai County (Yang, Z., et al. 1985), and the Dadunzi site in Yuanmou County (Zhang, X. 1985) (see locations in Figures 2.23).

X. Zhang has made a large contribution to the study of animal remains recovered from sites in Yunnan. In his (Zhang, X. 1987) article 'Yunnan Xinshiqi Shidai de Jiachu [Domestic Animals in Neolithic Yunnan]', the finds of domestic animal remains from the early Holocene Epoch to the Bronze Age in Yunnan were outlined, with some sites discussed in detail. This article collected all the reported data of animal remains found in Yunnan prior to publication, provided a brief illustration of the distribution of animal species from the early Holocene Epoch to the Bronze Age, and drew preliminary conclusions from discussion of the data.

Other articles by X. Zhang and his colleagues focused on the analysis of animal remains from specific sites in Yunnan. These include the Tangzigou site in Baoshan City

[1] The definition of domestication used in this book follows Bökönyi (1989:22) and is discussed more in chapter four.

(Zhang, X., et al. 1992) and the Dadunzi Site in Yuanmou County (Zhang, X. 1985). During these analyses, the descriptions, measurements, and abundance of animal specimens were presented, along with primary discussions. However, these articles still emphasised species lists and elaborate descriptions of the specimens in terms of morphology, and no age profile or kill-off patterns were provided or discussed.

Like other scholars in the last decades in China who specialise in studies on animal remains from archaeological sites, X. Zhang is actually a palaeontologist. This is probably why his goals and methods of studying animal remains differ from that employed by modern zooarchaeologists. His palaeontological descriptions focused on horns, antlers, skulls, mandibles, and teeth, with no limbs or vertebrae included. This indicates an obvious bias in specimen selection and collection during excavations, reflecting palaeontological rather than zooarchaeological interests.

Other researchers to publish more interpretive zooarchaeological papers include Z. Yang and colleagues (1985). They (Yang, Z., et al. 1985) published an article describing more than 20 taxa of vertebrate fossils, including *Homo sapiens,* recovered from Holocene sedimentary deposits in Qilu Lake, Tonghai County, Yunnan Province. Z. Yang and colleagues also have palaeontological backgrounds, thus no age profiles or kill-off patterns were given in this article. However, they did note the lack of analysis of limb bones by X. Zhang (1992), a topic that was covered in their paleontological descriptive research.

After the publication of X. Zhang and colleagues' (1992) article on faunal remains in 1992, few zooarchaeological studies in Yunnan were carried out and few published articles focused on faunal remains. Nevertheless, these early studies helped to build a database for comparative purposes in this study.

1.3 Date and Citation Style

Dates presented in this book are given as they were in the original publication. Earlier publications typically presented dates as uncalibrated or in years B.C. To avoid the complexities of modern calibration programs these older dates, which typically refer to a lengthy range or time period, and are generally widely understood and accepted in China, are retained here. The preference throughout has been to present dates in approximate years B.P.

References cited are presented according to the American Anthropological Association (AAA) 2009 style guide[2]. This style guide is based on the Chicago Manual of Style (15th edition). This means that where authors have the same surname, as is common in China, primary initials are used to distinguish between them. Where primary authors have the same first surname and the same initials, in-text references give the full name to allow differentiation of cited authors.

1.4 Structure of the Book

To answer the questions proposed in Section 1.1, the following chapters of this book will discuss several important issues regarding the human-animal relationships at the Haimenkou site. These issues include an investigation of animal exploitation patterns and the local animal domestication processes. The human subsistence economy based around faunal remains is another topic of discussions. Cultural communication or interaction with other people in prehistoric Yunnan and from other regions in China is another key issue that this research aims to expand on. Chapter two describes the environmental setting and archaeological background of the Haimenkou site. Chapter three outlines the basic procedures and limitations of the analyses carried out on the faunal assemblage from Haimenkou. Chapter four presents the analysis of the range and relative importance of identified taxa from Haimenkou. Chapter five presents the analysis of skeletal part representation and bone modification details for the site. Chapter six presents an analysis of kill-off patterns for domestic animals.

A general discussion and outline of conclusions that may be drawn from the zooarchaeological study of the Haimenkou site is presented in chapter seven. The origins of domestic animals at the site will be discussed, along with the exploitation patterns of animals from this site. The changes in human subsistence economy through time, and implications for understanding communications between Haimenkou and other regions in prehistoric China will also be discussed. Directions for further research are offered.

[2] URL: http://www.aaanet.org/publications/guidelines.cfm.

2

Environmental Setting and Archaeological Background of the Haimenkou Site

2.1 Introduction

The Haimenkou archaeological site dates from the late Neolithic Period to the middle Bronze Age. This chapter will describe the environmental setting and archaeological background of Haimenkou. It will provide an overview of the site, and then summarise the importance of Haimenkou for this study.

2.2 Site Location and Environmental Setting

The Haimenkou site is located near Jianhu Lake in Haimenkou Village, Jianchuan County, northwestern Yunnan Province, southwestern China. The geographic coordinates of the site are 26°28′ north latitude and 99°55′ east longitude (YPM 1995:775) (Figures 2.1 and 2.2).

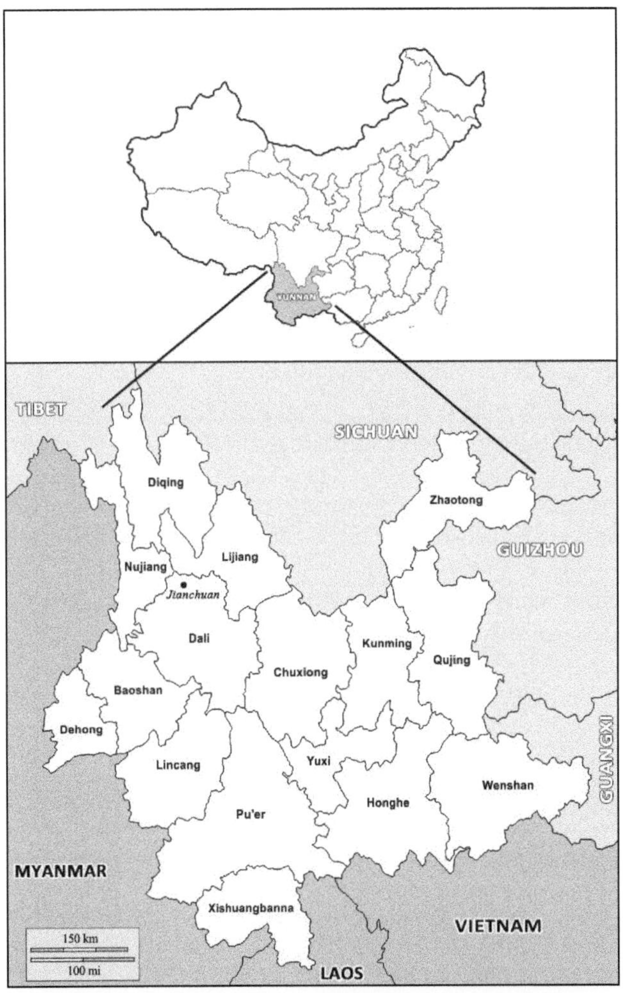

Figure 2.1 Locations of Yunnan Province and Jianchuan County (The original map is from URL: http://d-maps.com/carte.php?lib=yunnan_map&num_car=19490&lang=en)

Yunnan Province (Yunnan means 'south of the clouds') is located on the Yunnan-Guizhou Plateau in southwestern China. It covers a total area of 394,000 square kilometres. The current population of this province is forty million people, of which ethnic minority groups make up one third (Chou 1992:384). By 2007, the province had been divided into 8 prefecture-level cities, 8 autonomous prefectures, 12 municipal districts, 9 cities at the county level, 79 counties, and 29 autonomous counties[1].

Lying on the southern border of southwestern China, Yunnan Province adjoins several provinces. These include Guizhou Province and Guangxi *Zhuang* Autonomous Region (Guangxi Province is where many members of the *Zhuang* ethnic minority live) to the east, Sichuan Province to the north, and Tibet Autonomous Region to the northwest. Even today, Yunnan is considered by many Chinese a 'frontier province'. This is because it also borders countries such as Myanmar in the west and Laos and Vietnam to the south (Figure 2.1). Due to its pivotal position connecting these surrounding regions, Yunnan was important in antiquity as a path for human migration and cultural communication. A research focus of this book has been to examine cultural connections around Haimenkou site based on the faunal remains recovered.

As a plateau and mountainous region, Yunnan is part of the southward extension of the Qinghai-Tibetan Plateau. With an average elevation of around 2,000 metres, it shows large variations in elevation across the whole province. Mountains cover 48 percent of its land area, while 10 percent of the province is covered by plateaus, and six percent by basins and river valleys. In general, the province is high in the northwest and has lower elevations in the south and southeast.

There are a large number of waterways in Yunnan, the total length of which amounts to over 14,200 kilometres (Chou 1992:389). Although there are over six hundred rivers and streams in a variety of sizes in Yunnan, the major rivers only number 180. These mainly include the upper reaches of rivers running from north to south and emptying into the sea. The six main river systems of Yunnan include the Dulong (Irrawaddy), Nu (Salween), Lancang (Mekong), Jinsha (Yangtze), Yuan (Red), and Nanpan (Pearl) Rivers (Figure 2.3). The Haimenkou site, on which this study is based, is situated at the bottom of the 'San Jiang Bing Liu [Three Parallel Rivers of Yunnan]' region. These rivers comprise the Nu (Salween), Lancang (Mekong), and Jinsha (Yangtze) Rivers. Differing from other major

[1] Data from URL: http://www.360doc.com/content/10/0114/16/92853_13539157.shtml.

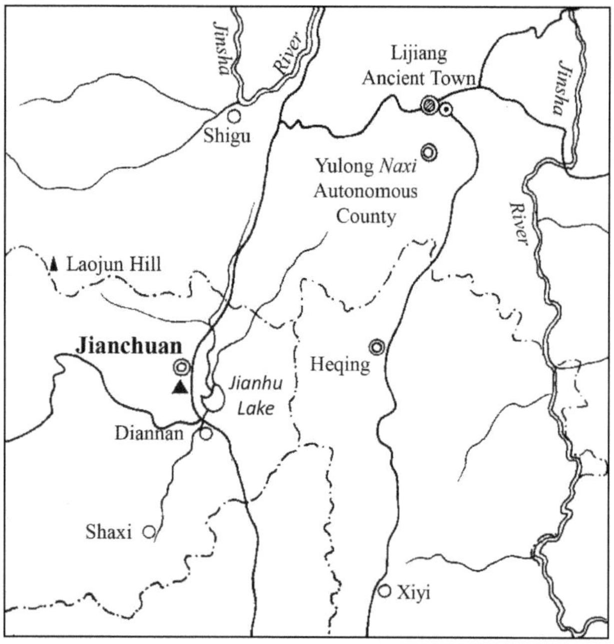

Figure 2.2 Locations of Jianchuan County and Haimenkou Site. (Modified from YPICRA, et al. 2009b: figure 1; reproduced with permission of Archaeology Press)
Note: The ▲ refers to the location of the Haimenkou site.

river systems in China that flow in a west-east direction, the rivers of Yunnan flow in a north-south direction. The north-south direction may have played a major role in the interregional communication among peoples from different regions in Yunnan and from other provinces in China, such as Sichuan, Tibet, Qinghai, and Gansu (see Figure 2.24).

Jianchuan County (26°12′–26°41′N, 99°33′–100°33′N) is located in northwestern Yunnan, northern Dali *Bai* Autonomous Prefecture, in the middle part of the Hengduan Mountain region (YPICRA, et al. 2009b:675). This is where the *Bai* ethnic minority group live. This county

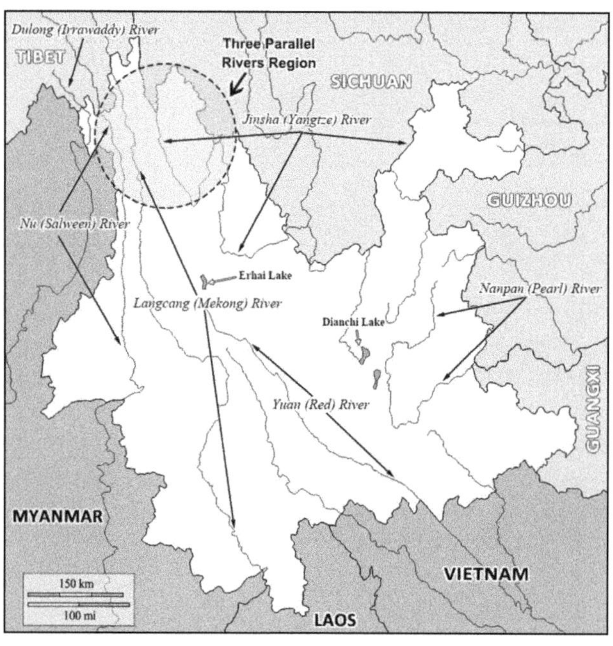

Figure 2.3 Rivers in Yunnan (The original map is from URL: http://d-maps.com/carte.php?num_car=19482&lang=en)

covers an area of 2,250 square kilometres, abutting Heqing County to the east, Eryuan County to the south, Lanping and Yunlong Counties to the west, and Lijiang County to the north (Figure 2.1) (YPICRA, et al. 2009b:675). With an altitude difference of more than 2,300 metres and an average elevation of around 2,190 metres, the topography in this area is extremely complex (YPICRA, et al. 2009b:675). The county has a south temperate monsoonal climate, with four distinct seasons. It is simultaneously hot and humid in summer (YPICRA, et al. 2009b:675).

Yunnan has been called a 'plant kingdom' due to its abundant plant resources which include around 15 thousand plant species (Chou 1992:392). It is also known as an 'animal kingdom' due to its rich wildlife including over 230 mammal species, more than 770 avian species, and 130 reptiles (Chou 1992:392). As a county in Yunnan, Jianchuan also has a large number of animal species. According to 'Jianchuan Xian Zhi [Jianchuan County Annals]' (CCAJCYP 1999:78), the main mammal species in Jianchuan include a number of carnivorous and omnivorous taxa. Large mammal taxa include leopard (*Panthera pardus*), wolf (*Canis lupus*), dhole (*Cuon alpinus*), Asian black bear (*Ursus thibetanus*), and wild boar (*Sus scrofa*). Smaller carnivores include leopard cat (*Prionailurus bengalensis*), masked palm civet (*Paguma larvata*), small Indian civet (*Viverricula indica*), large Indian civet (*Viverra zibetha*), raccoon dog (*Nyctereutes procyonoides*), fox (*Vulpes* sp.), red panda (*Ailurus fulgens*), and Siberian weasel (*Mustela sibirica*). Smaller omnivorous or herbivorous taxa in the county include gibbon (*Hylobates lar*), macaque (*Macaca* sp.), otter (*Lutra lutra*), deer (*Cervus* sp.), muntjac (*Muntiacus* sp.), musk deer (*Moschus* sp.), roe deer (*Capreolus* sp.), blue sheep (*Pseudois nayaur*), porcupine (*Hystrix brachyura*), bamboo rat (*Rhizomys* sp.), flying squirrel (*Hylopetes* sp.), squirrel (*Sciurus* sp.), house mouse (*Rattus* sp.), pangolin (*Manis* sp.), and bats (Vespertilionidae). In addition, there used to be tigers (*Panthera tigris*) in the county but they are now locally extinct (CCAJCYP 1999:78). The variety of mammal taxa in the county today probably indicates the rich resources of game animals the ancient Haimenkou inhabitants had access to.

The Haimenkou site, covering an area of about 50 thousand square metres (5 ha.), is situated south of the exit of Jianhu Lake, which is about one kilometre northwest of Haimenkou village, Jianchuan County (Figure 2.4). Jianhu Lake is a typical plateau rift freshwater lake, which covers an area of 7.5 square kilometres. The stream that flows from Jianhu Lake into Heihui River is finally incorporated into the Lancang (Mekong) River (YPICRA, et al. 2009a:594).

2.3 Excavations at the Site

2.3.1 History of Excavations at the Site

The Haimenkou site was first discovered on March 25, 1957, and then excavations were conducted over three

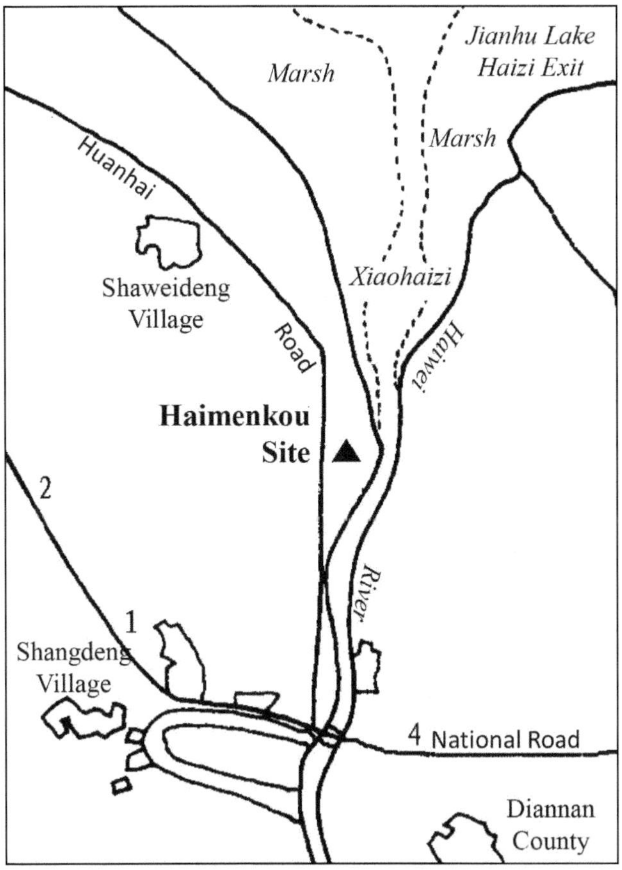

Figure 2.4 Location of Haimenkou Site (Modified from YPICRA, et al. 2009b: figure 2; reproduced with permission of Archaeology Press)

seasons in 1957, 1978, and more recently in 2008 (POYPM 1958; YPICRA, et al. 2009a; b; YPM 1995).

Archaeological excavations at Haimenkou took place from March to April of 1957 by the Preparatory Office of Yunnan Provincial Museum (POYPM), for the purpose of cooperating with local government on a river course construction project. More than three hundred wooden poles and beams (remains of ancient pile-dwellings) were

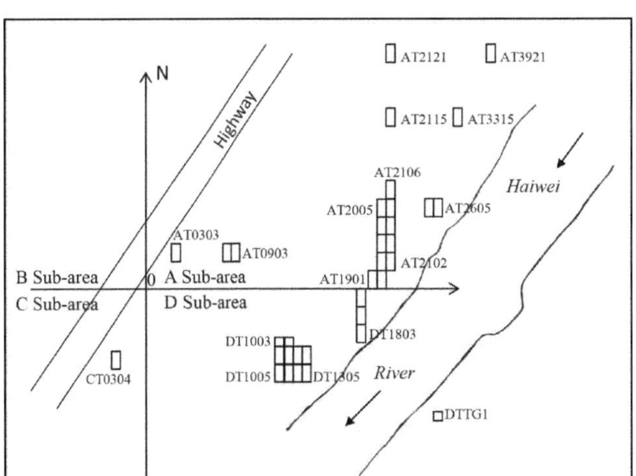

Figure 2.5 Plan of the 2008 Excavation Area at Haimenkou (Drawn by Rui Min, no scale available; used with his permission)

recovered in 1957, as well as nearly one thousand artefacts. These included pottery sherds, bronzes, tools of stone, wood, horn, and antler, and botanical and animal remains. This field season lasted only six days, leading to relatively coarse and incomplete results, such as a rough stratigraphic classification, an indistinct excavation area, and inadequate dating (POYPM 1958; YPM 1995).

In April 1978, the second season of excavation at Haimenkou was organised by the Yunnan Provincial Museum (YPM). During this season three excavation grids were arranged, covering an area of 215 square metres. Numerous artefacts and features (the number is not clear) were revealed and recorded. Nevertheless, there were still many questions remaining. Because the second season of excavation was conducted beside Jianhu Lake, the process of excavation was hindered by water seepage. Moreover, the archaeological team leader of the excavation passed away suddenly after this season so relevant data disappeared and the publication of the excavation report was postponed (YPM 1995:775).

Despite these setbacks, on the basis of the outcomes of the two earlier excavations, the significance of Haimenkou was gradually recognised (YPICRA, et al. 2009a:595). To further investigate the site, a third season of excavation was conducted in 2008. The latest excavation, carried out from January 8 to May 25, was led by the Yunnan Provincial Institute of Cultural Relics and Archaeology (YPICRA), and involved a joint team from the Cultural Relics Departments of Dali Prefecture and Jianchuan County (YPICRA, et al. 2009b). The whole site measured 350 metres from north to south and 150 metres from east to west, and was estimated to be over 50 thousand square metres (5 ha.). The central area was approximately 20 thousand square metres. The excavation regions were planned as quadrants, meaning that there were four sub-areas arranged clockwise following the letters A, B, C, and D (Figure 2.5). Twenty-five 5 m × 10 m grids, three 5 m × 5 m grids, and seven 5 m × 2 m grids were arranged. The whole excavation area covered 1,395 square metres. Advanced and multidisciplinary techniques were employed during this excavation season, including aerial photography, ETS (electronic total station) surveying and mapping, digital photography drawing, and flotation. More than three thousand artefacts and implements were recovered, including pottery sherds, stone implements, bone tools, and bronzes. At the same time, a large number of archaeological features were revealed, including house foundations, hearths, wooden columns and beams (remains of ancient pile-dwelling), gray stones (they may have been used in hearths), a human bone pit, and postholes. More than four thousand wooden columns and beams, which covered the central area of 20 thousand square metres (2 ha.), were also recovered (YPICRA, et al. 2009b:676).

2.3.2 The Stratigraphy

According to the first excavation report, the stratigraphy of Haimenkou was divided into three layers under limited circumstances in 1957 (POYPM 1958). Levels (1) and (2) were modern disturbed layers, 160 centimetres thick (POYPM 1958). Level (3) was only 8–10 centimetres thick, and incorporated pottery sherds, charcoal, and wooden columns (POYPM 1958).

Based on the second season report, published in 1995, stratigraphy at the site was revised and divided into four layers according to more intensive work and a larger excavation area. The stratigraphy identified by the second season of excavation at the site is described as follows (YPM 1995).

Level (1) was the modern disturbed layer caused by the construction of the river course. It was 40–50 centimetres thick, and was mixed with pottery sherds and stone implements.

Level (2) was the plough layer and was 50 centimetres thick. It contained a few pottery sherds and modern rubbish.

Level (3) was a gray sandy soil layer approximately 15–30 centimetres thick. It contained pottery sherds, stone implements, and bronze and iron objects. This level was thought to have been disturbed by Iron Age (after *ca.* 2100 B.P.) inhabitants.

Level (4) was a gray clay layer, 60–70 centimetres thick. It contained abundant pottery sherds, bronzes, stone implements, bone and antler tools, and a number of wooden columns and beams (these were 60–100 cm in length). In this level one beam was found with a bronze axe inset into it. This provided critical evidence for dating the site. This level contained evidence of the prehistoric Haimenkou culture.

Underneath this level the sediment was culturally sterile.

According to the latest reports published in 2009, the stratigraphy of Haimenkou has now been divided into ten layers by means of changes in soil composition and consistency (YPICRA, et al. 2009a; b). Figure 2.6 displays the profile of layers of DT1005 (T is an abbreviation for the Chinese term *Tanfang* meaning grid, and D refers to subarea D), which is characterised as follows:

Level (1) was the modern plough layer, 7–12 centimetres thick, with brown topsoil mixed with plant roots, charcoal, and modern ceramics.

Level (2) was the Ming and Qing layer (Ming and Qing are late dynastic periods of China, *ca.* A.D. 1400–1900). It was 7–12 centimetres below the surface and 5–17 centimetres thick, and was comprised of reddish-brown soil containing a few Ming and Qing ceramic sherds, tiles, and pottery sherds.

Level (3) was the Song and Yuan layer (Song and Yuan are dynastic periods of China, *ca.* A.D. 1000–1400), 12–29 centimetres below the surface and 6–13 centimetres thick. This layer of gray soil mixed with white grits and shells, and contained cord-impressed and squarely-stamped pottery sherds, tubular net sinkers, a few stone implements, and iron objects.

Level (4) was a Bronze Age layer, 28–36 centimetres below the surface and 20–23 centimetres thick. The sediment was black clay mixed with charcoal and shells, and contained pottery sherds, stone implements, animal bones, and a small quantity of bronze objects.

Level (5) was a Bronze Age layer, 48–50 centimetres below the surface and 20–45 centimetres thick. The sediment was greenish-gray clay and contained a large number of pottery sherds, some tools of stone and wood, animal bones, wooden columns, and a small quantity of bronze objects.

Level (6) was a Bronze Age layer, 75–90 centimetres below the surface and 10–28 centimetres thick. It was comprised of yellowish-brown soil mixed with sand and many wood fragments and contained pottery sherds, tools of stone and wood, animal bones, a small quantity of bronze objects, and wooden columns.

Level (7) was the early Bronze Age layer, 100–105 centimetres below the surface and 15–25 centimetres thick. The sediment was greenish-gray soil and contained many pottery sherds, tools of stone and wood, animal bones, carbonised rice, wheat and barnyard millet, and wooden columns.

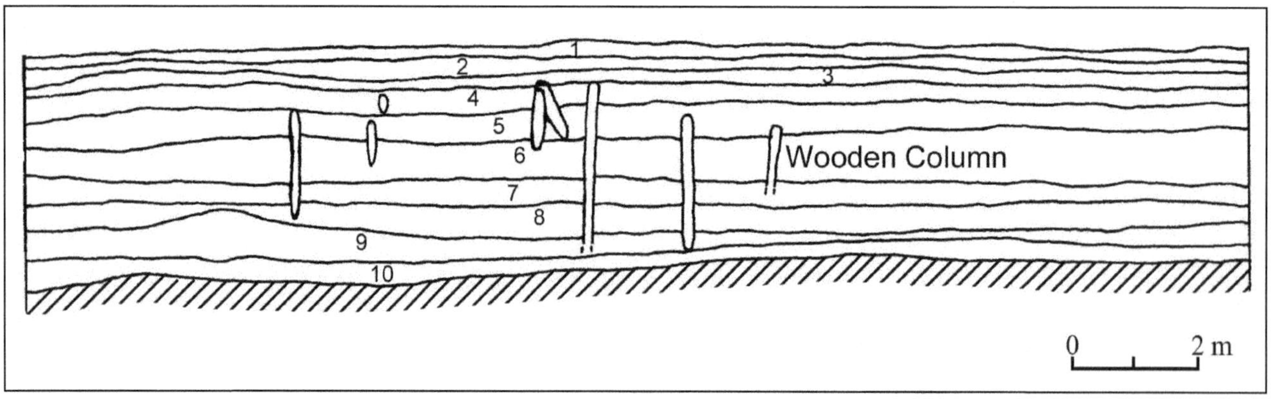

Figure 2.6 Stratigraphy of the East Wall of DT1005 in the Third Excavation (Drawn by Rui Min; used with his permission)

Level (8) was the late Neolithic Period layer, 120–130 centimetres below the surface and 5–30 centimetres thick. It was a greyish-brown soil mixed with charcoal and contained pottery sherds, tools of stone and wood, animal remains, and carbonised rice and millet.

Level (9) was the late Neolithic Period layer, 135–140 centimetres below the surface and 7–20 centimetres thick. This was yellowish-gray soil mixed with wood fragments and charcoal. This layer contained a small quantity of pottery sherds, tools of stone and wood, animal remains, and carbonised rice.

Level (10) was the late Neolithic Period layer, 155–160 centimetres below the surface and 10–20 cm thick. Sediment was yellowish-brown mingled with grit and bits of wood fragments and charcoal, as well as a few pottery sherds.

Underneath this level the sediment was culturally sterile.

The thickness of the entire archaeological deposits varies from 160 to 320 centimetres. In general, the northern area and the area near Jianhu Lake, which were relatively thick in Level (3), yielded more abundant cultural remains.

According to the archaeological features and artefacts revealed, the strata described above were classified into four groups from top to bottom, in accordance with four phases described as follows.

The first group containing only Level (3) remains was attributed to the Haimenkou 'Late Phase' due to the presence of iron artefacts. This phase will only be mentioned briefly as it was not the focus of the analysis of this book due to its late period.

The second group comprising Levels (4) and (5) was attributed to the Haimenkou 'Phase III' based on the presence of bronze and similar artefacts belonging to the early to middle Bronze Age in Yunnan.

The third group consisting of Levels (6) and (7) was attributed to the Haimenkou 'Phase II' based on the presence of artefacts belonging to the early Bronze Age in Yunnan.

The fourth group including Levels (8) to (10) was attributed to the Haimenkou 'Phase I' because of the presence of similar artefacts and an absence of bronzes. This phase belongs to the late Neolithic Period in Yunnan.

As described above, different stratigraphy was determined in each of the three excavation seasons. This was probably due to the different excavation conditions and areas for each excavation season (Table 2.1). The third season, however, provided the most detailed and accurate information due to its large excavation area and the advanced techniques employed. This book is based on the faunal assemblage recovered from this fieldwork season.

Table 2.1 Differences between Three Excavation Seasons

Season	Number of Grids	Excavation Area
1957	Unclear	Unclear, but probably small
1978	3	215 m^2
2008	35	1,395 m^2

Note: The excavation area from 1957 is unclear, but the site area was reported as 140×12 m^2. Data are taken from the archaeological field reports (POYPM 1958; YPICRA, et al. 2009b; YPM 1995).

2.3.3 Dating

On the basis of radiocarbon dating and dendrochronology calibration, a wooden column unearthed in 1957 was dated to 3115±90 B.P. (LIACAS 1972:57). Another wooden pole unearthed in 1978 from Level (4) in T4 was dated to 2595±75 B.P. (LIACASS 1990:667), calibrated to 2660±125 B.P. (YPM 1995:785).

According to the reports for the 2008 excavation season (YPICRA, et al. 2009a; b) the dating at Haimenkou was determined by the data from the previous two seasons (1957 and 1978) in combination with typological data. For example, a few pottery sherds with incised decorations unearthed from Phase I seemed to resemble sherds from other sites dating to the late Neolithic Period in the vicinity, including the Caiyuanzi site (*ca.* 4300 B.P.) in Yongren County, Yunnan (YPICRA, et al. 2003) (see location in Figure 2.23). The stone knives characterised by double-sided grooves and perforations from Phase I (Figure 2.7) were also recovered at the Karuo site (*ca.* 5000–4000 B.P.) in Changdu County, Tibet (CPAMTAR and DHSU 1985) (see location in Figure 2.24). Given the dates of these two sites, Phase I of Haimenkou may be typologically dated to around 5000–3900 B.P.

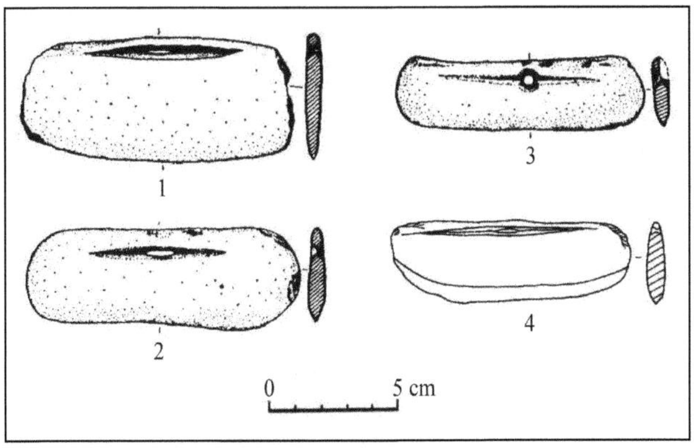

Figure 2.7 Comparisons between Stone Knives Characterised by Double-Sided Grooves and Perforations from Haimenkou and Karuo

1–3: Stone Knives from Karuo (After CPAMTAR and DHSU 1985: figure 61:4, 5, 22; reproduced with permission of Cultural Relics Press);

4: Stone Knife from Haimenkou (After YPICRA, et al. 2009b: figure 6:4, see also Figure 2.12:4; reproduced with permission of Archaeology Press).

In addition, due to the similarities in bronze objects, the date of Level (4) at Haimenkou could be inferred to be earlier than the date provided for the early stage of the Aofengshan grave site, which is also located in Jianchuan County (YPICRA 1990). The Aofengshan grave is a Bronze Age site that was radiocarbon dated to 2420±80 B.P., calibrated to 2450±90 B.P. (LIACASS 1984:652; YPICRA 1990:257).

As a result of the above data the approximate date range of the phases of Haimenkou can be summarised as follows: Phase I dates to *ca.* 5000–3900 B.P., Phase II dates to *ca.* 3800–3200 B.P., Phase III dates to *ca.* 3100–2500 B.P., and the Late Phase dates to the late dynastic periods (i.e., Song, Yuan, and Ming dynasties, A.D. 960–1644).

New data based on materials recovered from the 2008 season were released in 2011 (Li, K. and Min 2011:101) (Table 2.2). Consequently, the dates of Haimenkou have been revised on the basis of new radiocarbon dates and typological data as follows: Phase I dates to around 5000–3800 B.P., Phase II dates to around 3700–3200 B.P., Phase III dates to around 3100–2400 B.P., and the Late Phase dates to the late dynastic periods (i.e., Song, Yuan, and Ming dynasties, A.D. 960–1644).

As the earliest Bronze Age site discovered in Yunnan, Haimenkou sets the date for the Bronze Age on the Yunnan-Guizhou Plateau. As N. Wang (1992) proposed, the periodisation of Yunnan differs from the Central Plain region of China (the area on the lower reaches of the Yellow River), where the Bronze Age began around the 21st century B.C. (Bai 2002; Chang 1983:2; Zhang, Zhiheng 1995:6). K. Li and Min (2011) suggest the Bronze Age in Yunnan may be divided into three periods. That is, the early period from the 17th century to the 8th century B.C., the middle period from the 8th century to the 4th century B.C., and the late period from the 4th century to the 2nd century B.C. According to the division, the date of the Haimenkou site ranges from the late Neolithic Period to the middle Bronze Age.

2.3.4 Archaeological Features

The 2008 season excavation at Haimenkou revealed a large number of archaeological features, including house foundations, wooden poles, pits, postholes, and a human bone pit (not a grave). The largest pile-dwelling site discovered in China was also uncovered, and was represented by four thousand wooden columns and beams with mortise and tenon joint structures (YPICRA, et al. 2009a; b). The features uncovered from the site of the 2008 season are described as follows in chronological order.

2.3.4.1 Phase I

The archaeological features revealed in Phase I were mainly wooden columns. Less than one-fifth of the total of the wooden columns at Haimenkou and one house foundation were uncovered and dated to this phase. The house foundation is described below.

House Foundation

F1 (F is an abbreviation for the Chinese term Fangzi meaning house) was located at the north-central section of T1204 in sub-area D (i.e., DT1204) below Level (8). As a water-logged pile-dwelling construction, it had 45

Table 2.2 Carbon-14 Dates of Haimenkou in Jianchuan County, 2008

Lab Number	Sample	Original Number	Carbon-14 Date (B.P.)
BA081094	Seed	2008JHDT1304(5)	3000±35
BA081095	Plant Root	2008JHAT2121(5)	2200±35
BA081096	Carbonised Wheat	2008JHAT2002(6)	2435±35
BA081097	Carbonised Plant	2008JHDT1005(6)	3020±35
BA081098	Plant Fibre	2008JHDT1304(6)	3075±35
BA081099	Carbonised Rice	2008JHAT2003(6)	2930±35
BA081100	Carbonised Foxtail Millet	2008JHAT2003(6)	2940±35
BA081101	Carbonised Foxtail Millet	2008JHAT2003(7)	3550±40
BA081102	Plant Root	2008JHAT2505(7)	3205±35
BA081103	Charcoal	2008JHDT1205(8)	3605±40
BA081104	Charcoal	2008JHDT1005(9)	3345±35
BA081105	Wooden Stick	2008JHDT1004(9)	4210±35
BA081106	Plant Root	2008JHDT1003(10)	4485±35

Note: Based on K. Li and Min (2011: table 1). Data are from Laboratory of Quaternary Dating, Peking University, China.

round wooden columns (8–15 cm in diameter). Three column samples, Number 6, 7, and 8, were 95, 56, and 55 centimetres in length respectively, with chopped and pointed-bottoms. All of these columns were arranged in an orderly plan as a rectangle. The rectangle was divided into two smaller rectangles by a row of columns lying at the north-central part (Figure 2.8), probably for better support. This house foundation was 550 centimetres long and 220 centimetres wide. The floor was probably constructed of wood, the wall structure was of wattle and daub, and the roof was possibly made of thatch (YPICRA, et al. 2009a:597). Three grey stones were found at the southern end of house F1, and may have been used to form an indoor hearth and then scattered after the collapse of the house. The exact orientation and position of doorways is as yet unclear.

2.3.4.2 Phase II

The archaeological features revealed in Phase II included abundant wooden columns and beams and a human bone pit, as well as a house foundation. No further description of the human bone pit was provided in the original brief reports. Around two-fifths of the total of the wooden columns unearthed were dated to this phase.

Wooden Columns

The shapes of the wooden columns yielded in this phase were round, semi-round, triangular, and elliptical. The bottoms of these columns were chopped into a point, while some had small triangular mortises. For example, the columns found at the grid DT1003 (Figure 2.9) were 26–137 centimetres in length, between 4–20 centimetres in diameter, and with bottoms located from Levels (8) to (10).

Wooden Beams

A considerable number of wooden beams were revealed in phase II. The longest was 601 centimetres in length, 34 centimetres in width, and eight centimetres in thickness. It was unearthed below Level (6) at AT2004 with one end showing an arc-square hole, while the other end was broken by a column from a later period. Another example was that of a wooden beam, also recovered below Level (6) at AT2003. This beam was 278 centimetres long, 22 centimetres wide, and six centimetres thick, with one end having a square mortise.

2.3.4.3 Phase III

The archaeological features revealed in Phase III included abundant wooden columns, along with several hearths. About two-fifths of the total of the wooden columns were found in this phase. Four hearths were revealed, which are all located below Level (5).

Wooden Columns

The cross-section of the wooden columns unearthed from this phase were round, polygonal, rectangular, and triangular. Their bottoms were also chopped into a point, and some had mortises whose shapes were small triangles or large rectangles. There were many large columns recovered in this phase, and the largest one was 38 centimetres in diameter and almost 200 centimetres in length.

Hearths

The area of the smallest hearth found in this phase was less than one square metre, while the biggest covered over

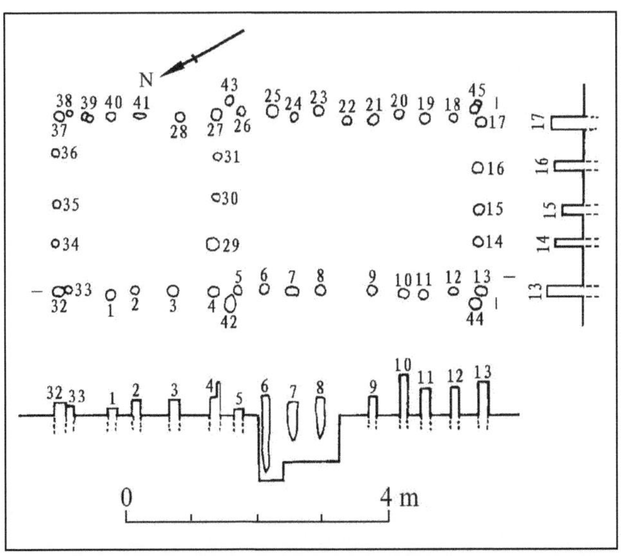

Figure 2.8 Plan and Section of House Foundation F1 (After YPICRA, et al. 2009b: figure 5; reproduced with permission of Archaeology Press)

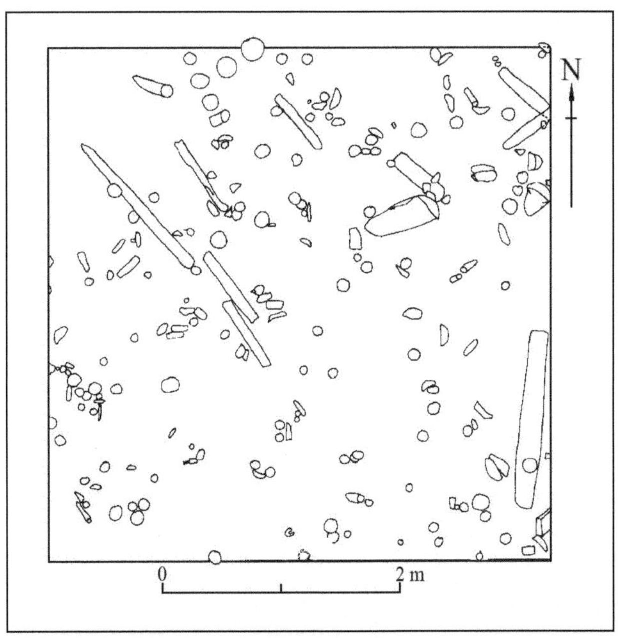

Figure 2.9 Plan of the Distribution of Wooden Columns at DT1003 (After YPICRA, et al. 2009b: figure 9; reproduced with permission of Archaeology Press)

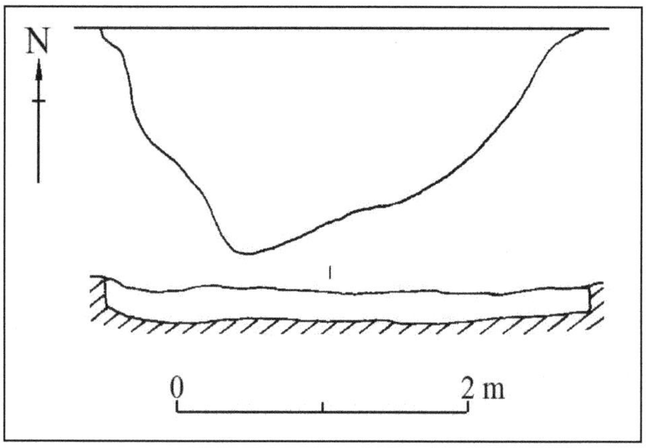

Figure 2.10 Plan and Section of Hearth HD3 (After YPICRA, et al. 2009b: figure 16; reproduced with permission of Archaeology Press)

ten square metres. For example, the hearth HD3 (HD is an abbreviation for the Chinese term *Huodui* meaning hearth), which is below Level (5) while 62–67 centimetres below the ground surface, had a generally round shape that was 360 centimetres in length, 155 centimetres in width, and 18–22 centimetres in thickness (Figure 2.10). It contained a lump of completely burnt clay, which was covered by a small amount of black-grey soil and a small number of pottery sherds.

2.3.4.4 Discussion and Summary

More than four thousand wooden columns and beams were revealed in the 2008 season, as well as two house foundations and hearths. From Phases I to II, the numbers of wooden columns and beams doubled. In Phase III, the number of wooden columns and beams was generally the same as in Phase II. This phenomenon may indicate an increase in the number of houses, as well as population growth through time. In addition, an obvious increase in the size of the wooden columns and beams was observed in chronological order. This augmentation may indicate more advanced techniques and tools for chopping and cutting wood (i.e., stone and bronze axes). This is in accordance with the changes occurring from the late Neolithic Period to the early Bronze Age at Haimenkou.

The house foundation F1 is similar to other Neolithic sites in Yunnan, such as the Caiyuanzi site (*ca.* 4300 B.P.) (YPICRA, et al. 2003) in Yongren County, and the Baiyangcun site (*ca.* 3800 B.P.) (LIACASS 1978:287; YPM 1981) in Binchuan County (see locations in Figure 2.23). At these sites the house foundations have a rectangular plan. They are similar to the house foundations recovered at Haimenkou, although they were ground surface constructions rather than pile-dwelling constructions. Another site with water-logged pile-dwelling constructions in Yunnan is the Yinsuodao site (*ca.* 5000–2400 B.P.) (YPICRA, et al. 2009c). This site is located near Erhai Lake (see location in Figure 2.23). These similar architectural styles may indicate cultural communication and/or ethnic relations between these sites in prehistoric Yunnan.

2.3.5 Artefacts

More than three thousand artefacts were recovered during the excavations from 1957 to 2008. These artefacts may be classified into groups based on raw materials, including pottery, bronze and iron objects, and tools of stone, wood, bone, and antler. The artefacts unearthed from the 2008 season are described as follows in chronological order (YPICRA, et al. 2009a; b).

2.3.5.1 Phase I

Pottery

The pottery collected in this phase was hand-made and fired at a high temperature. Sand-tempered black-grey pottery constituted the largest part of the assemblage, followed by sand-tempered grey pottery and polished black pottery (Figure 2.11).

The predominant type of pottery surface decoration was engraving or incising, with a variety of punctuated patterns also represented. The decorative designs included triangular, zigzag, crosshatched, wavy, and nipple-shaped patterns. These surface decorations are similar to other late Neolithic sites in Yunnan, such as Caiyuanzi (*ca.* 4300 B.P.) (see location in Figure 2.23) (YPICRA, et al. 2003).

Pottery unearthed in this phase was relatively small in size, including jars, basins, and bowls. Jars made up the largest portion of the assemblage, followed by bowls. In addition,

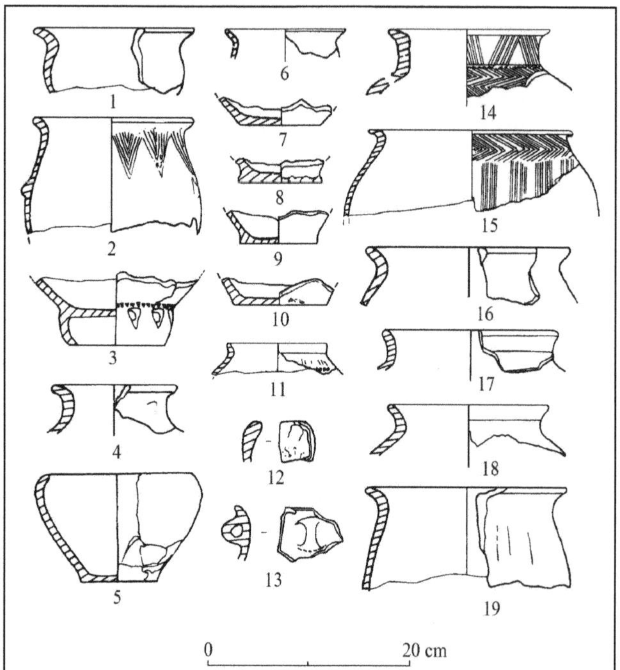

Figure 2.11 Pottery Unearthed from Phase I (After YPICRA, et al. 2009b: figure 7; reproduced with permission of Archaeology Press)

3: Ring-Footed Ware; 5, 12: Bowl; 7–10: Ware Bases; Others: Jars.

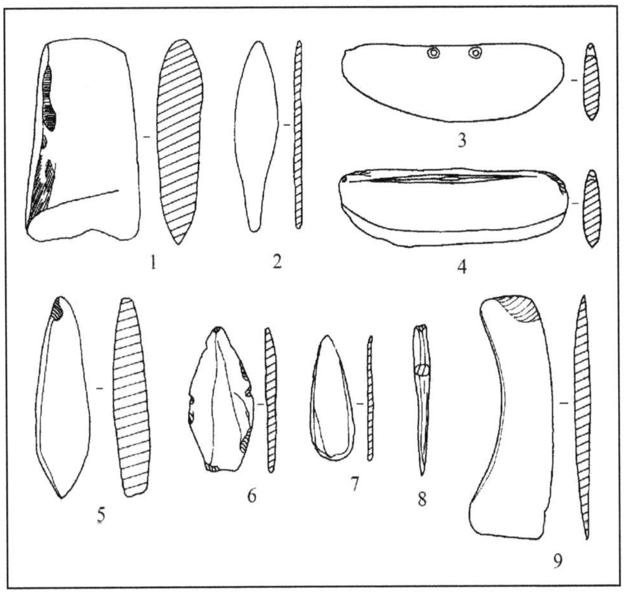

Figure 2.12 Tools of Stone and Bone Unearthed from Phase I (After YPICRA, et al. 2009b: figure 6; reproduced with permission of Archaeology Press)
1: Stone Axe; 2: Stone Arrowhead; 3, 4, 6, 7: Stone Knives; 5: Stone Chisel; 8: Stone Awl; 9: Bone Ornament (Scale: 1, 3, 4 = 1/5; others = 2/5).

a small number of ring-foot wares made of polished black pottery (Figure 2.11:3) were found in this phase and were absent in the following phases.

Stone Implements

Stone implements from this phase included axes, adzes, knives, chisels, awls, and arrowheads (Figure 2.12). Axes were trapezoid-shaped and most of them were more than ten centimetres in length, while adzes were generally quite small. Knives were half-moon or rectangle in shape, had single or double holes in the body, and were characterised by a double-sided and grooved pattern. Awls made up the majority of the stone implements, followed by knives, arrowheads, and axes. The frequency of each stone implement type is summarised in Table 2.3.

Table 2.3 List of Stone Implements from Phase I

Stone Implement	Number	Percentage*
Awl	58	54.7
Knife	16	15.1
Arrowhead	16	15.1
Axe	11	10.4
Chisel	5	4.7

Note: *The number of adzes was not present in the report (YPICRA, et al. 2009b), thus the percentages are not of the stone tools as a whole.

Others

A small number of tools of antler and bone as well as animal tooth ornaments were uncovered. The exact numbers were not provided in the report (YPICRA, et al. 2009b).

2.3.5.2 Phase II

Pottery

The pottery unearthed in this phase was all hand-made and was fired at a high temperature. Sand-tempered grey pottery dominated this phase, while black-grey pottery made up a lesser portion, followed by red-brown and grey-brown pottery. The polished black pottery found in Phase I was absent. Some of these wares were polished on the rim.

Decorative designs on vessel surfaces were reduced when compared to those of Phase I. Decorations present included punctuated spots and lines, nipple-shaped patterns, and red painting.

The pottery wares unearthed in this phase include jars, basins, bowls, *yi* ladles (a gourd-shaped ladle), with jars and bowls being the most common (Figures 2.13a–b). Pottery vessels with handles increased in number and other types of pottery became larger. The double-handled jars with red painting revealed from this phase may indicate some influence from the upper reach of the Yellow River (northwestern China) due to similarities in design (see Figure 2.13a:7, 18; see also Figure 2.13b:17, 19). In addition, 35 pottery net sinkers were revealed, which demonstrates the importance of fishing in the subsistence strategy of Haimenkou people.

Stone Implements

More than six hundred stone implements were uncovered in this phase. This is approximately six times more than in Phase I. They consisted of awls, axes, adzes, knives, chisels, arrowheads, and grinding tools (Figures 2.14a–b). Stone awls and adzes were dominant, followed by axes, arrowheads, and knives. The number of arrowheads increased in this phase. The frequencies of stone implement types are summarised in Table 2.4.

Bronzes

Small bronze objects began to appear in the later level of this phase (i.e., Level (6)). They include one chisel, one bell, one awl, and one knife, and were all small in size (Figure 2.15).

Others

Other artefacts were revealed in this phase, such as tools of wood, bone, and antler, and ornaments made of animal teeth (Figure 2.16). Wooden tools consisted of *si* spades (a spade-shaped farm tool used in ancient China), rakes, spoons, and oars. Bone and tooth artefacts included awls, spades, and ornaments. The exact number of wooden tools recovered was not given in the published field report (YPICRA, et al. 2009b). Details of the bone and tooth artefacts recovered are shown in Table 2.5.

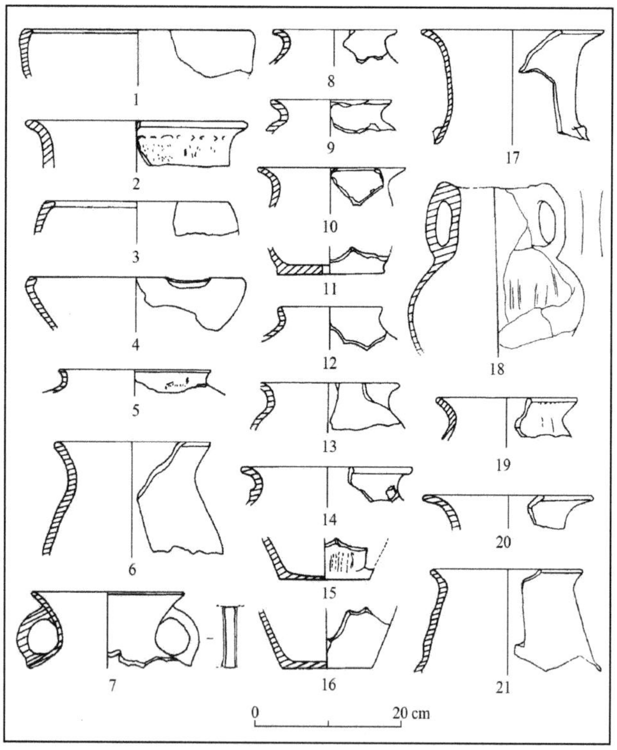

Figure 2.13a Pottery Unearthed from Phase II (Part 1) (After YPICRA, et al. 2009b: figure 12; reproduced with permission of Archaeology Press)

2: Basin; 4: *Yi* Ladle; 11, 15, 16: Ware Bases; Others: Jars.

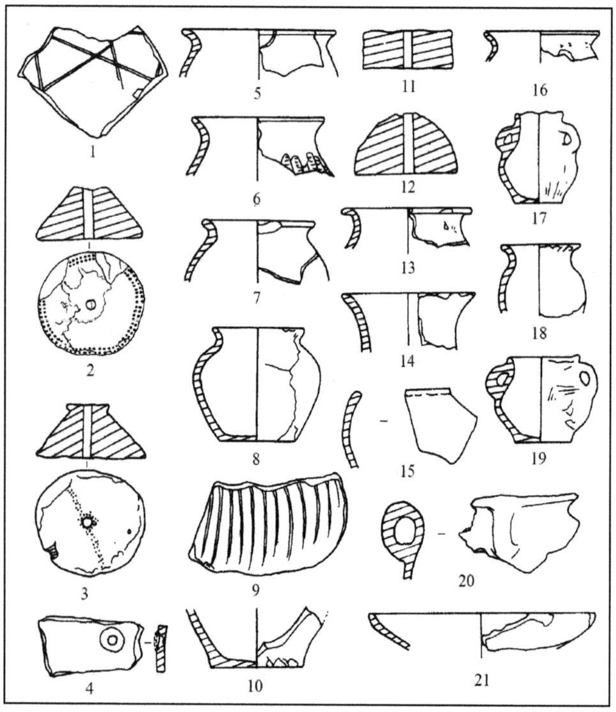

Figure 2.13b Pottery Unearthed from Phase II (Part 2) (After YPICRA, et al. 2009b: figure 13; reproduced with permission of Archaeology Press)

1, 4, 9: Sherds; 2, 3, 11, 12: Spindle Whorls; 10: Ware Bases; 21: Bowl; Others: Jars (Scale: 1–4, 9, 11, 12 = 2/7; others = 1/7).

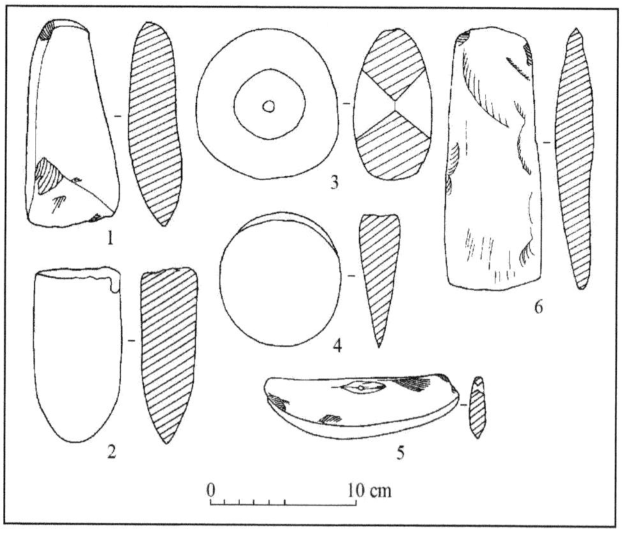

Figure 2.14a Stone Implements Unearthed from Phase II (Part 1) (After YPICRA, et al. 2009b: figure 10; reproduced with permission of Archaeology Press)

1: Adze; 2, 6: Axes; 3: Perforating Objects; 4: Grinding Tool; 5: Knife.

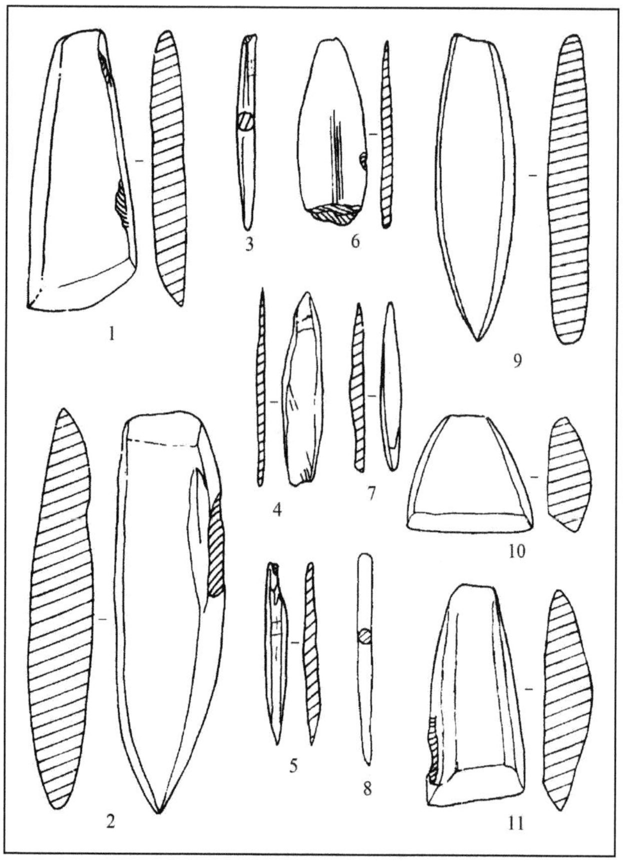

Figure 2.14b Stone Implements Unearthed from Phase II (Part 2) (After YPICRA, et al. 2009b: figure 11; reproduced with permission of Archaeology Press)

1, 10, 11: Adzes; 2, 9: Chisels; 3, 5, 8: Awls; 4, 6, 7: Arrowheads (Scale: 5, 7, 8 = 4/7; others = 1/1).

Table 2.4 List of Stone Implements from Phase II

Stone Implement	Number	Percentage*
Awl	193	31.2
Adze	166	26.8
Axe	79	12.8
Arrowhead	76	12.3
Knife	69	11.1
Chisel	22	3.6
Grinding tool	14	2.3

Note: *A small number of perforating objects were mentioned in the report but the exact number was not provided (YPICRA, et al. 2009b), thus the percentages are not of the whole range of stone tools.

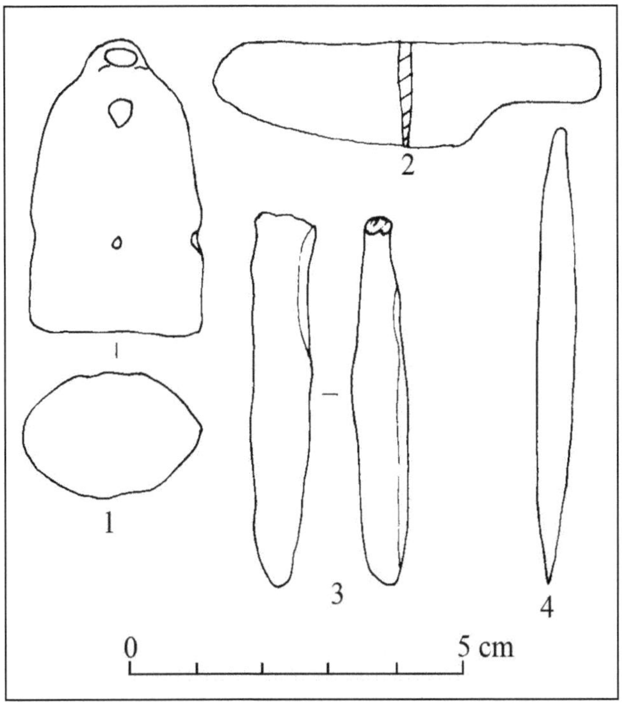

Figure 2.15 Bronzes Unearthed from Phase II (After YPICRA, et al. 2009b: figure 15; reproduced with permission of Archaeology Press)
1: Bell; 2: Knife; 3: Chisel; 4: Awl.

Table 2.5 List of Bone and Tooth Artefacts from Phase II

Artefact	Material	Number	Percentage*
Awl	Bone	40	50.0
Hairpin	Bone	24	30.0
Arrowhead	Bone	1	1.3
Spade	Bone	1	1.3
Ornament*	Tooth	14	17.5

Note: *There were a small number of bone ornaments, but the exact number was not provided in the report (YPICRA, et al. 2009b). Thus the percentages are not of the whole range of tools.

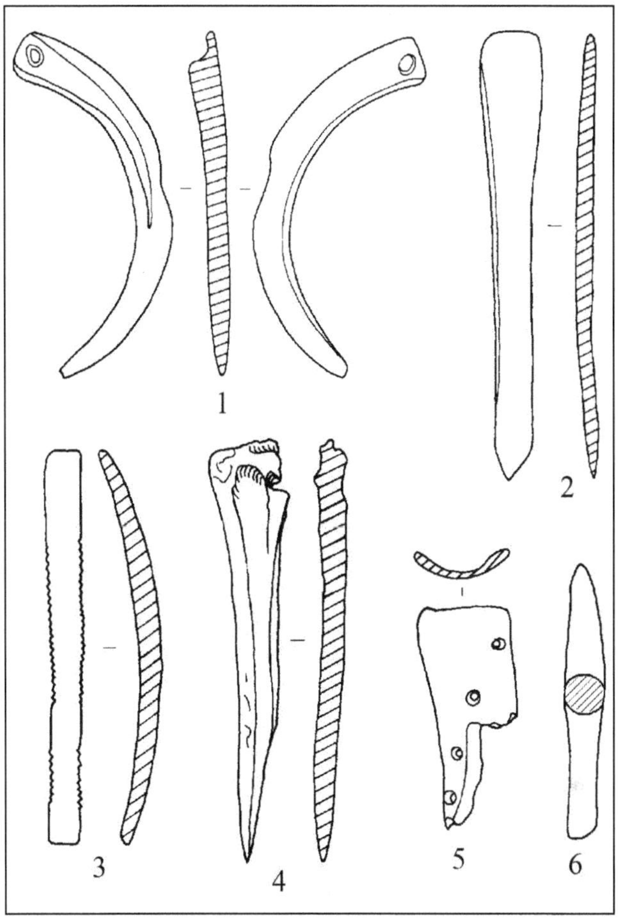

Figure 2.16 Bone and Tooth Artefacts Unearthed from Phase II (After YPICRA, et al. 2009b: figure 14; reproduced with permission of Archaeology Press)
1: Tooth Ornament; 2, 3: Bone Hairpin; 4: Bone Awl; 5: Bone Ornament; 6, Bone Arrowhead (Scale: 4, 5 = 1/2; others = 1/1)

2.3.5.3 Phase III

Pottery

The pottery unearthed in this phase was all hand-made, and was fired at a lower temperature than in previous phases. Sand-tempered grey pottery made up the majority of the wares, red-brown and grey-brown pottery increased in number, and black-grey pottery decreased. In addition, a few red-slipped wares were present.

Decorative designs on vessel surfaces were rare, but where they did exist nipple-shaped patterns were dominant.

Pottery wares unearthed from this phase included jars, basins, bowls, *yi* ladles (a gourd-shaped ladle), and some uncertain or irregular shapes, with jars making up the majority (Figure 2.17). The double-handled jars with red painting increased in number in this phase, which may indicate the increased influence from the upper Yellow River region (northwestern China). More than 30 net sinkers were also recovered, and some of these were made of pottery sherds.

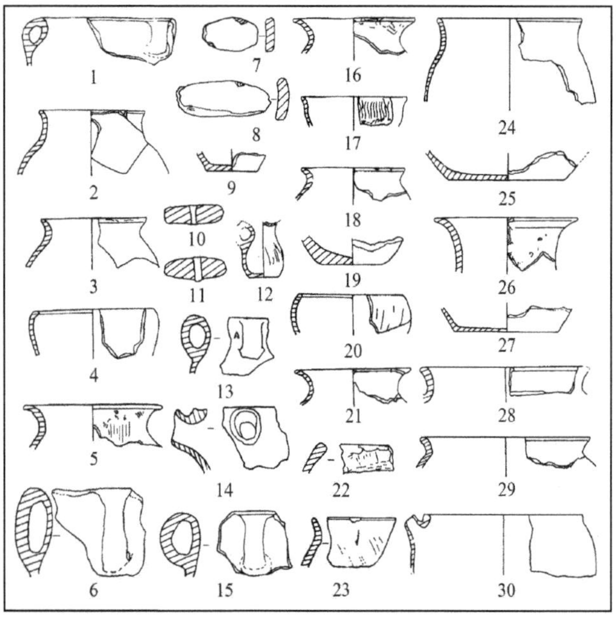

Figure 2.17 Pottery Unearthed from Phase III (After YPICRA, et al. 2009b: figure 18; reproduced with permission of Archaeology Press)

4, 17, 20: Bowls; 7, 8: Net Sinkers; 9, 19, 25, 27: Ware Bases; 10, 11: Spindle Whorls; 14: *Yi* Ladle; 22, 30: Urns; Others: Jars (Scale: 7, 8, 10, 11, 25, 27 = 1/5; others = 2/5).

Stone Implements

More than 250 stone implements were recovered in this phase. They consist of axes, adzes, knives, chisels, awls, arrowheads, grinding tools, and a casting mould (Figure 2.18). Stone awls were most common, followed by arrowheads. The details are provided in Table 2.6.

Bronzes

There were more bronze objects in this phase than in Phase II. Bronzes included three bracelets, two arrowheads, one awl, and one chisel (Figure 2.19). As in Phase II, these bronzes revealed in Phase III were also small objects, which may indicate an early stage of bronze manufacture.

Table 2.6 List of Stone Implements from Phase III

Stone Implement	Number	Percentage*
Awl	72	28.1
Adze	47	18.4
Knife	49	19.1
Arrowhead	43	16.8
Axe	36	14.1
Chisel	8	3.1
Casting Mould	1	0.4

Note: *Grinding tools were mentioned, but no exact number was provided in the report (YPICRA, et al. 2009b), thus the percentages are not of the stone implements as a whole.

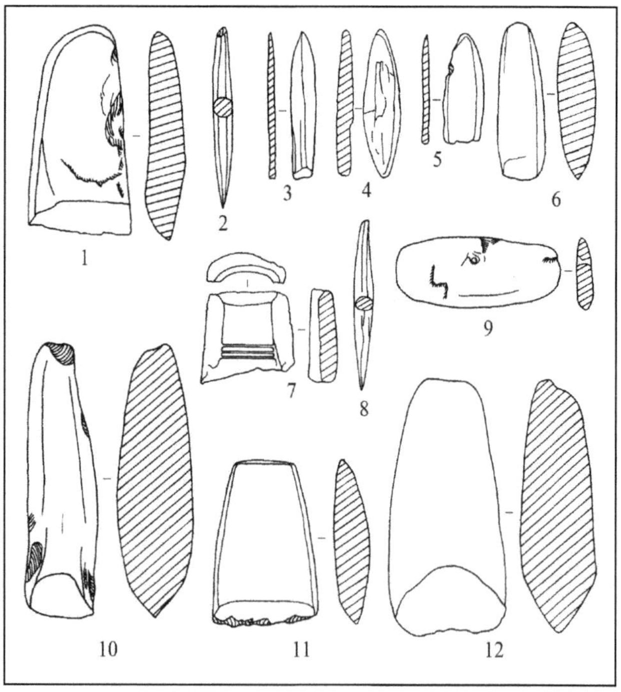

Figure 2.18 Stone Implements Unearthed from Phase III (After YPICRA, et al. 2009b: figure 17; reproduced with permission of Archaeology Press)

1, 11: Adze; 2, 8: Awls; 3, 4, 5: Arrowheads; 6, Chisel; 7: Mould; 9: Knife; 10, 12: Axes (Scale: 1, 9, 10, 12 = 1/5; 7 = 1/10; others = 2/5).

Bone Tools

Tools made of animal bones were also found in this phase, including a small number of bracelets (exact number is unavailable), 23 arrowheads, and four awls (see Figure 2.19).

2.3.5.4 Discussion and Summary

Pottery

Pottery dominated the artefact assemblages excavated from the three seasons and weighed a total of approximately four tons. Painted pottery, which was recovered from Levels (7) to (4) (i.e., Phases II and III, *ca.* 3700–2400 B.P.), was discovered for the first time from archaeological sites in Yunnan (YPICRA, et al. 2009a:598). Most of the painted pottery wares were double-handled jars painted in red (see Figure 2.13a:7, 18, see also Figure 2.13b:17, 19). Vessels unearthed from later layers were more vivid in colour, including red slip wares.

There were a number of archaeological cultures in northwestern China (Qinghai and Gansu Provinces) from the Chalcolithic Period to the Bronze Age. These included Qijia (*ca.* 4100–3700 B.P.) (IACASS 1991:274-286; 2003b:535-558), Siba (*ca.* 3900–3500 B.P.) (IACASS 1991:272; 2003b:558-562), Kayue (*ca.* 3600–2600 B.P.) (IACASS 1991:287-289; 2003b:562-566), Xindian (*ca.* 3400–2700 B.P.) (IACASS 2003a:500-506; Xie, D. 2002:174), Siwa

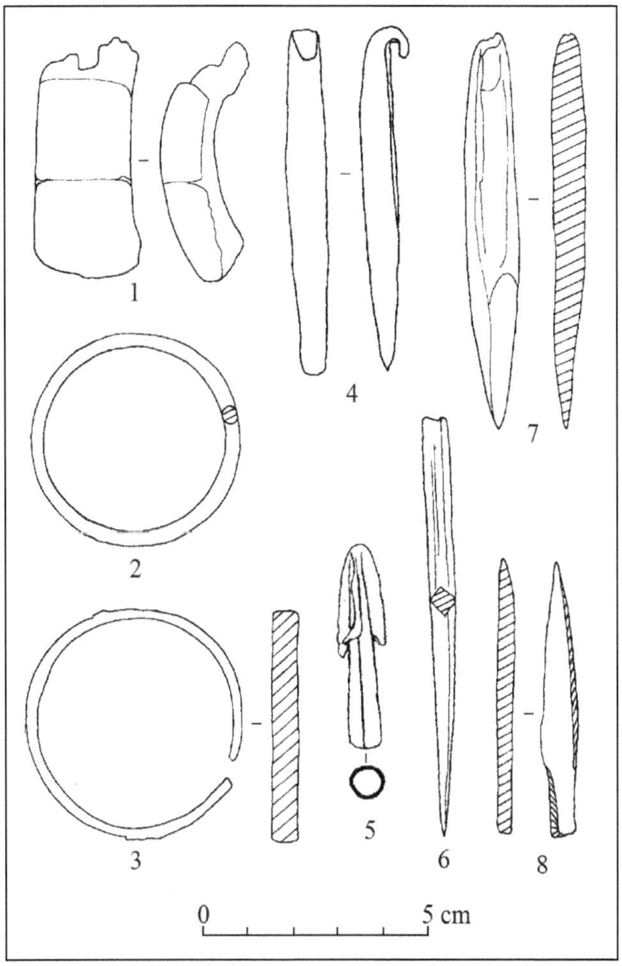

Figure 2.19 Artefacts Unearthed from Phase III (After YPICRA, et al. 2009b: figure 19; reproduced with permission of Archaeology Press)
1: Bone Bracelet; 2, 3: Bronze Bracelets; 4, Bronze Chisel; 5: Bronze Arrowhead; 7, Bone Awls; 8, Bone Arrowhead.

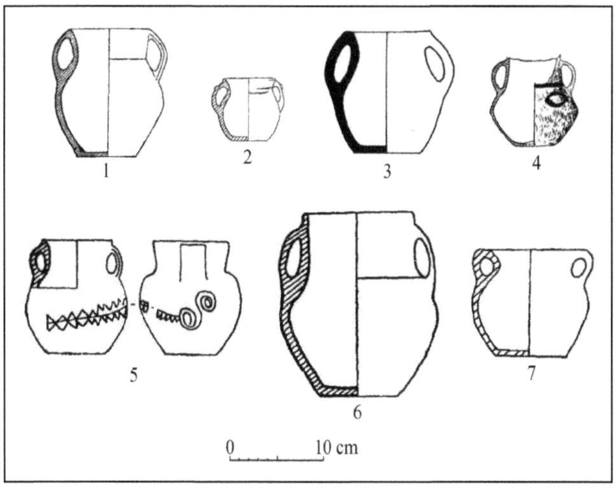

Figure 2.20 Double-Handled Pottery Jars from Sites in Northwestern China (Qinghai and Gansu Provinces)
1: Jar from Qinweijia (Qijia Culture) (After GATIACAS 1975: figure 15:14; reproduced with permission of Archaeology Press);
2: Jar from Donghuishan (Siba Culture) (After GPICRA and ALNCJU 1998: figure 57:10; reproduced with permission of Yongjie Xu);
3: Jar from Dahuazhongzhuang (Kayue Culture) (After HCM, et al. 1985: figure 15:9; reproduced with permission of Archaeology and Cultural Relics);
4: Jar from Jijiachuan (Xindian Culture) (After GSYRRAT 1962: figure 3:4; reproduced with permission of Archaeology Press);
5, 6: Jars from Lanqiao (Siwa Culture) (After GPAT, et al. 1987: figure 15:2, 5; reproduced with permission of Archaeology Press);
7: Jar from Hamadun (Shajing Culture) (After GPICRA 1990: figure 17:6; reproduced with permission of Archaeology Press).

(*ca.* 3400–2700 B.P.) (IACASS 2003a:507-510; Xie, D. 2002:190), Nuomuhong (*ca.* 3000–2800 B.P.) (Xie, D. 2002:204), and Shajing cultures (*ca.* 2900–2400 B.P.) (IACASS 2003a:510-515; Xie, D. 2002:215). In general, these archaeological cultures appear to be connected with some ancient Chinese ethnic groups, normally referred to as *Qiang*[2] in historical records (IACASS 2003a:506, 510, 515; 2003b: 562, 566; Xie, D. 2002-237; Yan 1978). The *Qiang* inhabited northwestern China (Qinghai and Gansu Provinces) from the Neolithic and Bronze Age Periods (Li, K. 1983:76; 1988; Li, K. and Xiao 1980:142; Yan 1978). Although it is unclear exactly when the name *Qiang* came to existence in prehistory, for the convenience of discussion it is used here to describe the peoples who were the carriers of archaeological cultures in the northwestern region concerned in the study. Double-handled jars were present at sites of all these culture. These jars are in a similar style to those recovered at Haimenkou (Figure 2.20). Therefore, the painted double-handled jars revealed at Haimenkou probably indicate the influence of the *Qiang* style from the region of the upper Yellow River Valley, especially Qinghai and Gansu Provinces (see locations of Qinghai and Gansu Provinces in Figure 2.24).

Also, the *Qiang* style double-handled jars were recovered from other sites in northwestern Yunnan (Figure 2.21), including Nagu (ATYPM 1983a) and Shidi (ATYPM 1983b) in Deqin County, stone graves in Zhongdian County (YPICRA 2005), Daxingzhen (ATYPM 1983c) in Ninglang County, and Aofengshan (YPICRA 1990) which is also located in Jianchuan County (see locations in Figure 2.28). These sites date to the early and middle Bronze Ages in Yunnan but do not occur earlier than the Western Zhou dynasty (*ca.* 3100–2800 B.P.) in the Central Plain region according to typological data (Yang, F., et al. 2010:211-218). They also show strong influences from the cultures in northwestern China (*Qiang* style cultures) based on artefacts including *Qiang* style double-handled jars.

In addition, these *Qiang* style double-handled jars were also widely recovered at sites in Sichuan Province (see location of Sichuan in Figure 2.24) (Xie, C. 2005), such as Zhaodian (*ca.* 2700 B.P.) in Wenchuan County (Ye and Luo 1999), Chengguan (*ca.* 2500–2000 B.P.) in Mao County (CPAMSP and CCMC 1983), and Jiashan (*ca.*

[2] There were different *Qiang* peoples mentioned in ancient texts.

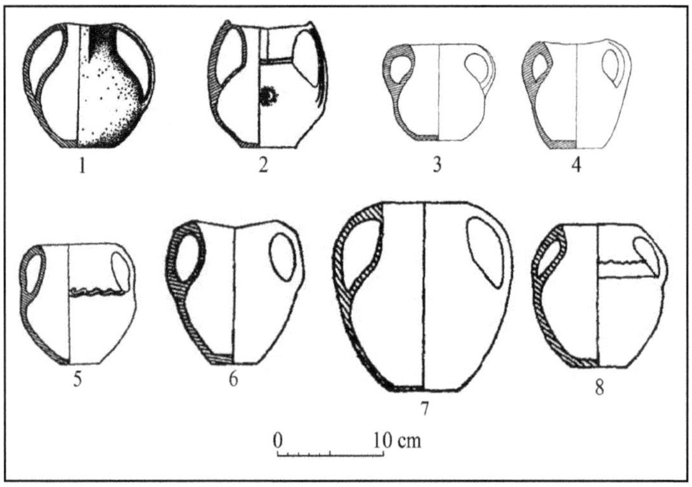

Figure 2.21 Double-Handled Pottery Jars from Sites in Northwestern Yunnan
1: Jar from Nagu (After ATYPM 1983a: figure 6:1; reproduced with permission of Archaeology Press);
2: Jar from Shidi (After ATYPM 1983b: figure 2:3; reproduced with permission of Archaeology Press);
3, 4: Jars from Zhongdian (After YPICRA 2005: figure 20:2, 6; reproduced with permission of Archaeology Press);
5, 6: Jars from Daxingzhen (After ATYPM 1983c: figure 6:1, 3; reproduced with permission of Archaeology Press);
7, 8: Jars from Aofengshan (After YPICRA 1990: figure 23:1, 2; reproduced with permission of Archaeology Press).

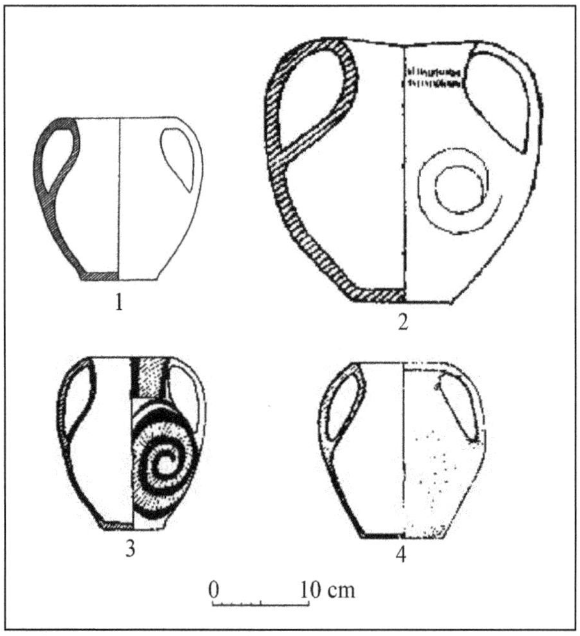

Figure 2.22 Double-Handled Pottery Jars from Sites in Sichuan
1: Jar from Zhaodian (After Ye and Luo 1999: figure 3:2; reproduced with permission of Archaeology Press);
2: Jar from Jiashan (After CRMATAP 1987: figure 14:3; reproduced with permission of Xueshu Xu);
3, 4: Jars from Chengguan (After CPAMSP and CCMC 1983: figure 10:1, 2; reproduced with permission of Cultural Relics Press).

2000 B.P.) in Li County (CRMATAP 1987) (Figure 2.22). Consequently, the locations of these sites from which *Qiang* style double-handled jars have been recovered in Yunnan, Gansu, Qinghai, and Sichuan Provinces suggest multiple expansion routes of the *Qiang* style cultures. These cultures may have spread to Yunnan from northwestern China. The routes may have included pathways through Tibet along the major rivers (the Nu, Lancang, and Jinsha Rivers) and through Sichuan along branches of the Yangtze River. Given the association with the dates of these sites and Haimenkou, it is probable that the cultural communication and/or population movement between Yunnan and northwestern China began at least from the early Bronze Age. It may have become more frequent throughout the Bronze Age.

Stone Implements

The stone awls dominated the assemblage with lower proportions of relatively large, long, stripe-like axes, varied-sized adzes, and relatively abundant arrowheads and knives (Table 2.7).

The stone implements represent a style prevalent around the upper reaches of the Nu, Lancang, and Jinsha Rivers during Neolithic times (YPICRA, et al. 2009a:598-599). Different scholars have divided Neolithic Yunnan into diverse cultural regions over several decades (Kan 1998; Li, K. and Xiao 1980; Wang, D. 1998; Wang, N. 1992). Nevertheless, the area near the upper reaches of the Nu, Lancang, and Jinsha Rivers (see Figure 2.23) is typically considered to be characterised by non-shouldered and non-stepped stone implements (Xu and Xiao 2004:209-223), such as the axes and adzes revealed at Haimenkou.

The relatively abundant adzes and knives may indicate the development of local agriculture at Haimenkou. In Chinese

Table 2.7 List of Stone Implements from Haimenkou

Stone Implement	Number	Percentage
Awl	323	32.9
Adze	213	21.7
Knife	134	13.7
Arrowhead	135	13.8
Axe	126	12.8
Chisel	35	3.6
Grinding tool	14	1.4
Casting Mould	1	0.1
Total*	981	100.0

Note: *As noted in Tables 2.2, 2.3, and 2.5, because some exact numbers of stone implements were not provided in original reports, Table 2.6 may only display an outline of the frequencies of stone implement types.

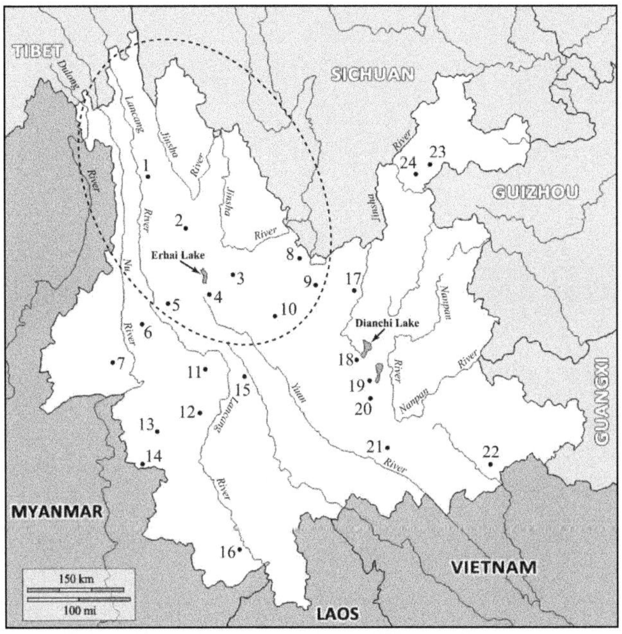

Figure 2.23 Map of Major Neolithic Sites in Yunnan (The original map is from URL: https: http://d-maps.com/carte.php?num_car=19482&lang=en)

1: Gedeng; 2: Haimenkou; 3: Baiyangcun; 4: Jinsuodao & Yinsuodao; 5: Xinguang; 6: Wanzipu; 7: Dahuashi; 8: Caiyuanzi & Mopandi; 9: Dadunzi; 10: Sunjiatun; 11: Manghuai; 12: Nabang; 13: Shifodong; 14: Dinglai; 15: Bingkuang; 16: Manbangnan; 17: Aqiao; 18: Shizhaishan; 19: Guangfentou; 20: Haidong & Huangjiaying; 21: Tangdian; 22: Xiaohedong; 23: Zhaxinchang; 24: Machang.

Note: The dashed circle refers to the area near the upper reaches of the Nu, Lancang, and Jinsha Rivers, which was characterised by non-shouldered and non-stepped stone implements.

prehistoric agriculture, an adze was used to loosen the soil, while a stone knife was employed to mow and harvest (Zhang, Zengqi 1990:182-183). Stone axes may also have been used to fell trees and make more space for cultivated land (Zhang, Zengqi 1990:182-183). In addition, the stone knives unearthed from Phase I are similar in style to their counterparts excavated from the Karuo site (*ca.* 5000–4000 B.P.) in Changdu County (see Figure 2.7), eastern Tibet (CPAMTAR and DHSU 1985:107-110). The Karuo site is located near the upper reach of the Lancang River (Figure 2.24). The artefacts at this site show similarities to both northwestern China and the area near the upper reaches of the Nu, Lancang, and Jinsha Rivers in Yunnan (Li, K. and Xiao 1980:141-142). These similarities help to date Haimenkou and may also indicate cultural communication and/or human migration from Qinghai and Gansu Provinces through Tibet to northwestern Yunnan along the Lancang River.

Bronzes

In the third excavation season bronze objects were uncovered in Level (6) (*ca.* 3200 B.P.). A total of 44 bronze and copper objects were found at Haimenkou in the three excavation seasons (Table 2.8). Most of them were small bronze tools and no large vessels were present. According to isotope and X-ray analyses applied on the 26 objects

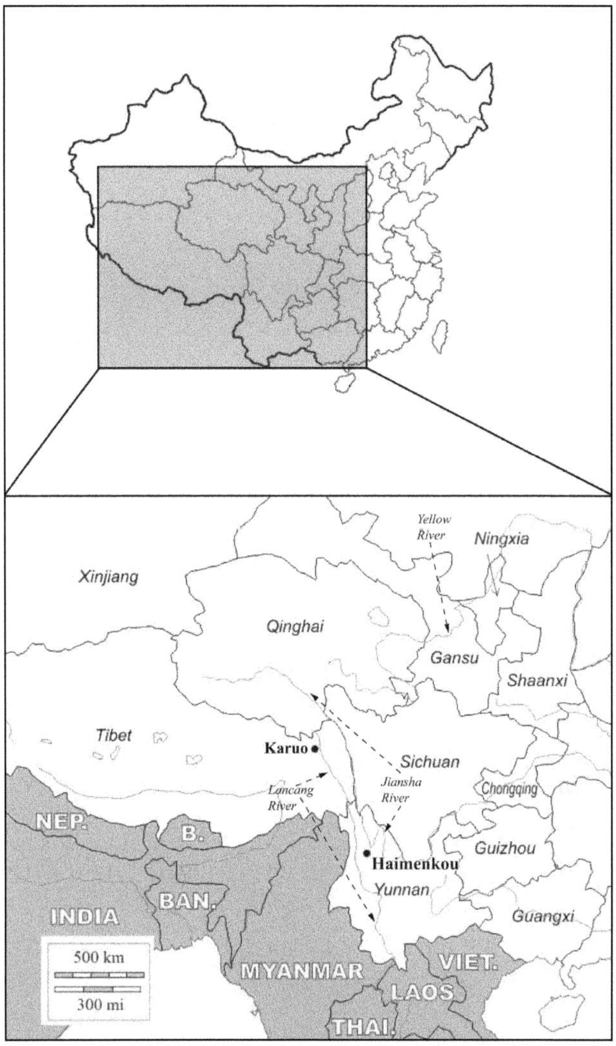

Figure 2.24 Locations of the Haimenkou and Karuo Sites, with the Provinces of Qinghai, Gansu, Sichuan, and Yunnan, and the Distribution of the Lancang, Jinsha, and Yellow Rivers
(The original map is from URL: http://d-maps.com/carte.php?lib=cina_cartina&num_car=11572&lang=it)

recovered from the first two seasons in 1957 and 1978, these bronzes were mainly tin bronze, with a few lead bronzes included (YPM 1995:787). The technologies used in their manufacture mainly consisted of single-mould casting, along with combined-mould casting and some forging. No evidence of pottery-mould casting was found at the site although two stone-moulds were discovered, one in each of the 1957 and 2008 seasons.

The patterns inside these moulds matched the bronze axes recovered in 1957, suggesting that these bronzes were locally produced. Furthermore, the bronze axes recovered at Haimenkou are similar to those from the Wanjiaba site (*ca.* 3000–2400 B.P.) in Chuxiong City, Yunnan (YPAT 1983) (Figure 2.25). The *yue* axes (a type of axe used in ancient China) found at Haimenkou also resemble those found at the Dabona site (*ca.* 2450 B.P.) (Li, J. 1965; LIACAS 1977:203) in Xiangyun County, Yunnan (YPAT 1964). The stone mould for the *yue* axe is similar to one

Table 2.8 List of Bronze Objects from Haimenkou

Bronze Object	Number (1957 & 1978)	Number (2008)**	Number Total	Percentage
Bracelet	6	3	9	24.3
Awl	6	2	8	21.6
*Yue** Axe	3	0	3	8.1
Chisel	1	2	3	8.1
Ornament	3	0	3	8.1
Knife	1	1	2	5.4
Arrowhead	0	2	2	5.4
Axe	1	0	1	2.7
Adze	1	0	1	2.7
Sickle	1	0	1	2.7
Fishhook	1	0	1	2.7
Shovel-Shaped Object	1	0	1	2.7
Broken Slice	1	0	1	2.7
Bell	0	1	1	2.7
Total**	26	11	37	100.0

Note: **Yue* is a type of axe used in ancient China.** Although the total number of bronze objects from the 2008 season was 18, only 11 items were present in the report (YPICRA, et al. 2009b), thus the percentages are not of the bronze objects as a whole.

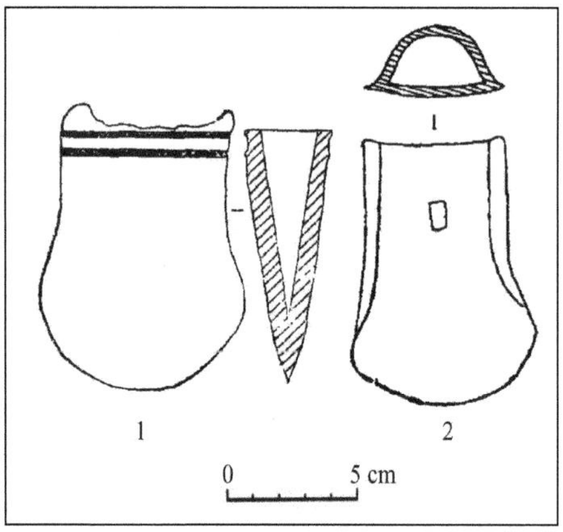

Figure 2.25 Comparison between Bronze Axes from Haimenkou and Wanjiaba
1: Bronze Axe from Haimenkou (After YPM 1995: figure 2:1; reproduced with permission of Archaeology Press);
2: Bronze Axe from Wanjiaba (After YPAT 1983: figure 25:7; reproduced with permission of Archaeology Press).

uncovered from the Aofengshan site (*ca.* 2500–2000 B.P.) (Yang, F., et al. 2010:199), located in Jianchuan County (YPICRA 1990) (Figure 2.26). These similarities seem to demonstrate the local production and development of bronze in Yunnan and to indicate cultural connections within the province (see locations of these sites in Figure 2.28). However, Cui and Wu (2008:60) suggest the characteristics of bronze metallurgy and foundry technologies at Haimenkou are similar to the technologies discovered in northwestern China (Qinghai and Gansu Provinces) in the early Bronze Age. Also, small tools and ornaments made of bronze which are similar in style and technology to those from Yunnan have been dated to sites in northwestern China much earlier than sites in Yunnan (IACASS 2003b:535-565; Sun and Han 1997) (Figure 2.27). Consequently, it is possible that bronze metallurgy was introduced to Yunnan from northwestern China and then developed into a distinct local style.

The bronze *yue* axes revealed at Haimenkou also show similarities with stone *yue* axes from the Manghuai site (ATYPM 1977) (see Figure 2.26) in Yun County, Yunnan. This site is located in the middle Lancang River Valley (see Figure 2.23:11). Some scholars (e.g., Kan 1998:21-22; Li, K. and Xiao 1980:139; Xu and Xiao 2004:397) suggest that the bronze *yue* axes found at Haimenkou evolved from the stone axes found at Manghuai. The axes from Manghuai, in the middle Lancang River Valley, have a distinct axe shape (see Figure 2.26:1). If this is so, communication trade or exchange may have occurred between Haimenkou and Manghuai in prehistoric Yunnan using the routes along the Lancang River.

Tables 2.7 and 2.8 show a clear contrast between the number of stone and bronze objects found at the site. There are far more stone implements than bronze objects, indicating that the use and manufacturing techniques of

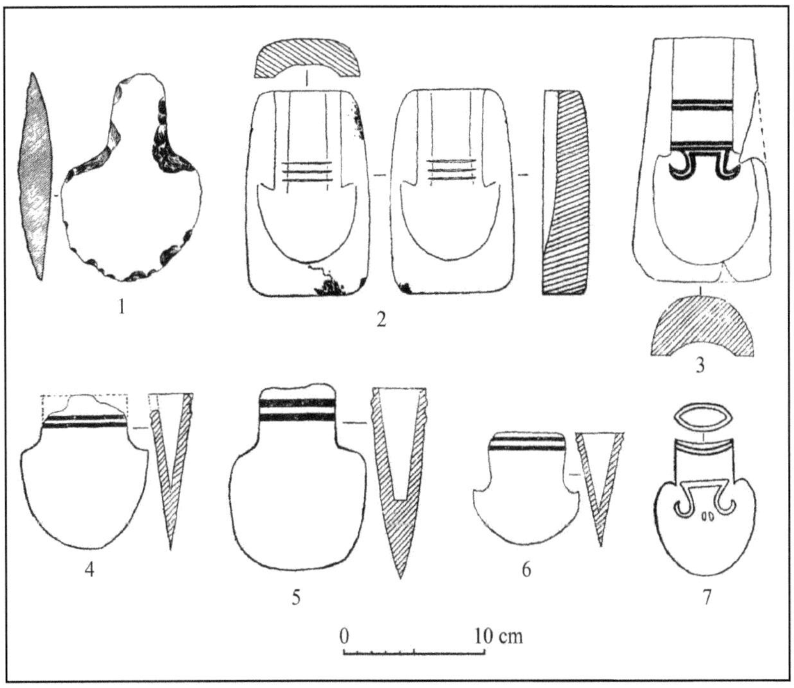

Figure 2.26 Comparisons between Stone and Bronze *Yue* Axes, and Stone-Moulds Revealed in Yunnan

1: Stone *Yue* Axe from Manghuai (After ATYPM 1977: figure 2:4; reproduced with permission of Archaeology Press);
2: Stone-Mould for *Yue* Axe from Aofengshan (After YPICRA 1990: figure 23:15; reproduced with permission of Archaeology Press);
3: Stone-Mould for *Yue* Axe from Haimenkou (After YPM 1995: figure 2:22; reproduced with permission of Archaeology Press);
4–6: Bronze *Yue* Axe from Haimenkou (After YPM 1995: figure 2:2,4,5; reproduced with permission of Archaeology Press);
7: Bronze *Yue* Axe from Dabona (After YPAT 1964: figure 7:1; reproduced with permission of Archaeology Press).

bronze tools were still at an early stage of development. Meanwhile, compared with the developed bronze production technologies in the contemporaneous Central Plain (Shang and Zhou dynasties, *ca.* 3600–2200 B.P.) (e.g., Chang 1983; Jiang, X. 2010), the bronze metallurgy and foundry technologies at Haimenkou apparently belonged to an earlier stage of development (YPICRA, et al. 2009a:599). Major sites belonging to the Bronze Age in Yunnan are shown in Figure 2.28. The area near the upper reaches of the Nu, Lancang, and Jinsha Rivers (see the dashed circle in Figure 2.28) also shows distinct signs of a local culture evolved from and sharing similar cultural connections to these Neolithic cultures (Wang, D. 1994; Xu and Xiao 2004:395-433).

Bone and Tooth Artefacts

Some artefacts made of bone and teeth were recovered from Haimenkou in the 2008 season. Although the exact numbers of some types are unavailable, an approximation is shown in Table 2.9. Awls dominate the bone artefacts, followed by hairpins and arrowheads. In contrast to stone and bronze objects, bone and tooth artefacts are more fragile and much less commonly preserved. Therefore they

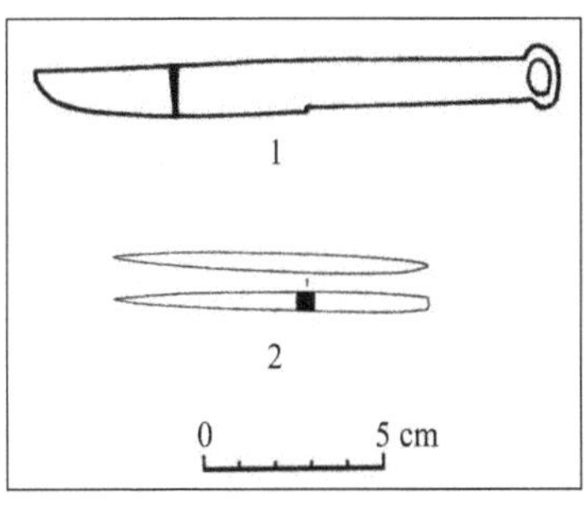

Figure 2.27 Small Bronze Tools from Sites in Northwestern China

1: Bronze Knife from Dahuazhongzhuang (Kayue Culture) (After HCM, et al. 1985: figure 16:10; reproduced with permission of *Archaeology and Cultural Relics*);
2: Bronze Awl from Qinweijia (Qijia Culture) (After GATIACAS 1975: figure 12:1; reproduced with permission of Archaeology Press).

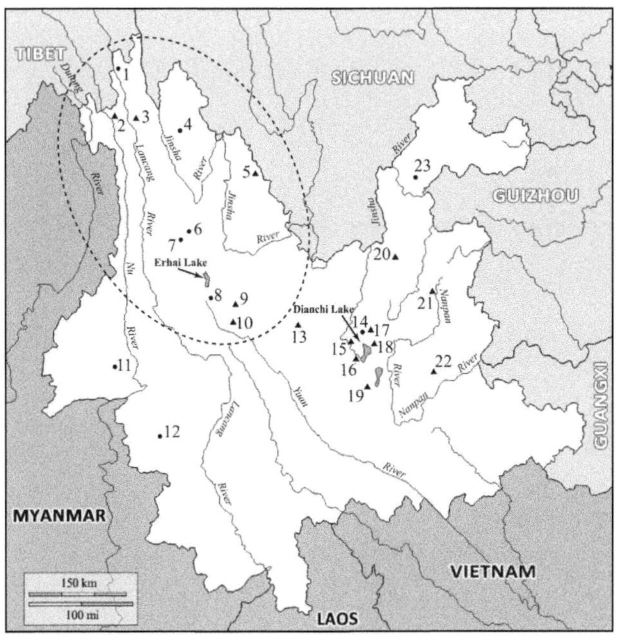

Figure 2.28 Map of Major Bronze Age Sites in Yunnan (The original map is from URL: http://d-maps.com/carte.php?num_car=19482&lang=en)

1: Nagu; 2: Yongzhi; 3: Shidi; 4: Zhongdian; 5: Daxingzhen; 6: Haimenkou; 7: Aofengshan; 8: Yinsuodao; 9: Dabona; 10: Juli; 11: Dahuashi; 12: Shifodong; 13: Wanjiaba; 14: Wangjiadun; 15: Taijishan; 16: Shizhaishan; 17: Yangfutou; 18: Tianzimiao; 19: Lijiashan; 20: Puchehe; 21: Batatai; 22: Shidongcun & Dayipu; 23: Yeshishan.

Note: the ● refers to early Bronze Age site, and the ▲ refers to middle and late Bronze Age site. The periodisation is based on K. Li and Min (2011:105).

are less commonly revealed at archaeological sites. Even so, there are still more bone and tooth tools than bronze artefacts. This may suggest that bone and tooth artefacts were very common tools used by the Haimenkou people in prehistory. Moreover, the bone bracelets recovered in Phase III (*ca.* 3100–2400 B.P.) (Figure 2.19:1) are similar to bone bracelets from the Dadunzi site (*ca.* 3300 B.P.) (LIACAS 1974:337) in Yuanmou County (Wang, N. 1992:18), which is also near the upper reaches of the Nu, Lancang, and Jinsha Rivers (see locations of these two sites in Figure 2.23). This similarity may indicate some communication between Haimenkou and Dadunzi in prehistory.

In addition, three oracle bones made of animal scapula were recovered in the 1978 season (YPM 1995:783). Oracle bones made of animal scapulae (sheep, goat, pig, deer, and cattle) have been widely recovered in sites of the Qijia culture (*ca.* 4100–3700 B.P.) in northwestern China (IACASS 2003b:553-555). They are still used among present ethnic minority groups in Yunnan and Sichuan Provinces, including the *Yi* and *Naxi* groups in Yunnan and the *Qiang* group in Sichuan (Lin 1963). These three ethnic groups, as well as the *Bai* group who live in Dali County today, are thought to have originated from the ancient ethnic *Qiang* groups (Kan 1998:129; Li, K. 1985:41) who are noted in ancient historical records. It is difficult to demonstrate archaeologically the ethnic identity of the people who inhabited Haimenkou. Nevertheless, the cultural influence from the upper Yellow River Valley is indicated by bone artefacts, as well as other evidence revealed at Haimenkou.

Wooden Tools

The exact number of wooden tools was not provided in the reports for the third excavation, although they were described as relatively abundant at the site (YPICRA, et al. 2009b). The abundance of these tools may be partly due to the excellent preservation condition at this waterlogged site. The *si* spade, a spade-shaped farm tool used in ancient China, is considered of particular interest by the excavators (YPICRA, et al. 2009a:599). The *si* spade is recognised as a farm tool from North China and was first unearthed in Level (7) (*ca.* 3700 B.P.) accompanied by some rakes (YPICRA, et al. 2009a:599). Its presence at the site, as well as the appearance of foxtail millet and wheat, suggests that agricultural elements from some North China cultures may have been transmitted to this area during the early period of the site (no later than Phase II, *ca.* 3700 B.P.) (YPICRA, et al. 2009a:599).

2.3.6 Faunal and Botanical Remains

2.3.6.1 Faunal Remains

Compared with the first and second excavation seasons, the faunal assemblage revealed during the 2008 season is similar to the faunal taxa recovered in these earlier excavations in terms of the range of taxa represented. Eight taxa were identified from the faunal assemblages in the previous excavations: sika deer (*Cervus nippon)*, sambar (*Rusa unicolor*), hog deer (*Axis* sp.), Reeve's muntjac (*Muntiacus* cf. *reevesi*), muntjac (*Muntiacus* sp.), water buffalo (*Bubalus bubalis*), dog (*Canis familiaris*), domestic pig (*Sus domesticus*), and wild boar (*Sus scrofa*) (YPM 1995:783). Bird remains were first noted in the reports for the 2008 excavation (YPICRA, et al. 2009a; b). In general, pig bones made up the largest portion of faunal remains from the Haimenkou site, followed by specimens of deer, large bovid, dog, and bird taxa. This is discussed in detail in chapters four, five, and six.

2.3.6.2 Botanical Remains

According to the second excavation report (YPM 1995), botanical taxa at Haimenkou included rice (*Oryza sativa*) and barnyard grass seeds (*Echinochloa* sp.). Wheat (*Triticum aestivum*) and foxtail millet (*Setaria italica*) were found in the 2008 season. More details were provided in the reports for the third season (YPICRA, et al. 2009a; b) as follows.

In Phase I, carbonised rice dominated. Rice was first present in Level (9) in a small quantity and increased in Level (8). A small amount of foxtail millet was first revealed from Level (8). However, there was more rice than millet present.

In Phase II, rice and foxtail millet made up the major part of the botanical remains. In addition to rice and foxtail millet, wheat was found for the first time from Level (7) at Haimenkou. The quantity of wheat was quite small, whereas rice and foxtail millet seeds increased. Some peach-stones and wild chestnuts were also revealed from this phase.

In Phase III, botanical taxa included rice, foxtail millet, wheat, barnyard grass, and other taxa, with rice and foxtail millet predominating and a relatively small quantity of wheat also present.

Table 2.9 List of Bone and Tooth Artefacts from Haimenkou

Bone Artefact	Number	Percentage
Awl	63	67.7
Hairpin	24	25.8
Arrowhead	5	5.4
Spade	1	1.1
Total	93	100.0
Tooth Artefact	**Number**	**Percentage**
Tooth Ornament	14	100.0

Note: As noted in Table 2.4, because some exact numbers of bone artefacts were not provided in original reports, Table 2.8 may just display an approximation of the frequencies of bone artefacts types.

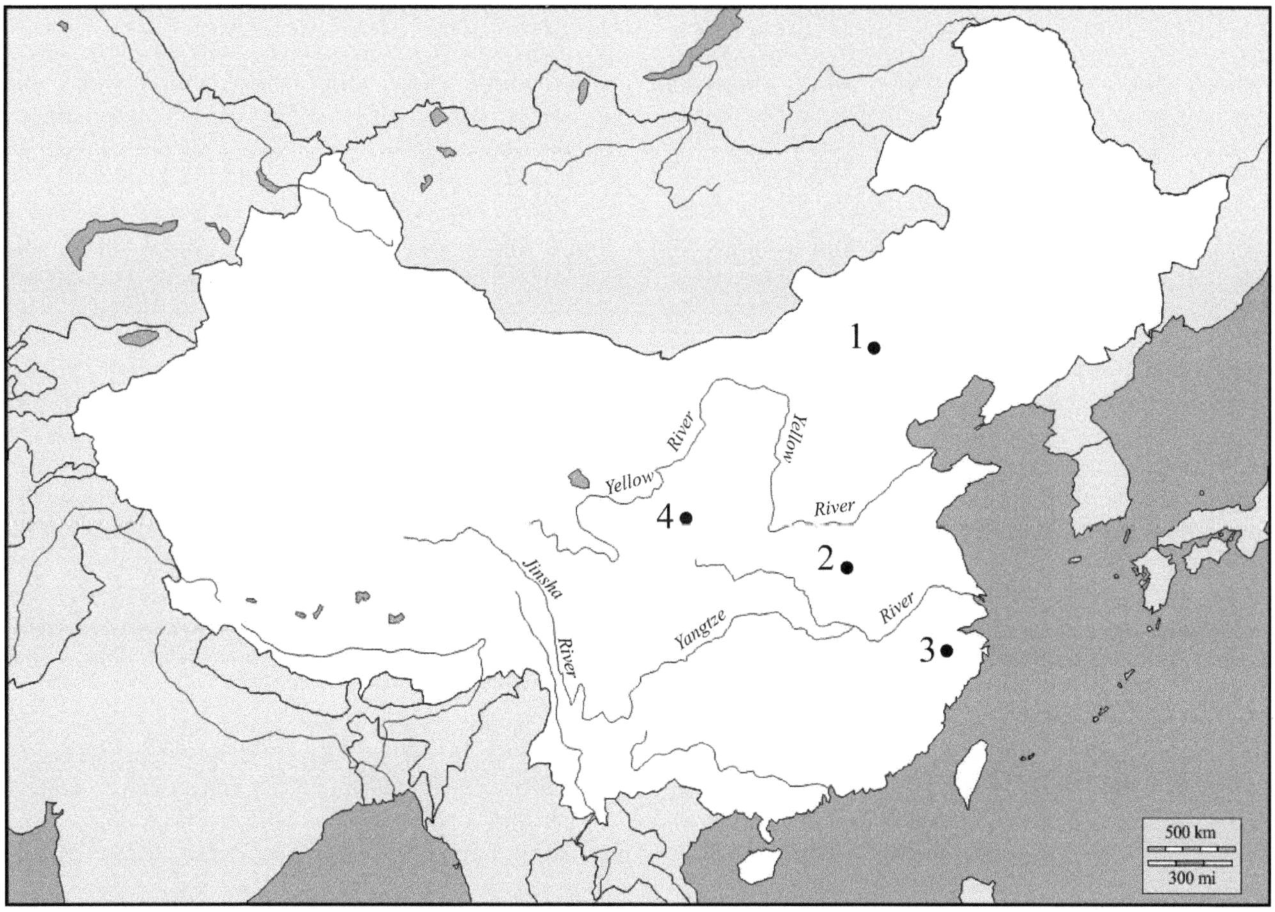

Figure 2.29 Locations of the Earliest Agricultural Sites in China
(The original map is from URL: http://d-maps.com/carte.php?lib=china_map&num_car=4644&lang=en)
1: Xinglonggou (Foxtail Millet and Broomcorn Millet);
2: Jiahu (Rice); 3: Shangshan (Rice);
4: Xishanping (Wheat)

According to Yan (2005), China has two original centres of agriculture. One was in the middle and lower Yangtze River Valley, with rice as the staple crop. The other was in North China, characterised by foxtail millet and broomcorn millet cultivation. More recently, Zhao (2011) has suggested that a third centre of agricultural origins may have occurred in the sub-tropical zone of far southern China and been characterised by taro and tuber consumption. Details on this latter are still sketchy so no further discussion of this possible third plant domestication centre will be provided. The earliest rice agriculture may have occurred at sites in the lower Yangtze River Valley around 11,000–9,000 years ago, such as at the Shangshan site in Zhejiang Province (d'Alpoim Guedes 2011; Jiang, L. and Liu 2006; Liu, L., et al. 2007). The Jiahu site (*ca.* 9000–7800 B.P.) (HPICRA 1999:515-519) in Henan Province, central China, also shows the characteristics of early rice agriculture (HPICRA 1999:462-463; Zhao 2011:S296). The Xinglonggou site (*ca.* 8000–7500 B.P.) (Liu, G. 2004; Liu, G., et al. 2004), which is located in the Liaoning Province, northeastern China, is one of the earliest sites with millet (both foxtail millet and broomcorn millet) cultivation in China (Zhao 2005:525; 2011:S301). Wheat is considered to have been introduced from West Asia around 4,500 years ago (Zhao 2005:527). To date, Xishanping in Gansu Province, dating to around 4,600 years ago, is the earliest archaeological site in China where wheat has been found (Jin 2007:17; Li, X., et al. 2007:1374).

Rice was cultivated at Haimenkou prior to the cultivation of millet and wheat, and it was evidently more abundant than millet and wheat. It is hard to indentify the origin of rice at Haimenkou due to the relatively late date of this site. Nevertheless, it is very possible that rice was the original crop at Haimenkou, and was the staple crop throughout the entire occupational history of the site (YPICRA, et al. 2009a; b). Foxtail millet was revealed at the Karuo site (*ca.* 5000–4000 B.P.) in Tibet (CPAMTAR and DHSU 1985:168). As mentioned previously, the Karuo site, located on the upper Lancang River (Figure 2.24), may have played an important role as an intermediate station for cultural communication and/or population movement between northwestern China and northeastern Yunnan. Foxtail millet was also recovered at a number of sites in Sichuan Province (d'Alpoim Guedes 2011), including Haxiu (*ca.* 5300–4700 B.P.) in Maerkang County (CRMATQAP, et al. 2007), and Yingpanshan (*ca.* 5300–4600 B.P.) in Mao County (Zhao and Chen 2011). Both rice and foxtail millet were present at the Baodun site (*ca.* 4700–4100 B.P.) in Xinjin County, Sichuan Province (d'Alpoim Guedes 2011; SJAE 1998). Therefore, the foxtail millet that was first found in Level (8) at Haimenkou may have been a crop introduced from the upper Yellow River Valley region through Tibet or Sichuan. It is possible that wheat may have not been introduced to Haimenkou until the beginning of Phase II (i.e., Level (7), *ca.* 3700 B.P.). Wheat may have been introduced by the same routes as that suggested for foxtail millet. In addition, the remains of rice, millet, and wheat were continuously present in the upper layers after their first recorded finding at the site. This suggests that the three crops were cultivated together throughout the Phases II and III at Haimenkou. This may also indicate the earliest practice of mixed-sowing and multi-cropping in Yunnan. Multi-cropping has been practised by ethnic minority groups in Yunnan since ancient times (Li, G. and Lu 1987:63-64).

2.3.6.3 Discussion and Summary

Botanical and faunal remains, combined with the abundance of arrowheads and net sinkers, indicate that the subsistence strategy at Haimenkou was a mixture of multi-cropping, gathering, hunting, and fishing. The influence of northwestern China is reflected by the presence of foxtail millet and wheat. Pig domestication may have occurred but this needs to be confirmed by the zooarchaeological analysis in this study (see chapters four and six).

2.4 Summary

Based on the data outlined above, the outcomes of the three excavations at the Haimenkou site may be preliminarily summarised as follows (YPICRA, et al. 2009a:599).

(1) The Haimenkou site is the most extensive water-logged settlement characterised by pile-dwelling constructions so far discovered in China. It provides a valuable case study of Chinese prehistoric settlement patterns.

(2) The Haimenkou site has distinctive cultural deposits, abundant material remains, and covers a long period of site-formation from the late Neolithic Period to the middle Bronze Age. These discoveries allowed the construction of a cultural sequence for prehistoric northwestern Yunnan and the revision of the cultural pedigree of southwestern China.

(3) There are a number of similarities between cultural remains found at Haimenkou and those at other sites in Yunnan during the late Neolithic Period and the Bronze Age. There are also some similarities between cultural remains found in Yunnan and some regions in northwestern and southwestern China. This may indicate cultural continuities and interactions during the late Neolithic Period and the Bronze Age in northwestern Yunnan and among Yunnan and other regions of China, especially Qinghai and Gansu Provinces in northwestern China. The cultural relics found at Haimenkou may also aid in investigating cultural communication and human migration on the prehistoric eastern Qinghai-Tibet Plateau and northwestern Yunnan-Guizhou Plateau.

(4) The botanical remains, including rice, foxtail millet, and wheat, recovered from the Haimenkou site may demonstrate that millet and wheat cultivation in the Yellow River Valley had extended to northwestern Yunnan sometime between the late Neolithic Period and the early Bronze Age. Additionally, the joint existence of rice, millet and wheat may be an indication of the earliest practice of a multi-cropping system.

(5) Bronze objects and stone moulds for bronze-casting were recovered from a distinct stratigraphic context. The morphological and technical similarities of bronzes between Haimenkou and early archaeological cultures in northwestern China suggest that bronze metallurgy at Haimenkou may have been introduced from the upper Yellow River Valley. However, a distinct local bronze culture may have then developed. The date of the bronzes found at Haimenkou also suggests that this site was the earliest Bronze Age site on the Yunnan-Guizhou Plateau. Northwestern Yunnan was an important region for the development of bronze culture and metallurgy in China.

(6) The abundant faunal and human remains together with other artefacts recovered from the site are likely to make a considerable contribution to studies of archaeology and anthropology. In view of the long period of human occupation at Haimenkou, the faunal remains revealed here may offer more information about the history of animal exploitation and domestication in China.

In the next chapter, the material for this zooarchaeological analysis, the methods of identification and recording, and the limitations of the analysis will be discussed.

3

Procedure of Faunal Analysis

3.1 Introduction

Before discussing the methods used in the Haimenkou faunal analysis, it is necessary to describe the material for analysis. Methods of identification and recording will be outlined and some limitations to the analyses discussed.

3.2 Material for Analysis

The material that forms the basis for this study was recovered during the third season of excavation at Haimenkou. This field season was held from January to May 2008. The excavation team was led by Professor Rui Min of the YPICRA. Since the author did not participate in the excavation and the identification and analysis did not start until 2010, onsite collection methods of faunal material are relatively unclear. According to the excavation reports which have been published (YPICRA, et al. 2009a; b) and personal correspondence with Professor Rui Min (March 30, 2011), flotation was employed during the third season. However, this approach was mainly used for collecting small botanical remains, while the faunal assemblage revealed from Haimenkou was collected by hand rather than by dry screening or water sieving. Numerous scholars have criticised the hand-collection method of recovering samples because of the resultant serious recovery biases (e.g., Casteel 1972; 1976; Clason and Prummel 1977; Payne 1972b; 1975). This issue will be addressed in more detail later. It appears that the majority of the faunal assemblage recovered from Haimenkou was composed of bones from large- and medium-sized animals. Small-sized animals, such as rodents, were infrequently represented in the faunal assemblage. This indicates that recovery bias has affected the faunal assemblage composition.

The faunal assemblage was not preserved very well. Damage resulting from the dry weather in Yunnan during the two years after the excavated area was uncovered was visible, including cracking and flaking. This negatively affected the identification of the faunal assemblage.

The identification of the bird bones from Haimenkou was carried out by Mr. Shuo Wang of the Key Laboratory of Evolutionary Systematics of Vertebrates, Institute of Vertebrate Palaeontology and Palaeoanthropology, Chinese Academy of Sciences (IVPPCAS), China. His work forms the basis for Section 4.3.3 (Birds) in chapter four of this book. Both Mr. Shuo Wang and the author carried out the measurement of the bird bones.

3.3 Identification and Recording

Each specimen of the faunal assemblage recovered from Haimenkou was identified to taxonomic level in as much detail as possible. However, it was not always possible to accurately identify all specimens recovered, thus faunal remains were identified to varying levels of family, genus, or species. Attributes of each specimen were recorded in detail. Recorded attributes included locus (where specimens were excavated), bone identification number, articulation condition, element, taxon, possibility of domestication (wild or domestic), certainty of identification, symmetry (left or right), sex determination, relative age, relative size, pathology, burning condition, modification (by humans or animals), origin of fragmentation, number of pieces, number of elements, weight, fragment size (the size of portion relative to the complete bone, for post-cranial bones only), portion present (proximal, distal, etc., for post-cranial bones only), state of fusion (for post-cranial bones only), comments, measurements, and any other details thought significant (see also Appendix 1).

These attributes were recorded based on the 'BONECODE' system developed by Meadow (1978) (see Appendix 1) and manipulated with Microsoft Excel. Eruption and wear stages of teeth were recorded following Payne (1973), Deniz and Payne (1982), and Silver (1970) for sheep and goats, and Grant (1972b), Silver (1970), and Schmid (1972) for pigs and cattle. Fusion stages for most animals were recorded following Silver (1970), Schmid (1972), and Amorosi (1989). The weight of the bones was measured to the nearest 0.1 gram by an electronic balance. Measurements were made with electronic digital callipers to the nearest 0.1 mm. Measurement definitions generally followed those of von den Driesch (1976) with personal advice from Professor Richard H. Meadow and Dr. Ajita K. Patel of the Peabody Museum, Harvard University (see details in Appendix 2).

Specimens that were identifiable to element and to taxon were labelled respectively with locus code and bone number, and described on Excel with a coding system based on Meadow's (1978) BONECODE. Most of the specimens that were not identifiable to a taxonomic level were collected together and categorised into a variety of groups. The groups were labelled according to body part: skull fragments, long bone fragments, ribs, vertebrae, indeterminate fragments, and so forth. Then, they were counted and weighed. While few of the unidentifiable specimens may still be identified to a taxonomical level, they could not be determined in this study due to time limitations and observer ability. Therefore, they were labelled and packed separately from other specimens and classified, for example, by animal size, such as 'SMALL' (smaller than rabbit-sized) or SM-MD (medium dog-sized to medium sheep-sized) (see details in Appendix 1). Further research needs to be carried out in the future on these unidentifiable specimens.

A variety of animal bone identification manuals and modern comparative collections were employed in the identification process during this study. The published atlases, including Schmid (1972) and Hillson (1992; 2005), were used during identification. Additional atlases and articles were also used for specific animal species, including Boessneck (1993), Payne (1985), and Prummel and Frisch (1986) for identification of sheep or goat bones. These additional references will be noted when the specific taxon is discussed.

A modern comparative collection was made and shipped to the temporary zooarchaeology laboratory at Dali City in Yunnan from the Zooarchaeology Laboratory of the Henan Provincial Institute of Cultural Relics and Archaeology (HPICRA), Henan Province, China. This collection comprised complete and partial skeletons of both domestic and wild animal taxa, including wild boar, cattle, water buffalo, sheep, goat, sika deer, Reeve's muntjac, dog, and cat. A partial chicken skeleton and an additional partial goat skeleton were purchased from the local market. Some other specimens, especially those from wild animals such as bear, deer, monkey, and some medium and small sized carnivores, were identified according to the faunal skeleton collection of the Reference Collection Department of Kunming Institute of Zoology, Chinese Academy of Science (KIZCAS), in Yunnan Province, China. The collection of the IVPPCAS was referred to for bird identification.

3.4 Problems Arising before Analysis

Some problems arose before the analysis was undertaken and are discussed below.

3.4.1 Clarifications of Terminology

Some clarifications of terminology are made here to ensure the standard application of specific terms used in this book.

In this book, the term 'specimen' is defined as a bone or tooth, or fragment thereof, from an archaeological or palaeontological site. The term 'element' refers to a single complete bone or tooth in the skeleton of an animal (Grayson 1984:16; Lyman 1994a:289). Thus, a complete femur or tibia is both an element and a specimen, but a proximal femur or distal tibia would only be called a specimen since it is just a part or a fragment of the original complete element.

In addition, based on the definition of Reitz and Wing (2008:9), the term 'sample' means a variety of faunal specimens from one species or different taxa, which are in a certain relationship with others before being excavated. Generally the whole set of specimens revealed from a unique archaeological provenience unit, such as a pit or a posthole, may be recognised as a sample. All samples from a single time period, coming from the same stratum, and those from a single site constitute a 'collection' (Reitz and Wing 2008:9). All the collections unearthed from a single site, representing the whole time span from when the site was first occupied to the day it is excavated, comprise an 'assemblage' (Reitz and Wing 2008:9).

3.4.2 Changes in the Nature of the Assemblage

Every faunal assemblage that was recovered from a site is a data resource for zooarchaeological study. Nevertheless, before processing an assemblage, some concepts should be considered. It is easy to understand that the samples gathered from archaeological sites, the faunal assemblages, are the remains of animals. The use of the word 'remains' indicates that much information has been lost during the transitive stages from live animals to disassociated and fragmentary bone specimens. After the specimens are excavated and sampled, zooarchaeologists must complete the time-consuming process of identification, recording, analysis, and later, publication (Meadow 1980). Klein and Cruz-Uribe (1984:3) identified five stages that outline the changes animals go through to become samples and pointed out the factors affecting the transition based on Clark and Kietzke (1967) and Meadow (1980). The five stages are listed below.

(1) Life assemblage: the community of live animals in their natural state.

(2) Death assemblage: the carcasses collected by people, carnivores or any other agent of bone accumulation.

(3) Deposited assemblage: the carcasses or parts of carcasses that come to rest at a site.

(4) Fossil assemblage: the animal parts that survive at a site until excavation or collection.

(5) Sample assemblage: the portion of fossil assemblage that is excavated or collected.

Along this chain of five stages, four transitions occur. According to Ringrose (1993:123-124), the four transitions are as follows. The first transition, from life assemblage to death assemblage, may be related to human behaviours such as hunting and domestication; the behaviours of other animals, such as carnivores or predators; the impact of the natural environment, including forest fires and accidents such drowning; or a natural death, from old age or disease. The second transition, from death assemblage to deposited assemblage, concerns behaviours from both humans and other animals and from natural forces in some circumstances. This transition, if made by humans, presents a process of first and secondary animal utilisations, involving food consumption, the extraction of secondary products (e.g., milk, wool, and bone glue), and the transportation or discarding of bone remains. Scavenging by humans may take place in this period. During this change, other animals may also play a role as consumers or transportation, for example, wolves or dogs may eat bodies of dead animals and bury the leftovers for their next meal. In addition, natural forces may achieve the same results as human or animal behaviours, for example, transportation by flood waters.

The third transition, from deposited assemblage to fossil assemblage, may have been affected by the second transition, since the changes and damages that occur in the second transition will probably influence the preservation condition of specimens during their long term underground. The third transition may be considered an interval between the discarding of animal remains and their burial. During this transition, for instance, weathering and trampling may play a crucial role in damaging and destroying the majority of assemblages. Consequently, only a small portion of deposited assemblage survive this transition and reach the next stage where they may suffer other damage including decay, leaching, or crushing. The last transition, from fossil assemblage to sample assemblage, is the only one that is under the control of archaeologists and zooarchaeologists. If a site undergoes an extensive and elaborate excavation in which all deposited animal bones are recovered and revealed, the faunal assemblage could be identical to the deposited assemblage. However, this is rarely the case. Due to a variety of limitations, including time and funding, a complete recovery excavation is commonly regarded as an ideal but unachievable condition. Hence, the choices made by excavators and the approaches and methods applied by them to retrieve material are critical for more effective research. These choices determine the contents and condition of the assemblage.

The stage-to-stage transitions and factors that affect the transformation of faunal data are illustrated in Figure 3.1. From this figure, there appear to be two sets of factors affecting the entire temporal journey of faunal data from living assemblage to published data (Davis 1987:23; Meadow 1980). The set of factors on the left side which cannot be controlled by the zooarchaeologists involve events that have already happened before the faunal data underwent a transition to become a fossil assemblage. The factors displayed on the right side of Figure 3.1 may be controlled by zooarchaeologists. This includes the operations related to the procedures comprising planning, excavating, sampling, and further investigation. Of all the factors which may be under the control of zooarchaeologists, the recovery bias, which commonly occurs in excavations, may lead to serious errors in data analysis and final interpretation (e.g., Cannon 1999; Casteel 1972; 1976; Clason and Prummel 1977; James 1997; Nagaoka 1994; Payne 1972b; 1975; Shaffer 1992; Shaffer and Sanchez 1994; Stahl 1996). However, this factor has not received sufficient attention in the circumstances of archaeological excavations in China (e.g., Chen, Q. 1993; Huang, W. and Ji 1963; Huang, Y. 1996; Jia and Zhang 1989; Li, M. 1991; Li, Y. and Han 1978; Lu, H. and Zhou 1990; Qi 1977; Wei, et al. 1989; Zhang, X. 1985; Zhang, X., et al. 1992; Zhou 1981; 1994). This was unfortunately the case at Haimenkou (YPICRA, et al. 2009b; YPM 1995). More details of the recovery bias will be presented in the next section.

3.4.3 Recovery Bias

A number of possible factors may affect a faunal assemblage and result in a lack of specimens as discussed above. Of these factors, the recovery bias, which is one of the factors that can be controlled by zooarchaeologists, will be further discussed below.

As Clason and Prummel (1977:171) stated, the faunal remains unearthed from archaeological sites can be assigned into two classes, namely micro-elements and macro-elements. Micro-elements refer to the bones of small mammals (smaller than a rat), small birds (smaller than a crow), reptiles, amphibians, small fish, and the shells of small molluscs. Macro-elements refer to the bones of large mammals, large birds, large fish, and the shells of large molluscs (Clason and Prummel 1977:171-172). Although this classification seems relatively rough and arbitrary, it can be helpful when examining the efficiency and precision of screen sieving methods.

During normal zooarchaeological field work, a classification consisting of micro- and macro-remains may make more sense. Based on the definition by Clason and Prummel (1977:171-172), micro-remains comprise micro-elements and the small fragments of macro-elements, such as trivial splinters of the shafts of long bones from large bovids. The macro-remains, which are easily recoverable with the naked eye, include complete and relatively complete macro-elements and certain micro-elements, such as a complete femur of a small bird (Clason and Prummel 1977:171-172). Therefore, the class of micro- or macro-remains is actually a contrastive concept, without a clear boundary, and the status of remains assigned to these two classes may be confused. In practice, without the aid of scientific equipment and objective methods, faunal remains are usually placed in a class depending on the excavators' working experience, the archaeological contexts, or the sufficiency of time for collection. Given the limits presented above, animal remains assigned into the micro-remains class are usually overlooked or discarded more frequently during excavation. An obvious tendency to pick up macro-remains may be observed among many sites across the world. Consequently, an insufficient collection of micro-remains can be expected, especially of the tiny teeth, bones, or bone fragments of rats and small fishes.

The recovery bias described above has been explained by experiments presented by several scholars (e.g., Cannon 1999; Casteel 1972; 1976; James 1997; Payne 1972b; 1975; Shaffer 1992; Shaffer and Sanchez 1994). Payne (1972b; 1975), for instance, published several papers which drew the attention of archaeologists to this serious problem. He designed and carried out a series of experiments at different sites, and described the methods used and the results obtained. Based on these results, he (Payne 1972b; 1975) suggested that both the collection of micro-remains and macro-remains could be underestimated without the application of sieving methods. Compared

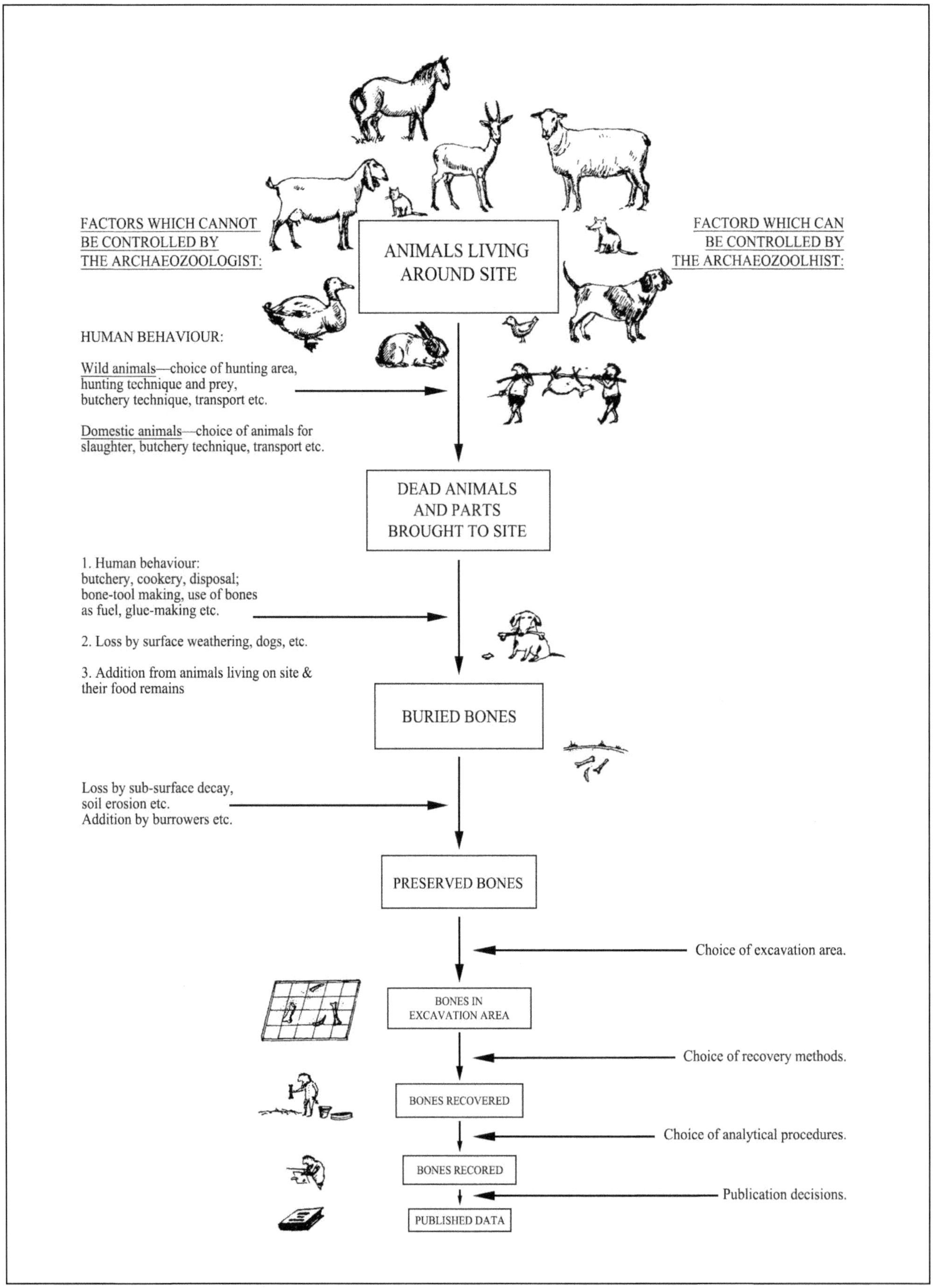

Figure 3.1 Factors Which May Affect Zooarchaeological Data
(After Davis 1987: figure 1.1; drawn by Sara Midda; reproduced with permission of Simon J. M. Davis)

with the results from hand-collection, the water-sieving method essentially changed the relative abundance among taxa. For example, the NISP (number of identified specimens) of cattle was doubled during water-sieving when compared to hand collection. In addition, the number of pig specimens missed in hand-collection but recovered in water-sieving was as much as ten times those retrieved by hand-collection, and for sheep or goat it was more than 20 times that collected by hand (Payne 1972b:59). Apart from the effect on taxonomic relative abundance, Payne (1972b:61) also emphasised that the recovery bias may affect relative skeletal part representation, which may influence the investigation of issues such as butchery patterns.

Similarly, according to Casteel's (1972:383) study, the discrepancy resulting from recovery bias could have 'amounted to a 100 percent loss of all fish remains from the units tested and later compared with materials which had been washed, screened, and separated in the laboratory using low-power magnification.' On the basis of other scholars' research (e.g., Fitch 1966; Sparks 1961; Thomas 1969), Casteel (1972:386) suggested that 'these data indicate that the efficiency of various mesh sizes and the efficacy of using sorting by eye in the field or binocular microscope in the laboratory in order to recover faunal material are dependent upon the class of animal being dealt with, the size of individuals, their species, and the relative size of the skeletal item(s) desired'.

In the faunal assemblage recovered from Haimenkou, only a few fish remains were present. This is obviously an uncommon phenomenon at a water-logged site beside a lake, as many pottery net sinkers and several bronze fishhooks from the Bronze Age were recovered (YPICRA, et al. 2009b; YPM 1995). Therefore, it is understandable to attribute the dearth of the fish remains and other micro-remains to inadequate sampling technologies or taphonomic effects. Accordingly, during the procedure of data analysis and interpretation, a recovery bias was carefully considered to exclude possible errors caused by the insufficient recovery of small remains and to avoid false or misleading results and analyses.

3.5 Summary

In this chapter, the factors determining the condition of the faunal assemblage and the process of identification and recording were described. Due to the use of a hand-collection rather than dry screening or water sieving method of faunal recovery during excavation, recovery bias may have affected the faunal assemblage to some extent in this study. This probably resulted in the absence of some samples of small-sized animals, as well as some micro-remains from large- or medium-sized animals. In the analyses in the following chapters, this should be born in mind.

In chapter four the range of identified animal taxa and their relative importance will be presented and discussed.

4

Range and Relative Importance of Identified Taxa

4.1 Introduction

Identifying the range and relative importance of taxa from archaeological sites is fundamental to any zooarchaeological investigation and analysis. Important data can be derived from this approach, such as skeletal part representation, age profiles, butchering patterns, and survival rates. The range of identified taxa provides a primary database for the site, including the diversity of animal taxa and past climates. Meanwhile, all detailed information is derived from more specific data such as the relative abundance of species, which may reflect the subsistence strategies of humans. In this chapter, the range and relative importance of identified taxa from Haimenkou will be presented and discussed.

Figure 4.1 gives locations of the archaeological sites mentioned in this chapter outside Yunnan Province in China. The late Palaeolithic site of Tangzigou (*ca.* 7000 B.P.) is the only site in Yunnan that is located on this map (the Neolithic Period in Yunnan began from *ca.* 6000 B.P.). Other Neolithic sites in Yunnan were depicted on maps in chapter two.

4.2 Methods of Quantification

Dexter Perkins noted that 'the primary objective of faunal analysis of material from an archaeological site (or from a palaeontological site) is to establish the relative frequency of each species' (Perkins, Dexter 1973:369). Numerous methods and approaches to quantification have been invented, developed, and continuously modified for the measure of relative abundance of diverse species excavated from archaeological sites. Discussions regarding these methods are still ongoing (e.g., Bökönyi 1970; Cannon 2001; Casteel 1977; 1978; Cruz-Uribe and Klein 1986; Fieller and Turner 1982; Grayson 1973; 1979; Horton 1984; Lyman 2008; Marshall, F. and Pilgram 1993; Payne 1972a; Ringrose 1993; Thomas 1969; Uerpmann 1973; Vermeij and Herbert 2004; Watson 1979; White, Theodore 1952; 1953a; b; Zeder

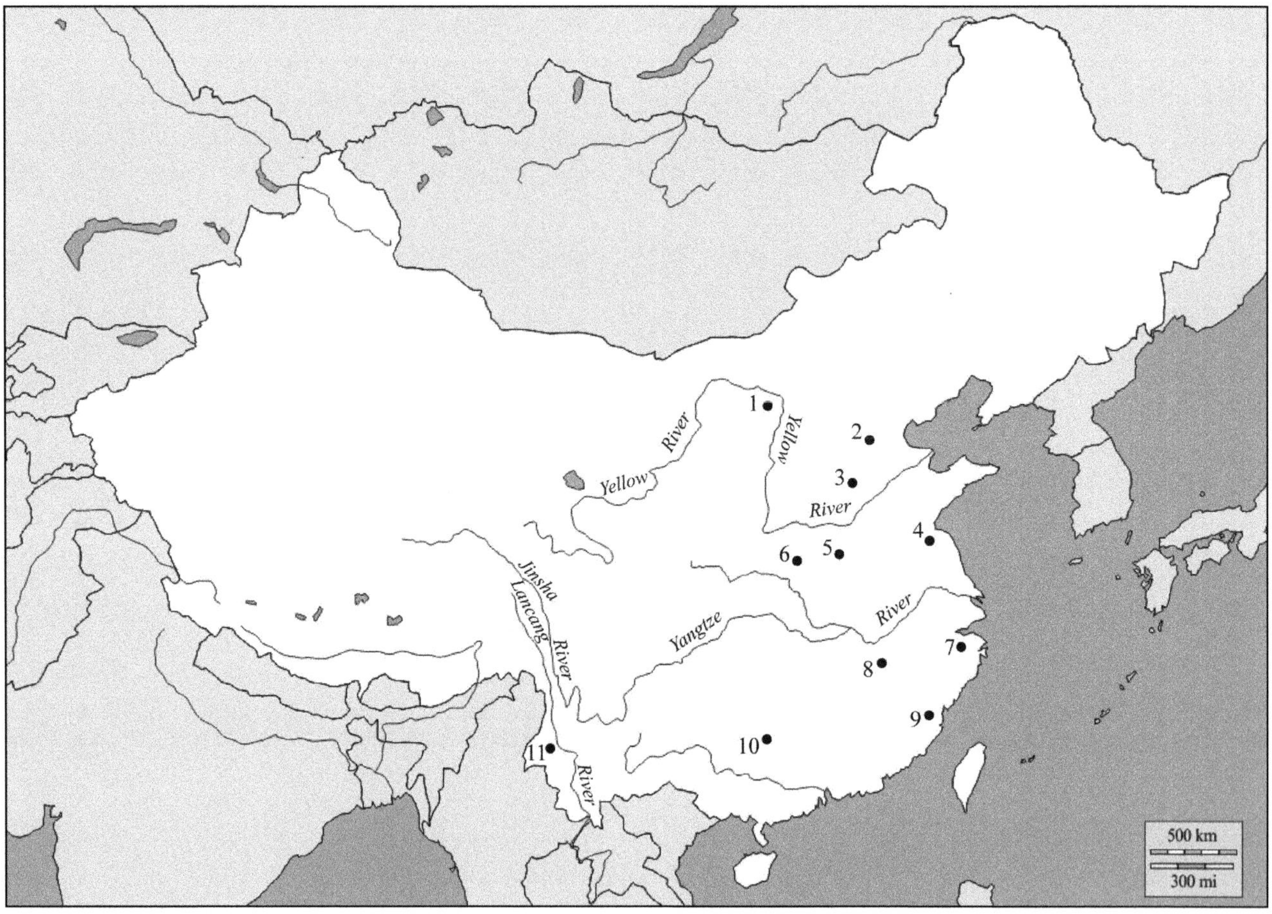

Figure 4.1 Map of Sites in China Mentioned in Chapter 4 (The original map is from URL: http://d-maps.com/carte.php?lib=china_map&num_car=4644&lang=en)

1: Zhukaigou; 2: Nanzhuangtou; 3: Cishan; 4: Wanbei; 5: Jiahu; 6: Xiawanggang; 7: Hemudu; 8: Xianrendong; 9: Tanshishan; 10: Zengpiyan; 11: Tangzigou. Note: Among the sites in Yunnan mentioned in this chapter, only the location of the late Palaeolithic site Tangzigou (No. 11) was presented on the map. See other sites in Yunnan mentioned in the text in figures in chapter two.

and Arter 2008). It should be mentioned here that the measures resulting from these methods are not absolute ones. That is, the number of the animals present at ancient sites may never be calculated accurately. The nature of the taxonomic abundance that can be obtained by archaeologists is 'relative'. In other words, 'each measure is usually meant to provide information on the abundance of each taxon in relation to the other taxa in the collection' (Grayson 1979:200). Of the methods used to estimate abundance, the most popular two are employed in this analysis. They are NISP (number of identified specimens) and MNI (minimum number of individuals). The method of weighing animal bones to determine their relative abundance, while not used so frequently, will also be applied as a supplement for comparison purposes. To better understand their usage in this book, these three methods are reviewed briefly below.

4.2.1 NISP

In addition to displaying a list of the taxa revealed from a site, the simplest way to reveal the properties of a faunal assemblage is to quantify them. The number of identified specimens (NISP) of each taxon is one of the most basic approaches to quantification. It has been widely applied in faunal analyses for several decades. Although some scholars prefer the MNI method to NISP (e.g., Chaplin 1971; Klein and Cruz-Uribe 1984), they generally also use NISP together with MNI in faunal quantification.

According to Klein and Cruz-Uribe (1984:25) and Lyman (2008:28), the NISP method has two prominent merits. Firstly, it can be calculated when the basic identification of samples is done, without further numerical manipulation. Secondly, it is cumulative. That means every time there are additional specimens, the new NISP can be calculated without complications. Thus, a NISP measure may be added to data from years or decades ago without major difficulty. It represents the maximum number of bones counted within an archaeological assemblage.

However, the NISP method still contains some serious limitations. Over several decades, a number of scholars have criticised its flaws (e.g., Bökönyi 1970; Chaplin 1971; Grayson 1973; 1979; 1984; Payne 1972b; Ringrose 1993; Shotwell 1955; Uerpmann 1973). Different scholars have used different ways to calculate NISP. Some scholars choose to count every identifiable fragment of bone (e.g., Chaplin 1971:65), whereas others collect all the fragments originating from the same element and count them as one (e.g., Reitz and Wing 2008:168). Therefore, the sample size from a single site may be exaggerated or decreased by different investigators. In addition to the various methods of calculation, NISP has inherent disadvantages. For example, NISP varies among taxa because bones and teeth show different frequencies in different taxa (Grayson 1984:21; Klein and Cruz-Uribe 1984:25; Lyman 2008:29). For example, a pig has 16 metapodials while a sheep only has four. These differences may exaggerate the abundance of some taxa while reducing that of others. Also, NISP cannot deal with articulated elements (Grayson 1984:22; Lyman 2008:30). For instance, zooarchaeologists must decide whether each tooth in a mandible is counted as an individual specimen, or whether the mandible and all the attached teeth are counted as one. Faunal remains are also subject to fragmentation effects and this can make for significant variation in NISP. Unless zooarchaeologists specify in what way they have counted, the NISP can be skewed for further studies.

Due to these and other disadvantages of the NISP method, it is not suitable for application for the sole measurement of taxonomically relative abundances. For this reason, the use of NISP along with MNI is necessary, because in some aspects the weaknesses in NISP may convert into advantages when the MNI method is utilised.

4.2.2 MNI

It is widely accepted that the initial use of MNI (minimum number of individuals) in archaeology was by Theodore White (1952; 1953a). However, it was used by palaeontologists for over 20 years (e.g., Howard 1930; Stock 1929) prior to its introduction into the zooarchaeological field. Over the years, over 16 definitions have been given for the MNI method (Lyman 2008:40). For example, Shotwell (1958:272) defines MNI as the 'number of individuals which are necessary to account for all of the skeletal elements (specimens) of a particular species found in a site'. On the basis of this definition, Ringrose (1993:126) makes a similar but more precise definition as 'the smallest number of individual animals needed to account for the specimens of a taxon found in a location'.

Since it was first used for the analysis of archaeology sites to estimate the number of individuals represented by the animal bones, the MNI method has developed a range of variations in terms of the differentiated procedures used to arrive at the final MNI measures (e.g., Bökönyi 1970; Casteel 1977; Chaplin 1971; Cruz-Uribe and Klein 1986; Fieller and Turner 1982; Grayson 1979; White, Theodore 1953a; Zeder and Arter 2008). Of these variations, the simplest one, which was also the earliest one employed and described by Theodore White (1953a), could be illustrated as follows: 'The method ... is to separate the most abundant element of the species found (usually the distal end of the tibia) into right and left components and use the greater number as the unit of calculation'. It is notable that the term 'element' used by Theodore White, is in fact representative of both 'element' and 'fragment' of an element. For example, if there are four left distal tibiae, three right proximal tibiae, and five right distal femurs, all belonging to cattle, revealed from the same archaeological unit such as a pit, then the MNI value of cattle in this pit is five.

According to Theodore White (1953a:397), the MNI method 'may introduce a slight error on the conservative side because, without the expenditure of a great deal of time with small return, we cannot be sure all of the lefts match the rights'. Theodore White (1953a:397) did not agree to 'use the total components of the most abundant element and divide by two' since this manipulation could lead to 'greater error because of the possible differential distribution of the kill' unless the site is completely excavated, which is not usually the case. In contrast to Theodore White, Bökönyi (1970), Chaplin (1971), and other scholars (e.g., Zeder and Arter 2008) preferred to spend 'a great deal of time' pairing specimens found at a site or a unit at the site. Bökönyi (1970), for example, suggested a variation on the MNI method concerning different age groups of specimens, dividing them based on epiphyseal fusion stages and bone sizes. Similarly to Bökönyi, Chaplin (1971:70-75) also recommended considering evidence for age, sex, and size when matching each specimen against the others, to determine whether two specimens (or more) could be from the same individual animal. However, according to Klein and Cruz-Uribe (1984:26-27) this matching process has some inherent problems. An example is the criterion 'size', since differences of bone size among adults are relatively subtle in most species. Given this, 'matching' by size therefore requires subjective or even arbitrary decisions on whether a particular left specimen and a particular right specimen come from the same individual. Aside from paying attention to pairing specimens, Klein and Cruz-Uribe (Cruz-Uribe and Klein 1986; Klein and Cruz-Uribe 1984) have taken another approach to the estimate of MNI. Using a computer program, they recorded bone fragments as fractions of complete elements, and then added the fractions to arrive at the final estimate. This method of computation is convenient when facing an assemblage of a large size or with great fragmentation, where pairing seems impractical.

It seems that MNI is a strong analytical tool where the NISP is weaker. As pointed out by Ringrose (1993:127), 'the basic principle of the MNI is to avoid "counting the same animal twice".' Accordingly, MNI may shed light on the solution of NISP disadvantages in those ways detailed above. Nevertheless, it is not unassailable and has its own disadvantages, and two of these flaws may call for more attention. Firstly, MNI values may exaggerate the importance of rarely represented taxa or taxa represented by low NISP values (Lyman 2008:46). For example, at a site the NISP of species A is one, while the NISP of species B is 10. Nevertheless, both MNIs of them are one, and thus the importance of species A is exaggerated by MNI. Secondly, different aggregates of specimens comprising a total collection will produce different MNI values (Lyman 2008:46). Agreeing with Grayson (1984), Lyman (2008:57-66) noted that the most serious problem with MNI is this one. That is, MNI is not cumulative, unlike NISP, thus it needs recalculating every time a new bag of specimens arrives. This causes problems when specimens are processed at different times as this will result in different outcomes. For instance, if MNIa is the sum of two MNIs which are calculated from Levels (4) and (5) at Haimenkou respectively and MNIb is found by grouping the specimens from Levels (4) and (5) and calculating them as a whole, MNIa will be greater than or equal to MNIb and in most cases MNIa is greater than MNIb. Currently there seems no really effective remedy for this problem. However, despite the existence of various problems that emerge during the process of estimating MNI, this method is still an advantageous way to estimate the relative abundances particularly when paired with NISP.

4.2.3 Weight Method

The weight method aims to assess the amount of meat each taxon contributed to ancient people's diets (Reitz and Wing 2008:239-237), thus it is an approximate measure of subsistence. According to Chaplin (1971:67-68), in this method 'the weight of bone from each species is multiplied by a factor to give the amount of meat represented per species'. The fundamental assumption underlying is that 'there is a constant relationship (within an acceptable range of error) between the total dressed carcass weight and the weight of the component bones' (Chaplin 1971:67-68). Therefore, this method is also termed the 'meat weight' method by scholars such as Grayson (1979).

It is reasonable to advocate this method in respect to taxonomic contribution to the diet. If, for example, one femur from a cow and another one from a rabbit were both excavated at a site, the NISP and MNI of these two taxa will be the same (i.e., one). However, it is not difficult to infer that the meat amount provided by a cow is dramatically greater than that provided by a rabbit. From this perspective, the estimate of various amounts of meat from different taxa is of considerable value.

The weight method, however, is problematic because its basic assumption may be questionable and unsubstantiated. For example, Casteel (1978) examined this method by utilising data for domestic pigs. He illustrated the relation between bone weights and meat weights of individuals, suggesting that it was over simplified and could not be graphed or described as linear. This result shows that the predictions made by the weight method appear to be in error by between 28 and 2,224 percent (Casteel 1978:71). For this reason the weight method of identifying taxonomic importance maybe unreliable (Casteel 1978:71; see also Chaplin 1971:69).

Nevertheless, the weight method will still be employed in this book as supplementary data for analysis and comparison.

4.2.4 Application of the Methods

In this book, the three methods for quantification overlap in certain ways. For NISP, the fragments of one element that were damaged prior to the excavation, in other words, during pre-depositional or post-depositional periods but before modern times, were collected together but counted as separate specimens. The fragments of one element that were damaged after the excavation, resulting from the excavation or transportation, were collected together, glued, and counted as one specimen. This process is time consuming, but it reveals the actual fragmentation and condition of bones as well as the actual number of specimens. Although some specimens from Haimenkou could not be identified even to family level, they were still recorded and counted. They will be discussed in this chapter to assist in the assemblage analysis. For example, some specimens could only be identified as 'small carnivore', which means they may have belonged to any one of a number of species including fox, dog, and cat. Other specimens that were not very diagnostic were unable to be identified to even a level as basic as a small carnivore. These relatively undiagnostic fragments (mostly shaft fragments), such as ribs and long bone fragments, were treated as questionable specimens, and were collected and labelled together and not included in the NISP count. Some trivial fragments, such as dozens of very small skull fragments where damage was probably caused by modern acts including excavation, although they were possibly able to be identified to certain taxonomic levels, were still counted as unidentifiable fragments.

MNIs were calculated in the following way. As introduced in chapter two, the strata of Haimenkou were divided into four phases, including Phase I, Phase II, Phase III, and the Late Phase. These phases all represent different chronological units. These are the analytical units used in this book. As explained above, one of the most serious problems of the MNI method is the aggregation effect. This effect is the difficulty caused by compounding or summing up the results of bone numbers from different layers, features, or units. Therefore, MNI was calculated for each analytical unit. For long bones, usually only extremities (proximal or distal ends) were used in MNI calculations, but the shafts were employed in the estimation of MNI when epiphyses were absent from an analytical unit.

In addition, since the specimens found in each phase were not excessively abundant, the MNI method was carried out by matching separate fragments. Loose teeth were matched with maxillae or mandibles without teeth, and epiphyses were matched with relevant diaphyses (shafts). Consequently, the loose teeth that were matched with the relevant maxilla or mandible or the epiphyses (or epiphysis) that could perfectly match with the corresponding diaphysis were treated as one element, resulting in an MNI value of one. The proximal epiphysis and the proximal diaphysis that could not match with each other were treated as two elements with the MNI value of two. However, if moderate or heavy weathering was present, even when the epiphysis did not match the diaphysis precisely, they were still assigned to one element in order to rule out the possibility that they came from the same individual.

Compared with the detailed methods of NISP and MNI applied in the book, the weight method of calculating proportions of animal bones present is relatively simple to carry out. Every identifiable fragment was weighed and recorded respectively, while other fragments, such as indeterminate fragments, long bone fragments, and ribs were collected based on provenience units (grids) and weighed as separate groups.

4.3 Results and Analysis

A total of around 2,800 specimens (bone fragments) weighing more than 63.5 kilograms (63,651.2 grams) were recovered during the third excavation season (2008) at the Haimenkou site. These were analysed in this study. Of the bone fragments, a total of 1,791 specimens, approximately 54 kilograms (54095.1 grams), were identified to taxonomic levels (species, genus, or family level, or a less diagnostic category such as 'small carnivore').

Table 4.1 displays the number and weight of both indentified and unidentified specimens (fragments). Table 4.2 shows the details of the number and weight of unidentified specimens. Figure 4.2 is derived from Table 4.1 and illustrates the proportional sample sizes of animal remains in each phase based on NISP and weight. This shows the distinct concentrations of animal abundances during Phases II and III, the phases representing the early and middle Bronze Ages.

Table 4.3 provides a summary of the range of identified taxa based on NISP, MNI, and bone weight, while Tables 4.4a–c provide more details than Table 4.3 by phase. The results of relative proportions provided in Tables 4.4a–c are visually displayed in Figures 4.3a–c. From these tables, the variations of taxonomic relative proportions can be compared in chronological order. Of all of the four phases, Phases I–III are the focus of the analysis and discussion, while the Late Phase is provided as supplementary data for analysis because it can only represent late periods near modern times (i.e., Song, Yuan, and Ming dynasties, A.D. 960–1644).

Table 4.1 Brief Distributions of Sample Sizes at Haimenkou by Phases

Phase	NISP	% NISP	WGT	% WGT	UID-NF	UID-WGT	T-NF	T-WGT
Phase I	103	5.8	2690.5	5.0	—	—	—	—
Phase II	766	42.8	25883.6	47.8	—	—	—	—
Phase III	905	50.5	24995.8	46.2	—	—	—	—
Late Phase	17	0.9	525.2	1.0	—	—	—	—
Total	1791	100.0	54095.1	100.0	1009	9556.1	2800	63651.2

Key: NISP = number of identified specimens; WGT = weight; UID-NF = number of unidentified fragments; UID-WGT = weight of unidentified fragments; T-NF = total number of fragments; T-WGT = total weight; — = no data available in this category.
Note: The number and weight of unidentified fragments are summarised by grid but not by phase.

Table 4.2 List of Unidentified Specimens at Haimenkou

Category	Number	%	Weight	%
Homo sapiens	20	2.0	340.7	3.6
Medium Mammal	10	1.0	142.7	1.5
Small Mammal	3	0.3	28.3	0.3
Fish	7	0.7	2.8	+
Modified Bone	20	2.0	103.2	1.1
Specimen with Questions	128	12.7	985.9	10.3
Unidentified Fragment	821	81.4	7952.5	83.2
Total	1009	100.0	9556.1	100.0

Key: + = less than 0.05%.

Table 4.3 Range and Relative Abundance of Identified Taxa from Haimenkou

Taxon	Common Name	NISP	% NISP	MNI	% MNI	WGT	% WGT
Sus domesticus	Pig	931	52.0	46	30.7	15353.7	28.4
Canis familiaris	Dog	25	1.4	6	4.0	164.1	0.3
Ovis & Capra	Sheep & Goat	65	3.6	7	4.7	1026.8	1.0
Bos gaurus	Gaur	387	21.6	11	7.3	23500.3	43.4
Rusa unicolor	Sambar	23	1.3	8	5.3	4147.9	7.7
Axis porcinus	Hog Deer	2	0.1	2	1.3	288.0	0.5
Large Cervid	Large Deer	147	8.2	10	6.7	5747.4	10.6
Muntiacus muntjak	Red Muntjac	21	1.2	9	6.0	1304.8	2.4
Medium Cervid	Medium Deer	108	6.0	9	6.0	1638.3	3.0
Muntiacus reevesi	Reeve's Muntjac	2	0.1	2	1.3	24.2	+
Moschus berezovskii	Forest Musk Deer	1	0.1	1	0.7	1.4	+
Small Cervid	Small Deer	11	0.6	4	2.7	95.0	0.2
Sus scrofa	Wild Boar	3	0.2	1	0.7	174.7	0.3
Macaca sp.	Macaque	3	0.2	2	1.3	43.7	0.1
Ursus sp.	Bear	7	0.4	3	2.0	393.6	0.7
Vulpes vulpes	Red Fox	1	0.1	1	0.7	7.8	+
Prionailurus bengalensis	Leopard Cat	1	0.1	1	0.7	6.9	+
Paguma larvata	Masked Palm Civet	3	0.2	2	1.3	14.7	+
Small Felid	Small Felid	3	0.2	2	1.3	15.4	+
Small Carnivore	Small Carnivore	8	0.4	4	2.7	35.8	0.1
Lepus sp.	Hare	7	0.4	2	1.3	24.5	+
Rhizomys sp.	Bamboo Rat	1	0.1	1	0.7	1.0	+
Hystrix brachyura	Porcupine	1	0.1	1	0.7	6.1	+
Rattus sp.	Rat	7	0.4	3	2.0	5.9	+
Bird	Bird	21	1.2	11	7.3	58.7	0.1
Unio douglasiae	Mussel	2	0.1	1	0.7	14.4	+
Total		1791	100.0	150	100.0	54095.1	100.0

Key: See Tables 4.1 and 4.2.

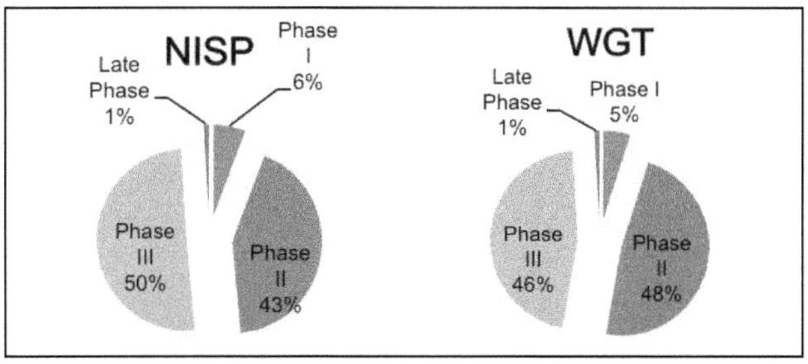

Figure 4.2 Distributions of Sample Sizes at Haimenkou by Phase
Key: See Table 4.1.

4.3.1 Domestic Mammals

To date, there is no consensus on the definition of 'domestication' of animals, although many of the more recent definitions share some similarities (e.g., Bökönyi 1989; Ducos 1978; Meadow 1989). According to Bökönyi (1989:22), the essence of animal domestication may be described as 'the capture and taming by man of animals of a species with particular behavioural characteristics, their removal from their natural living area, and breeding community and their maintenance under controlled breeding conditions for mutual benefit.'

Jarman and Wilkinson (1972:83) have distinguished domestic species from their wild counterparts based on two criteria. The first involves the physical features of the animals, including their size, morphology, and colour. The second involves the demography and ecology of the exploited animal populations and evidence of the way in which they were harvested. To be specific, the second criterion includes the investigation of distribution, sex ratio, mortality, survival pattern, and the ecological fitness of the animals to the local biotope. In addition to these criteria, artefactual and artistic data also provide a further source of potential evidence. However, this evidence is often ambiguous and difficult to interpret (Jarman and Wilkinson 1972:83). Mason (1984a:vii) has proposed another set of criteria to identify domestication. According to his criteria, a domestic animal in its most developed form should have four principal characteristics: its breeding is under human control; it provides a product or service useful to man; it is tame; and it has been selected away from the wild form. These characteristics cover the basic criteria that were applied during the identifying and distinguishing process of the study of animal remains. Although other criteria may be invoked by different authors, they are similar to those presented above (e.g., Bökönyi 1989; Clutton-Brock 1999; Davis 1981; Ducos 1978; 1989; Meadow 1989). Although the criteria have been summarised, revised, and supplemented, in practice it is still complicated to answer questions regarding which species is domesticated and which is not, especially when feral and semi-domesticated conditions may also exist. Additionally, while not all domestic animals can be distinguished under the criteria, exceptions do exist and certain cases should be handled with specific methods.

In this study, pig (*Sus domesticus*), dog (*Canis familiaris*), and sheep or goat are identified as domesticated taxa at Haimenkou, with details given in the sections below. As shown by Table 4.3, they amount to more than half of the total identified faunal specimens from Haimenkou based on NISP, nearly two-fifths based on MNI, and almost 30 percent based on bone weight. It is clear that domestic animals predominate in importance in the faunal assemblage from Haimenkou. The domestication of gaurs (*Bos gaurus*) cannot yet be determined based on available data from the 2008 season excavation. The gaur is assigned to the category of wild animals in this study, and the details concerned will be described later.

4.3.1.1 Artiodactyla

Suidae

Pig (*Sus domesticus*)

The *Sus* bones contribute more than half of the total indentified specimens based on NISP, and about 30 percent based on both MNI and weight. This rich abundance indicates the status of the pig as the most important animal in the faunal assemblage from Haimenkou.

It is difficult to elucidate the origins of modern domestic pigs since their wild ancestors, the wild boars (*Sus scrofa*), are generally distributed throughout Eurasia and North Africa (Epstein 1984:146; Higgs and Jarman 1972:4; Mayer, et al. 1998:39). The widespread distribution of its wild ancestor may imply there are multiple centres of domestication of pigs as also advocated by modern genetic analysis (Giuffra, et al. 2000; Larson, et al. 2005).

The earliest evidence of pig domestication worldwide was found at Çayönü, a Neolithic settlement in Anatolia (Near East), dating to around 9,000 years ago (Reed 1969). In China, the earliest domestic pig (*Sus domesticus*) is dated to around 8,500 years ago at the Jiahu site in Wuyang County, Henan Province (Cucchi, et al. 2011; Flad, et al. 2009; Luo and Zhang 2008). As the most important domestic animal in China (Yuan and Flad 2002:724), the pig has been found at archaeological sites dating from the early Neolithic Period across China (e.g., Cheng 1988; CPAMTAR and DHSU 1985; Li, Y. and Han 1959; Luo and Zhang 2008; Qi 1988; Wei, et al. 1989; Zhou 1981). In the case of Yunnan, the earliest domestic pigs appeared at Baiyangcun in Binchuan County (YPM 1981), dating to the late Neolithic Period (*ca.* 3800 B.P.) (LIACASS 1978:287). Pig bones have been revealed at most of the prehistoric sites in Yunnan and dominate the faunal assemblages from a large number of these sites. For example, according to the analysis of the Dadunzi site (*ca.* 3300 B.P.) in Yuanmou County (YPM 1977), the proportion of pig remains is more than one third of the whole number of identified specimens (Zhang, X. 1985).

Most of the *Sus* sp. remains from Haimenkou were identified as domesticated (*Sus domesticus*) while only three specimens were determined to be wild boar based on their exceptionally larger sizes. The *Sus* sp. specimens that were

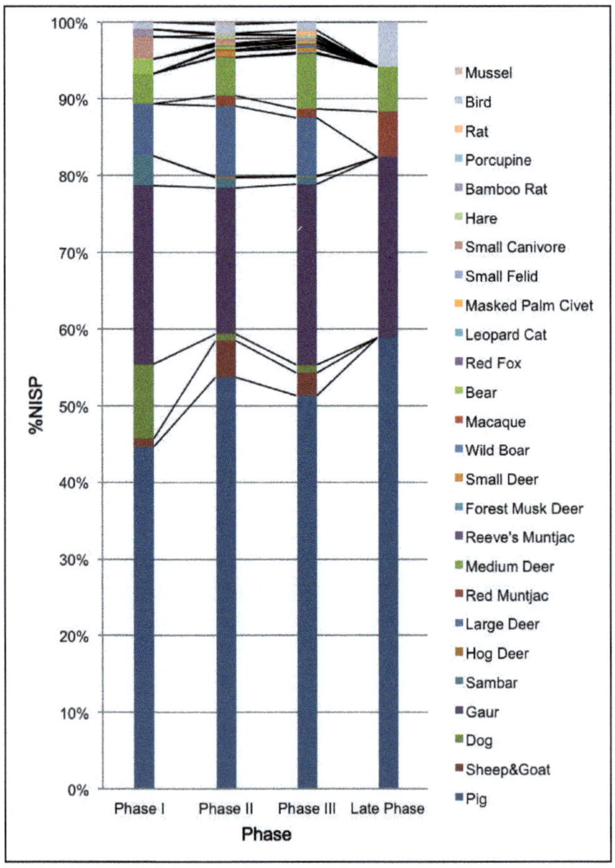

Figure 4.3a Relative Proportions of Identified Taxa Based on NISP

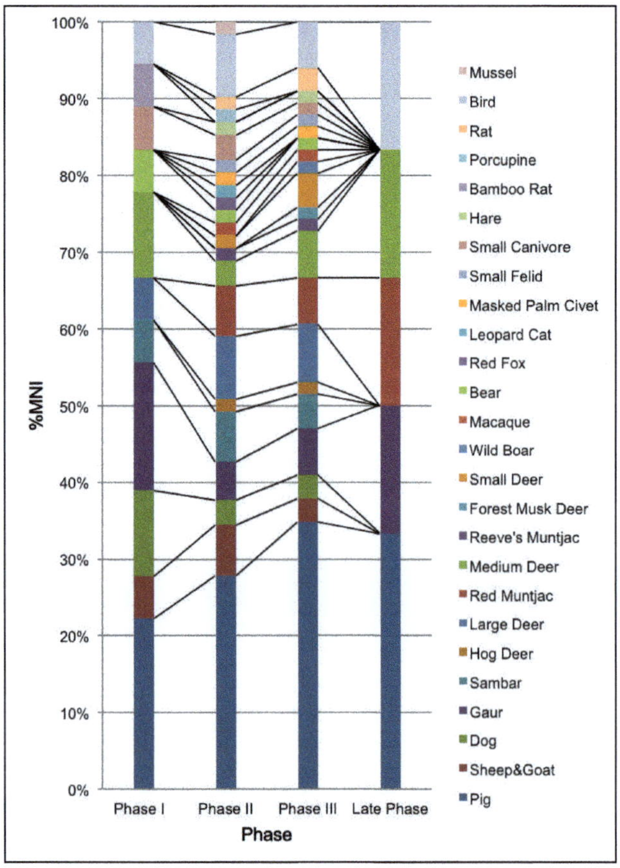

Figure 4.3b Relative Proportions of Identified Taxa Based on MNI

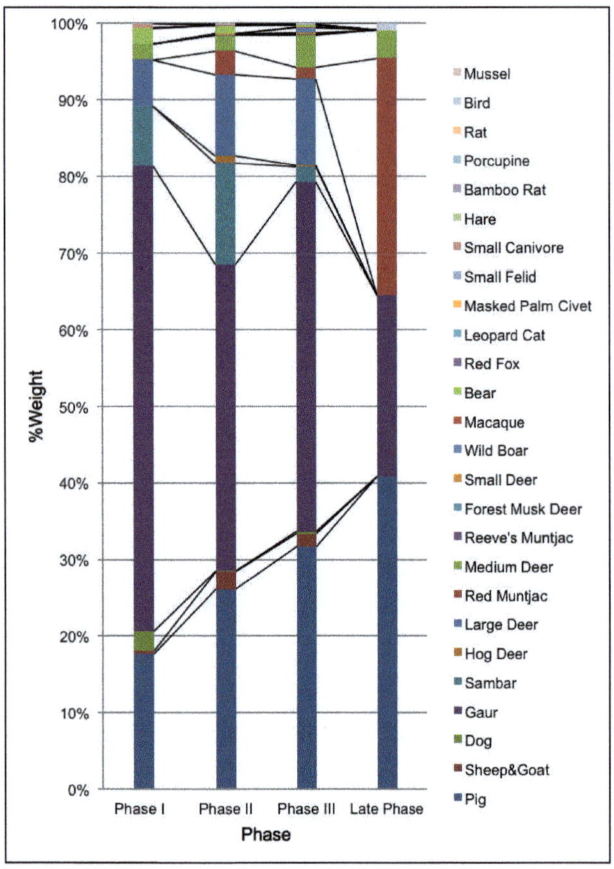

Figure 4.3c Relative Proportions of Identified Taxa Based on Weight

identified as domesticated will be analysed in this section, while the wild boar specimens will be discussed in Section 4.3.2.1.

The reasons that most of the *Sus* sp. remains were identified as domesticated are varied. Firstly, they have a much smaller body size when compared with both ancient and modern wild boars. Secondly, their kill-off patterns suggests that pigs were mainly slaughtered at a juvenile age. In fact, most were slaughtered intensively within a limited time span (see detailed discussion in chapter six).

As shown by Figures 4.4a–c, NISP, MNI, and weight of pig remains all present similar trends during Phases I to III (the Late Phase is not further discussed in this study). The numbers of pig bones increased from Phases I to III based on NISP, MNI, and weight, and the proportions grew slightly with fluctuations. It is clear that this abundance is concentrated in Phases II and III. This suggests a marked group size increase between Phases I and II. Thus increasingly intensive pig husbandry may have occurred at the transition from the late Neolithic Period to the early Bronze Age. In addition, especially from the proportions based on MNI and weight, it appears that a gradual increase in the numbers of pigs occurred between Phases I to III. This may indicate that the relative importance of pigs increased during human occupation at Haimenkou.

Over several decades, decreased body size has been employed as a criterion to distinguish the remains of

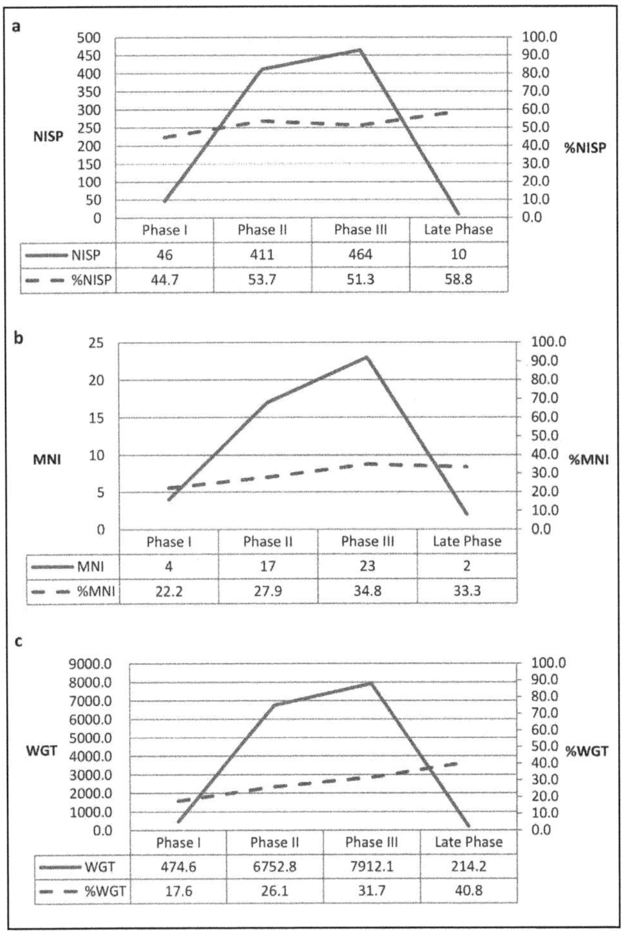

Figure 4.4a–c Numbers and Relative Proportions of Pig Based on NISP, MNI, and Weight

domestic pigs from their wild forms (e.g., Bökönyi 1974; Bökönyi 1989; Clutton-Brock 1981; 1999; Davis 1981; Ducos 1978; 1989; Flannery 1961; 1983; Jarman and Wilkinson 1972; Meadow 1989; Stampfli 1983; Zeuner 1963). Detailed observation on the *Sus* sp. remains from Haimenkou indicates that most of the bones and teeth are significantly smaller than wild boars. For the purpose of this comparison, a modern wild boar skeleton from the Zooarchaeology Laboratory of the HPICRA and also the three specimens of the wild boar recovered at Haimenkou were used as comparative reference material. A considerable proportion of the *Sus* bones appear immature, including at least one epiphysis that has not fused with its diaphysis (see details in chapter six). According to von den Driesch (1976:4), measurements taken to indicate age should only be taken from the bones or teeth of a fully grown animal. In young animals, an increase in size has to be considered, and consequently the corrected measurement is not precise enough to guarantee an accurate estimation of size. Thus, the measurement of bones that are taken from immature animals may only be justified in the case of exceptional size or in the study of particular age groups (e.g., Boessneck and von den Driesch 1973; von den Driesch and Boessneck 1974). Therefore, although most of the bones of the *Sus* sp. were measured and recorded (see Appendix 2), they are taken as supplementary data and will not be discussed in this book.

Teeth are the most durable and hard tissues in the body of an animal and are usually relatively abundant at archaeology sites (Silver 1970:289-290). When teeth are fully erupted their size can be measured and used to reflect body size. Thus, the teeth of pigs may provide a better choice than bones to investigate a decrease in pig body size. The reduction in the length of pig molars, with smaller sizes indicating domesticated forms, indicates that the body size of pigs decreased after domestication (e.g., Bökönyi 1974; Davis 1981; Flannery 1961; 1969; 1983; Mayer, et al. 1998; Stampfli 1983). For example, Flannery (1969:309) distinguished domestic pigs from their wild forms on the basis of the significant reduction in the length of the cheek tooth row, and especially by a reduction in the length of the mandibular third molar. Based on his method, in this book the data of the sizes (both breadth and length) of pig maxillary and mandibular third molars from Haimenkou were used as an indication of domestication.

The *Sus* sp. skeleton employed as a reference in this study was a female wild boar. It was modern sample Number 99 from the Zooarchaeology Laboratory of the HPICRA. Its left maxillary and mandibular molars were measured and used as a control group. With the exception of the canine, pig teeth are less dimorphic than limb bones (Flannery 1983:168), therefore the sex of Number 99 is not an issue here. Flannery's (1969:309) control group included 21 mandibular third molars of wild boars from Southwest Asia, which ranged from 38.8 to 49.3 millimetres in length, with a mean of 41.3 millimetres. The length of the mandibular third molar of Number 99 is 38.9 millimetres which is just within the standard thresholds. This indicates that it is a standard wild boar which may be usefully compared with the specimens from Haimenkou.

Due to the limited sample size of pigs from Haimenkou, only 13 mandibular third molars were measurable and employed in this book. Of these, 10 were from mandibles and three were loose teeth. All the specimens were from Phases II and III (see details in Table 4.5a, see also in Appendix 2). The measurements (length and breadth) of mandibular third molars of modern sample Number 99 and the specimens from Haimenkou are plotted in Figure 4.5a.

As shown by Figure 4.5a, no specimen of mandibular third molars from Haimenkou is larger than modern sample Number 99. Most specimens cluster between 25.7–31.4 millimetres in length and range from 13.0 to 15.3 millimetres in breadth. Therefore, there is a strong possibility that the specimens from Haimenkou were domestic pigs. However, there is no evident chronological size decrease, which may be due to the absence of specimens from Phase I as well as the relatively limited sample size.

Only four maxillary third molars were measurable and able to be used in this book. Of these, one was from maxillary alveolus and the other three were loose teeth. Specimens were from Phases I, II, and III (see details in Table 4.5b, see also in Appendix 2). The measurements of maxillary third molars of modern sample Number 99 and the specimens from Haimenkou are plotted in Figure 4.5b.

Table 4.4a NISPs of Identified Taxa from Haimenkou

Taxon	Common Name	Phase I	Phase II	Phase III	Late Phase
Sus domesticus	Pig	46	411	464	10
Ovis & Capra	Sheep & Goat	1	37	27	0
Canis familiaris	Dog	10	6	9	0
Bos gaurus	Gaur	24	146	213	4
Rusa unicolor	Sambar	4	10	9	0
Axis porcinus	Hog Deer	0	1	1	0
Large Cervid	Large Deer	7	71	69	0
Muntiacus muntjak	Red Muntjac	0	10	10	1
Medium Cervid	Medium Deer	4	38	65	1
Muntiacus reevesi	Reeve's Muntjac	0	1	1	0
Moschus berezovskii	Forest Musk Deer	0	0	1	0
Small Cervid	Small Deer	0	6	5	0
Sus scrofa	Wild Boar	0	0	3	0
Macaca sp.	Macaque	0	1	2	0
Ursus sp.	Bear	2	3	2	0
Vulpes vulpes	Red Fox	0	1	0	0
Prionailurus bengalensis	Leopard Cat	0	1	0	0
Paguma larvata	Masked Palm Civet	0	1	2	0
Small Felid	Small Felid	0	1	2	0
Small Carnivore	Small Carnivore	3	4	1	0
Lepus sp.	Hare	0	4	3	0
Rhizomys sp.	Bamboo Rat	1	0	0	0
Hystrix brachyura	Porcupine	0	1	0	0
Rattus sp.	Rat	0	1	6	0
Bird	Bird	1	9	10	1
Unio douglasiae	Mussel	0	2	0	0
Total		103	766	905	17

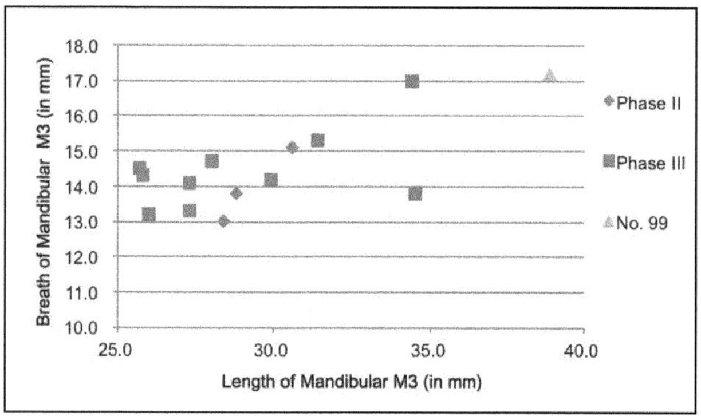

Figure 4.5a Measurements of Mandibular Third Molars of *Sus* sp. from Haimenkou Compared with Modern Sample No. 99

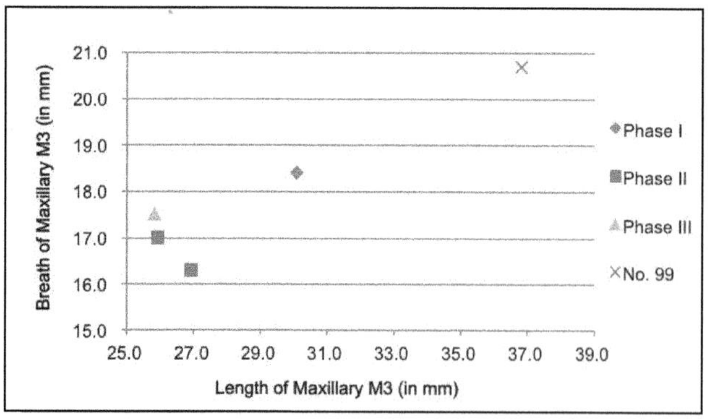

Figure 4.5b Measurements of Maxillary Third Molars of *Sus* sp. from Haimenkou Compared with Modern Sample No. 99

Table 4.4b MNIs of Identified Taxa from Haimenkou

Taxon	Common Name	Phase I	Phase II	Phase III	Late Phase
Sus domesticus	Pig	4	17	23	2
Ovis & Capra	Sheep & Goat	1	4	2	0
Canis familiaris	Dog	2	2	2	0
Bos gaurus	Gaur	3	3	4	1
Cervus unicolor	Sambar	1	4	3	0
Axis porcinus	Hog Deer	0	1	1	0
Large Cervid	Large Deer	1	5	5	0
Muntiacus muntjak	Red Muntjac	0	4	4	1
Medium Cervid	Medium Deer	2	2	4	1
Muntiacus reevesi	Reeve's Muntjac	0	1	1	0
Moschus berezovskii	Forest Musk Deer	0	0	1	0
Small Cervid	Small Deer	0	1	3	0
Sus scrofa	Wild Boar	0	0	1	0
Macaca sp.	Macaque	0	1	1	0
Ursus sp.	Bear	1	1	1	0
Vulpes vulpes	Red Fox	0	1	0	0
Prionailurus bengalensis	Leopard Cat	0	1	0	0
Paguma larvata	Masked Palm Civet	0	1	1	0
Small Felid	Small Felid	0	1	1	0
Small Carnivore	Small Carnivore	1	2	1	0
Lepus sp.	Hare	0	1	1	0
Rhizomys sp.	Bamboo Rat	1	0	0	0
Hystrix brachyura	Porcupine	0	1	0	0
Rattus sp.	Rat	0	1	2	0
Bird	Bird	1	5	4	1
Unio douglasiae	Mussel	0	1	0	0
Total		18	61	66	6

In Figure 4.5b, no specimen of maxillary third molars from Haimenkou is larger than the maxillary third molar of modern sample Number 99. This is consistent across length and breadth, with the disparity even more widespread than in Figure 4.5a. Since specimens from all three phases are present in Figure 4.5b, this condition may suggest the domesticated characteristics of the *Sus* sp. at Haimenkou appeared at the beginning of human occupation. Based on these specimens an obvious reduction in size over time is observable, with the maxillary third molar from Phase I the largest among all the specimens from Haimenkou and the one from Phase III the smallest. This may indicate that body-size reduction in the process of pig domestication continued after pigs arrived at Haimenkou with their keepers. However, in the mandibular third molar assemblage this trend is not evident. The small sample size limits the conclusions that can be drawn.

Consequently, according to the analysis of the third molars of pig from Haimenkou, it is probable that the third molars belonged to domestic pigs. However, the relatively limited sample size may require future investigation based on more abundant specimens. At present there is still a wide distribution of wild boars in Yunnan and also in other regions of China (Smith, A. and Xie 2008:452), so it is difficult to conclude where the ancestors of these domestic pigs came from. Nevertheless, it is probable that these pigs were domesticated before moving to Haimenkou with their keepers and it seems that the process of domestication continued alongside the Haimenkou people's occupation of the site.

Table 4.4c Weight of Identified Taxa from Haimenkou

Taxon	Common Name	Phase I	Phase II	Phase III	Late Phase
Sus domesticus	Pig	474.6	6752.8	7912.1	214.2
Ovis & Capra	Sheep & Goat	10.7	600.0	416.1	0.0
Canis familiaris	Dog	68.7	22.8	72.6	0.0
Bos gaurus	Gaur	1633.7	10348.6	11393.5	124.5
Cervus unicolor	Sambar	210.6	3435.9	501.4	0.0
Axis porcinus	Hog Deer	0.0	230.9	57.1	0.0
Large Cervid	Large Deer	162.9	2751.2	2833.3	0.0
Muntiacus muntjak	Red Muntjac	0.0	793.2	349.3	162.3
Medium Cervid	Medium Deer	54.3	515.9	1049.1	19.0
Muntiacus reevesi	Reeve's Muntjac	0.0	4.1	20.1	0.0
Moschus berezovskii	Forest Musk Deer	0.0	0.0	1.4	0.0
Small Cervid	Small Deer	0.0	50.5	44.5	0.0
Sus scrofa	Wild Boar	0.0	0.0	174.7	0.0
Macaca sp.	Macaque	0.0	17.4	26.3	0.0
Ursus sp.	Bear	56.7	268.2	68.7	0.0
Vulpes vulpes	Red Fox	0.0	7.8	0.0	0.0
Prionailurus bengalensis	Leopard Cat	0.0	6.9	0.0	0.0
Paguma larvata	Masked Palm Civet	0.0	8.5	6.2	0.0
Small Felid	Small Felid	0.0	4.6	10.8	0.0
Small Carnivore	Small Carnivore	15.4	11.0	9.4	0.0
Lepus sp.	Hare	0.0	14.7	9.8	0.0
Rhizomys sp.	Bamboo Rat	1.0	0.0	0.0	0.0
Hystrix brachyura	Porcupine	0.0	6.1	0.0	0.0
Rattus sp.	Rat	0.0	0.9	5.0	0.0
Bird	Bird	1.9	17.2	34.4	5.2
Unio douglasiae	Mussel	0.0	14.4	0.0	0.0
Total		2690.5	25883.6	24995.8	525.2

In summary, the pig is the most dominant species at Haimenkou due to its richest abundance. At present this animal is identified as domesticated based on the analysis of its body size and kill-off patterns. Specific details of the kill-off patterns will be discussed in chapter six.

Bovidae

Sheep (*Ovis aries*) and Goat (*Capra hircus*)

Domestic sheep and goat are considered to be two of the earliest domestic animals worldwide. Sheep are thought to have been domesticated in the Fertile Crescent region of West Asia around 10,000–9,000 years ago. (Higgs and Jarman 1972:3; Ryder 1984:66). Goats probably underwent this transition at the same period (Mason 1984b:89; Zeder and Hesse 2000).

Based on archaeological records, the domestication of sheep and goats in China appears to have taken place much later than in West Asia (Flad, et al. 2009:466; Yuan, et al. 2007:28). There seems to be a possibility that the domestic sheep or goat remains found in China were introduced species (Flad, et al. 2009:466), The subsequent

Table 4.5a Measurements (in mm) of Mandibular Third Molars of *Sus* sp. from Haimenkou and Modern Sample No. 99

Location	Layer	Phase	BON	SYM	M3-B	M3-L
08JHAT2002	(6)	II	1244	left	13.0	(28.4)
08JHAT2003	(6)	II	1364	left	13.8	28.8
08JHDT1802	(6)	II	540	left	15.1	30.6
08JHAT2003	(5)	III	1363	left	14.2	29.9
08JHAT2505	(5)	III	1816	left	17.0	(34.4)
08JHDT1802	(5)	III	539	left	14.3	(25.8)
08JHDT1105	(5)	III	429	left	14.1	27.3
08JHDT1802	(5)	III	122	right	14.5	(25.7)
08JHAT2002	(4)	III	1246	left	13.3	27.3
08JHAT1901	(4)	III	1967	left	13.8	34.5
08JHAT2001	(4)	III	1206	left	15.3	31.4
08JHAT2002	(4)	III	1246	right	13.2	26.0
08JHAT2103	(4)	III	1724	right	14.7	(28.0)
Modern Sample: Female Wild Boar			No. 99	left	17.2	38.9

Key: BON = Bone Identification Number; SYM = Symmetry (Right/Left); M3 = Third Molar; B = Breadth; L = Length.
Note: Definitions of measurements follow von den Driesch (1976:38-41).
The numbers in parentheses in column 'M3-L' refer to approximate estimates, as some specimens were slightly damaged meaning precise measurement could not be obtained.

Table 4.5b Measurements (in mm) of Maxillary Third Molars of *Sus* sp. from Haimenkou and Modern Sample No. 99

Location	Layer	Phase	BON	SYM	M3-B	M3-L
08JHAT2002	(8)	I	1262	Left	18.4	30.1
08JHDT1305	(6)	II	1095	left	17.0	25.9
08JHAT2002	(6)	II	1236	right	16.3	26.9
08JHDT1801	(4)	III	316	left	17.5	25.8
Modern Sample: Female Wild Boar			No. 99	left	20.7	36.8

Key: See Table 4.5a.
Note: Definitions of measurements follow von don Driesch (1976:38-41).

first process of sheep domestication may have occurred in Qinghai and Gansu Provinces in China from around 5600–5000 B.P. (see locations of Gansu and Qinghai in Figure 2.24) (Flad, et al. 2009:466; Yuan, et al. 2007:28). At present, the ancient DNA analysis of sheep remains from several archaeological sites scattered in northern China dating between 4500–2500 B.P. indicates multiple centres of origin for domestic sheep in China (Cai, et al. 2010). Consequently, domestic sheep and goats may have been introduced into China around 5,600 years ago, while the possibility of indigenous multiple centres of origin still cannot be ruled out.

In Chinese, the character '*yang*' can represent both sheep and goats. In a number of Chinese archaeological reports only the term '*yang*' was mentioned when sheep or goat remains were present (e.g., Li, Y. and Han 1959:176-177; YPM 1981:365; Zhang, X. 1985:33). To clarify this condition, in this book the caprine animals that cannot be clearly recognised as sheep or goat will be labelled as 'sheep/goat'. According to the limited archaeological and palaeontological data, sheep/goat (*Ovis/Capra)* remains were not found in Yunnan until the excavation at Dadunzi (YPM 1977; Zhang, X. 1985). This site dates to around 3,300 years ago (LIACAS 1974:337) and thus corresponds to the flourishing period of Haimenkou (i.e., Phases II and III, *ca.* 3700–2400 B.P.). To date, during the same prehistoric period in Yunnan Province only three sites yielded animal remains identified as sheep/goat (*yang*). These sites included Baiyangcun (*ca.* 3800 B.P.) in Binchuan County (LIACASS 1978:287; YPM 1981), Dadunzi (*ca.* 3300 B.P.) in Yuanmou County (LIACAS 1974:337; YPM 1977), and Haimenkou in Jianchuan County. Along with the archaeological material of the Bronze Age, new evidence of species domestication emerged, such as the bronze sheep figures unearthed at the Dabona site, Xiangyun County (*ca.* 2450 B.P.) (Li, J. 1965; LIACAS 1977:203; YPAT 1964:612). This corresponds approximately with the end of Phase III of Haimenkou (*ca.* 3100–2400 B.P.) (Li, K. and Min 2011:103). Baiyangcun, Dadunzi, Dabona, and Haimenkou are all located in the same region near the upper reach of the Nu, Lancang, and Jinsha Rivers, northwestern Yunnan (see Figures 2.23 and 2.28), and may represent a general culture evolved locally from the Neolithic Period (Wang, D. 1994; Xu and Xiao 2004:395-433). This indicates the presence of sheep or goats in Yunnan from the late Neolithic Period to the Bronze Age.

Table 4.6a NISP, MNI, and Weight of Sheep and Sheep/Goat

Phase	Taxon	Common Name	NISP	MNI	WGT
I	*Ovis aries*	Sheep	1	1	10.7
II	*Ovis aries*	Sheep	15	2	318.7
	Ovis/ Capra	Sheep/ Goat	22	4	281.3
III	*Ovis aries*	Sheep	10	2	157.9
	Ovis/ Capra	Sheep/ Goat	17	2	258.2
Late	*Ovis aries*	Sheep	0	0	0.0
	Ovis/ Capra	Sheep/ Goat	0	0	0.0

Key: See Table 4.1.

Table 4.6b NISP, MNI, and Weight of Sheep & Goat

Phase	Taxon	Common Name	NISP	MNI	WGT
I	*Ovis & Capra*	Sheep & Goat	1	1	10.7
II	*Ovis & Capra*	Sheep & Goat	37	4	600.0
III	*Ovis & Capra*	Sheep & Goat	27	2	416.1
Late	*Ovis & Capra*	Sheep & Goat	0	0	0.0

Key: See Table 4.1.

The sheep/goat specimens recovered from Haimenkou, as shown by Table 4.6a, were identified as sheep (*Ovis aries*) when certain, or 'sheep/goat' (*Ovis/Capra*) when unsure. Distinguishing bones of sheep from goats is always a vexing problem for zooarchaeologists, especially under conditions which include a high degree of fragmentation. Several relevant references were used for differentiating these two species (Boessneck 1970; Payne 1985; Prummel and Frisch 1986; Zeder and Pilaar 2010), but absolute certainty may never be achieved in practice.

As shown in Table 4.6a, the caprine remains recovered at Haimenkou were identified as sheep with confidence or sheep/goat when uncertain. Therefore, due to the absence of diagnostic goats in the assemblage from Haimenkou and the presence of bronze sheep figures from other sites in the later period of Yunnan (e.g., Dabona, *ca.* 2450 B.P., see Li, J. 1965; LIACAS 1977:203; YPAT 1964:612), it is probable that all these sheep/goat bones found at Haimenkou did actually belong to sheep. For this reason they will be combined as one category, 'sheep & goat' (*Ovis & Capra*), in following analyses and discussion (Table 4.6b).

Sheep & goat remains do not dominate any phase of Haimenkou, and this is similar to the findings from the Baiyangcun and Dadunzi sites where sheep/goat remains were also recovered. At Haimenkou, the size and features of these caprine bones are nearly the same as those of modern domestic sheep (see details of size in Appendix 2), suggesting they were domesticated taxa. There is no evidence for their wild ancestors in the area near the upper reach of the Nu, Lancang, and Jinsha Rivers, so it is probable that the sheep & goat remains were those of domesticated animals which were introduced to this region. Their relatively low numbers in the faunal assemblage from Haimenkou may indicate a relatively low status in human subsistence economy. Nevertheless, the evidence is enough to indicate that these sheep & goats were domesticated.

Although the abundance of sheep & goats is concentrated in Phases II and III, the trend, differing from pigs, does not increase from Phases II to III (see Figures 4.6a–f). In other words, the amount and relative abundance of sheep & goats both decreased in Phase III. This may result from a number of causes, such as inadequate sample collection of sheep & goat remains due to the relatively small size of their elements. This trend may also indicate an unsuitable husbandry pattern for sheep & goats, which may have reduced the group size.

As reviewed in chapter two, painted double-handled pottery jars and wheat began to appear during Level (7) of Phase II at Haimenkou. The pottery and botanical evidence reflects an influence from the upper Yellow River Valley region in the area of Qinghai and Gansu Provinces, northwestern China. At the same time, as mentioned above, the first domestic sheep were present in Qinghai and Gansu Provinces at about 5,600 years ago (Flad, et al. 2009:466; Yuan, et al. 2007:28). The presence of sheep remains at Haimenkou may therefore indicate influence from or

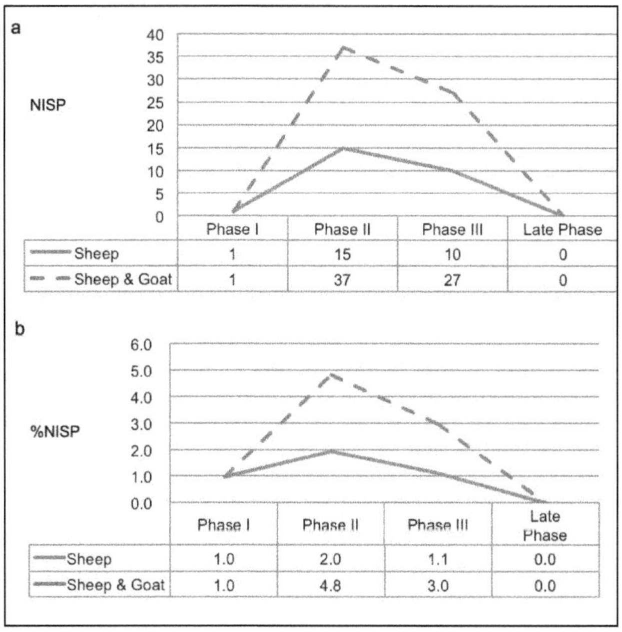

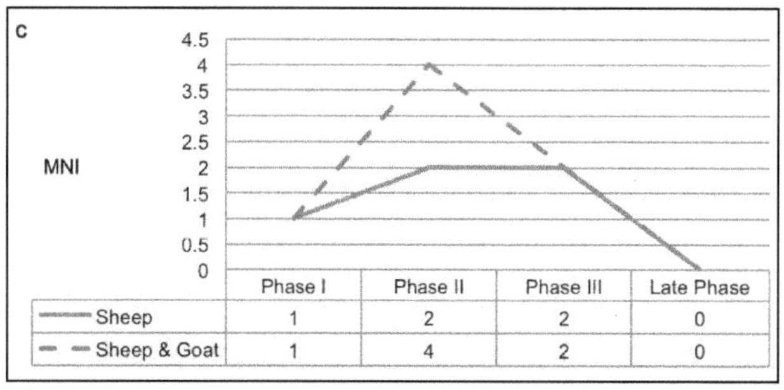

Figure 4.6a–c NISPs, % NISPs, and MNIs of Sheep and Sheep & Goat

connections with this same region of northwestern China. There was only one specimen of sheep in Level (8), so it is possible that this sole sheep specimen may have in fact belonged to Phase II and been mis-identified as coming from an earlier layer during excavations. Consequently, it is probable that sheep were introduced to Haimenkou at the same period (Phase II, *ca.* 3700–3200 B.P.) as the painted pottery style and wheat cultivation from the region of Qinghai and Gansu Provinces in China (see Figure 2.24). If sheep came to Haimenkou as introduced animals from Phase II, the unfamiliarity of the Haimenkou people to this exotic taxon may have led to the reduction of the group size evident in Phase III.

4.3.1.2 Carnivora

Canidae

Dog (*Canis familiaris*)

The dog (*Canis familiaris*) is said to be the only domestic

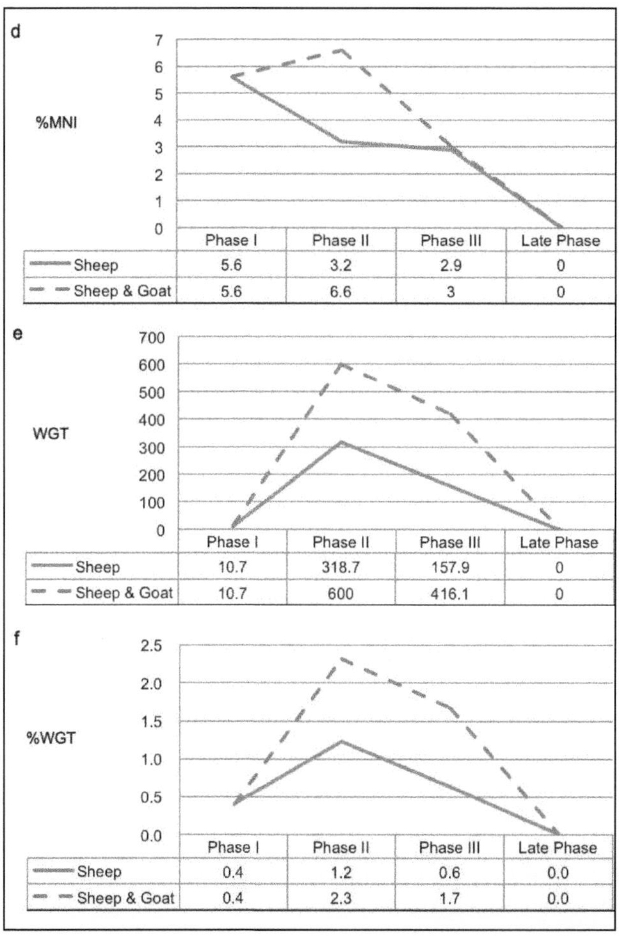

Figure 4.6d–f %MNIs, Weight, and %Weight of Sheep and Sheep & Goat

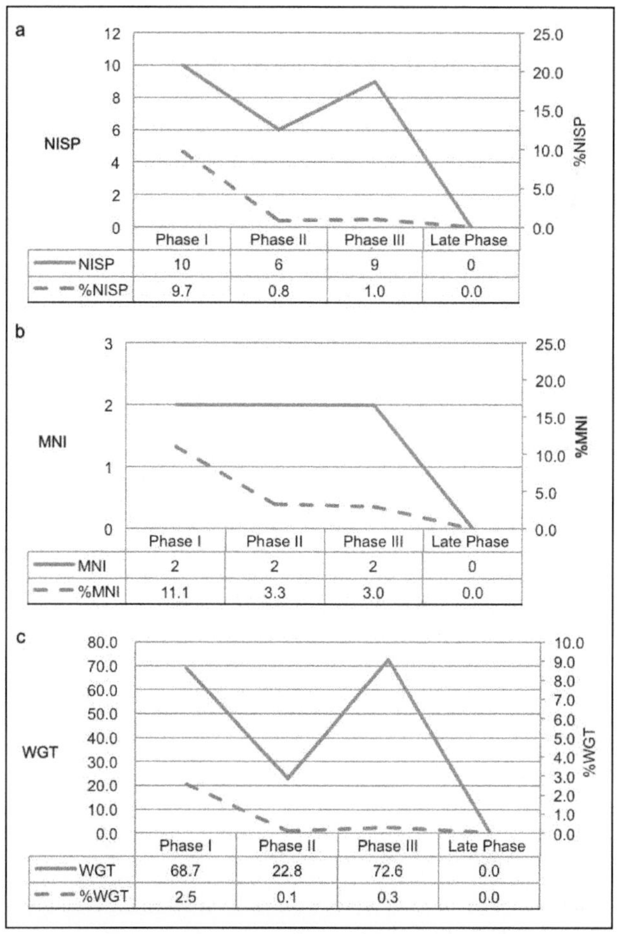

Figure 4.7a–c Numbers and Relative Proportions of Dog Based on NISP, MNI, and Weight

species that was distributed across Eurasia and the Americas before the 15th century, the period when transoceanic travel was developed (Leonard, et al. 2002:1613). To date, the earliest domesticated dog recovered came from the Razboinichya Cave in the Altai Mountains of southern Siberia (Ovodov, et al. 2011). It was dated to around 33,000 B.P. (Ovodov, et al. 2011). It is considered to have undergone part of the domestication process by this time, but this appears to have been disrupted by the Last Glacial Maximum (*ca.* 26,500–19,000 B.P.) (Ovodov, et al. 2011). More evidence concerning dog domestication in the late Glacial and early Holocene periods (*ca.* 14,000–9000 B.P.) has been revealed throughout the world, including in North America, northern Europe, and Asia (Clutton-Brock 1970; 1984; 1999; Lawrence 1967; Nobis 1979; 1981; Norbert 1987). Therefore, circumstantial archaeological evidence suggests multiple origins of domestic dogs, and this proposition agrees with genetic evidence (Leonard, et al. 2002; Savolainen, et al. 2002; Vilà, et al. 1997).

In China the earliest dog was reported at the Nanzhuangtou site in Xushui County, Hebei Province, and dated to around 10,510–9700 B.P. (Yuan and Li 2010). However, in Yunnan the dog remains recovered during archaeological excavations represent a much later period than in Hebei. In Yunnan, dogs were first unearthed at Baiyangcun (*ca.* 3800 B.P.) (LIACASS 1978:287) in Binchuan County (YPM 1981; Zhang, X. 1987).

The dog remains contributed about 10 percent of the total identified bone fragments at the Dadunzi site (*ca.* 3300 B.P.) (LIACAS 1974:337) in Yuanmou County (YPM 1977). At Dadunzi site they were the most abundant faunal remains after the remains of pigs and cattle (Zhang, X. 1985). Compared with the abundance of dogs from Dadunzi, the amount of dog remains excavated at Haimenkou is relatively low. The proportion of dog is slightly more than one percent of the identified specimens based on NISP, but accounts for four percent and three percent based on MNI and weight respectively. Its fluctuating distributions during Phases I to III based on NISP, MNI, and weight are shown in Figures 4.7a–c: where MNI in Phases I to III is fairly equal, while both NISP and bone weight show a dramatic fall-off in Phase II. Given the small size and fragile condition of these dog specimens, as well as the small sample size of dogs, this pattern may have been caused by a recovery bias due to random collection of specimens. The pattern may not truly reflect the fluctuations in the dog population at Haimenkou in antiquity.

It is difficult to discuss dog remains from Haimenkou in any detail because of the very limited sample size. However, possible domestication was exhibited by the size and morphological characteristics of the dog specimens. These dogs are similar in size to the red fox (*Vulpes vulpes*), also found at Haimenkou (see Table 4.7b). The red fox is obviously smaller than a modern Chinese rural dog (modern sample Number 46) from the comparative reference collection, and also much smaller than the dog remains from other Neolithic sites in China (Table 4.7a). From the skeletal characteristics, their mandibles can be easily distinguished from those of red fox and other similar sized canines. Therefore, dogs from Haimenkou are considered a domesticated breed with a reduced body size.

In addition to its occurrence at the site of Haimenkou, the small-sized dog remains recovered may have been more widely distributed in Yunnan. The dog remains uncovered at Dadunzi (*ca.* 3300 B.P.) (LIACASS 1978:287; YPM 1981; Zhang, X. 1985) may also have belonged to this small-sized breed, as their mandibular molars share similarities in size with their counterparts from Haimenkou (Table 4.7b). The dog remains recovered at Haimenkou were also similar to a small-sized dog breed depicted in bronze figures at the Shizhaishan site (*ca.* 2500–2100 B.P.) located in eastern Yunnan (see Figure 2.28:16) (Jiang, Z. 1999; Zhang, Zengqi 1998:157).

The small-sized dog breed found at Shizhaishan is likely to have been an agile animal which may have been used as a sheepdog or hunting partner rather than for food (Zhang, Zengqi 1998:157). According to ethnological data, a few decades ago in Yunnan the local ethnic minority groups still kept dogs for hunting, and eating dogs was forbidden (Li, G. and Lu 1987:172). Therefore, the dogs recovered at Haimenkou and Dadunzi may have played a similar role in hunting to those at Shizhaishan. However, as the analysis in chapter five shows, cut marks found on one dog radius (BON 685 from Level (8), Phase I) may indicate canine butchery and consumption by people (see Figure 5.15). In addition, traces of burning were found on some dog specimens (see details in chapter five). Hence, the possibility of dog meat consumption at Haimenkou cannot be definitely excluded. However, aside from this radius bearing cut marks, there is no strong evidence to suggest that dogs at Haimenkou were systematically consumed or bred for meat.

All these small-sized dogs may be related to a unique small-sized dog breed that was delivered to the Shang dynasty (*ca.* 3700–3100 B.P.) as tribute by inhabitants of the middle Lancang River Valley, southwestern Yunnan (Fang 2001:33; Huang, H. 2006:333-336). This tribute was recorded in an ancient Chinese book, *Yizhoushu* [the Superfluous (Chapters of the) Book of Zhou]. During the Ming and Qing dynasties (*ca.* A.D. 1400–1800), there was also a small-sized dog breed in the middle Lancang River Valley in southwestern Yunnan which was paid to the royal court as tribute (Fang 2001:33). T. Zhang (1868–1936) (1985:333), a famous Chinese philosopher and literary critic, considered this dog breed in Ming and Qing dynasties to be the same dog breed noted in the *Yizhoushu*. In addition, at present this small-sized dog breed can still be found in the same regions in southwestern Yunnan (Rui Min, personal communication, March 30, 2011). If the small-sized dog found at Haimenkou is related to the unique small-sized dog breed mentioned in historic records and occurring in present times in southwestern Yunnan, it may suggest the possibility of connections

Table 4.7a Measurements (in mm) of Dog Mandibles from Haimenkou Compared with Other Chinese Neolithic Sites and Modern Sample No. 46

Site	Date (B.P.)	TL	L	P1-M3-L
Haimenkou	ca. 5000–2400	96.4	96.0	62.8 (58.5-65.0) n=3
Nanzhuangtou	ca. 10,500–9700	—	—	79.4
Jiahu	ca. 9000–7800	—	—	72.6
Cishan	ca. 8000	110.25 (104.0-116.5) n=2	105.0 (99.0-111.0) n=2	67.0 (63.0-71.0) n=2
Wanbei	ca. 6000	—	—	71.0
Jiangzhai	ca. 6700	—	—	70.0
Zhukaigou	ca. 4200–3500	—	135.0	69.4 (67.6-71.2) n=2
Tanshishan	ca. 5500–4000	—	—	66.3(65.6-67.0) n=2
No. 46	Modern	134.1	135.1	69.3

Key: TL = total length (the length from the condyle process to infradentale); L = length (the length from the angular process to infradentale); P1-M3-L = length of the cheek-tooth row (first premolar–third molar); n = number; — = no data available in this category.
Note: Definitions of measurements follow von don Driesch (1976:60-61).
Data of other Neolithic sites are from: Nanzhuangtou (Yuan and Li 2010), Jiahu (HPICRA 1999), Cishan (Zhou 1981), Wanbei (Li, M. 1991), Jiangzhai (Qi 1988), Zhukaigou (Huang, Y. 1996), and Tanshishan (Qi 1977); No. 46 is a female Chinese rural dog, modern sample from the Zooarchaeology Laboratory of the HPICRA.

Table 4.7b Measurements (in mm) of Dog and Fox Mandibles from Haimenkou Compared with Dadunzi in Yunnan and Modern Sample No. 46

SIT	LAY	BON	ANM	SYM	TL	L	P1-M3-L	M1-L	M1-B	M2-L	M2-B
HMK	(5)	1240	Dog	left	—	—	65.0	19.9	8.1	8.2	6.6
	(4)	1661	Dog	left	—	—	64.8	18.7	7.3	8.6	6.5
	(4)	1662	Dog	right	96.4	96.0	58.5	18.1	7.5	7.8	5.6
DDZ	—	7	Dog	left	—	—	—	19.3	7.7	—	—
	—	7	Dog	right	—	—	—	19.3	7.6	—	—
	—	10	Dog	—	—	—	—	16.9	6.8	7.4	6.8
	—	24	Dog	—	—	—	—	18.8	7.4	8.0	—
HMK	(6)	1614	Fox	left	—	—	55.0	—	—	—	—
Reference		No. 46	Dog	left	134.1	135.1	69.3	20.5	8.5	7.5	6.2

Key: LAY = layer; HMK = Haimenkou; DDZ = Dadunzi; ANM = animal identification; see other abbreviations in Tables 4.5a and 4.6a.
Note: Definitions of measurements follow von don Driesch (1976:60-61).
Data of the Dadunzi site are from Zhang (1985: table 2).

between Haimenkou and the middle Lancang River Valley in prehistory. This possibility has also been discussed in chapter two, based on the recovery of bronze *yue* axes at Haimenkou which show similarities with the stone *yue* axes from the Manghuai site (ATYPM 1977) located in the middle Lancang River Valley, southwestern Yunnan. This site is about 300 kilometres from the Haimenkou site (see Figure 2.23:11). Consequently, there may have been some contacts between Haimenkou and the region where Manghuai is situated in prehistoric Yunnan, and this communication may have occurred along the Lancang River.

In summary, it may be inferred that the dog found at Haimenkou belonged to a unique small-sized dog breed, which possibly existed in Yunnan from the Neolithic Period to modern times. Furthermore, this dog breed may indicate some connections between peoples who lived in Haimenkou, northwestern Yunnan, and in the middle Lancang River Valley, southwestern Yunnan. Similar to pigs, dog remains with apparently domesticated characteristics were found in Phases I to III. Therefore, it is possible that dogs may have accompanied their owners from the beginning of human occupation at Haimenkou and played an important role in hunting. The scarcity of dog remains form Haimenkou may be due to a recovery bias during the excavation.

4.3.2 Wild Mammals

The numbers of wild mammal species remains dominate all the wild animal remains at Haimenkou. A total of 16 wild mammal taxa have been found in the faunal assemblage from Haimenkou. Among the wild mammals, gaurs contribute most of the NISP, MNI, and weight, followed by large cervids.

4.3.2.1 Artiodactyla

Bovidae

Gaur (*Bos gaurus*)

One important question raised by researchers in China relates to the introduction of water buffaloes to China.

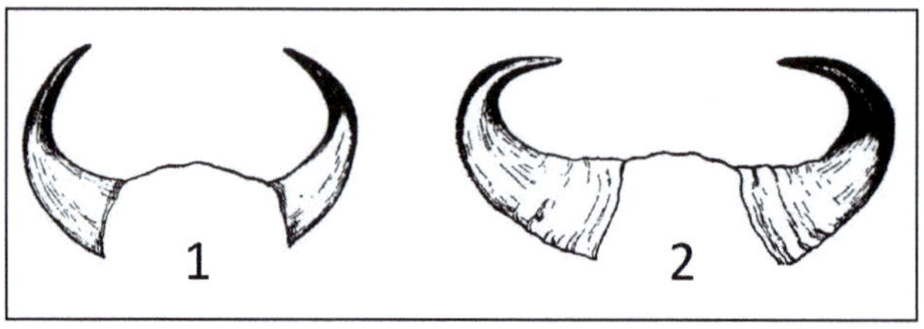

Figure 4.8a Horns of Gaur in Xishuangbanna, Yunnan, China (After Shou and Cai 1958:113; reproduced with permission of Science China Press)
1: Female; 2: Male.

Several routes of water buffalo introduction have been proposed (see Yang, Dongya, et al. 2008), so their presence or absence at a cultural centre such as Haimenkou is an important research question. However, no fragments of water buffalo horn cores were found at Haimenkou and no specimens show distinct features belonging to water buffalo bones. As all the cattle bones exhibit a consistent similarity, the possibility of the existence of water buffaloes (*Bubalus* sp.) in the faunal assemblage from Haimenkou has been excluded. Related taxa that may have been present in Yunnan in antiquity include gaur (*Bos gaurus*), gayal (*Bos frontalis*), and zebu (*Bos indicus*). Among these *Bos* species, to date only gaur remains have been identified in archaeological data from the late Palaeolithic Period to the Bronze Age in Yunnan Province. Despite this, the gayal and the zebu are evident in artefactual and artistic data from the Bronze to Iron Ages (see Figures 4.11 and 4.12). Based on the extremely robust skeleton together with the distinct water-buffalo-like, but oval-section shaped horn cores (Figure 4.8b), the *Bos* species recovered at Haimenkou were identified as gaur (*Bos gaurus*) (Figure 4.8a). This is a type of wild cattle that still lives in the Xishuangbanna *Dai* Autonomous Prefecture, southern Yunnan (Smith, A. and Xie 2008).

In China, cattle are traditionally considered domestic animals. This belief dates from an ancient period and is based on both historic records and archaeological evidence (Chen, W. 2002:68-71; ESBBC 1986:2-5; Yuan 2001). The most common domestic cattle type is called 'yellow cattle' (*huangniu*), and it belongs to the species *Bos taurus*

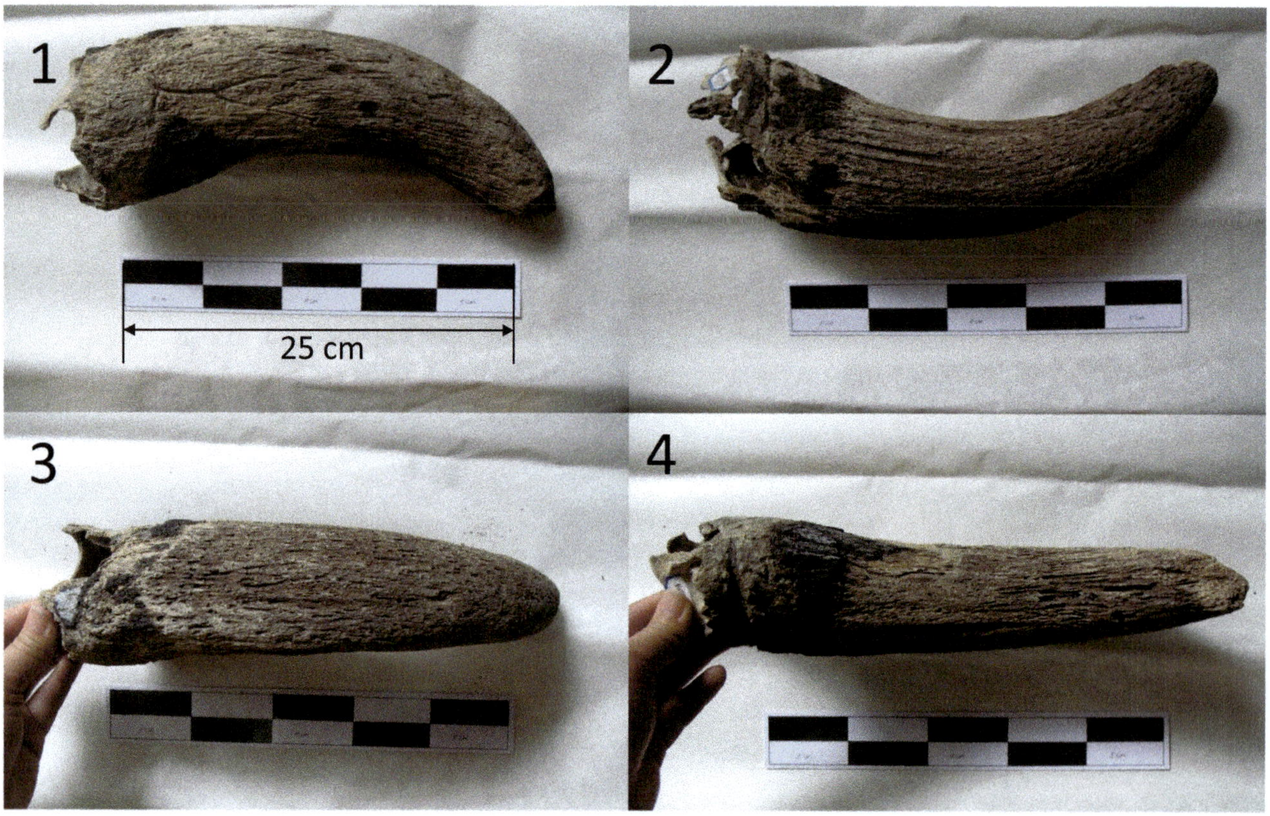

Figure 4.8b Horn Core (Right Side) of Gaur from Haimenkou (Photos by Juan Wang)
1: Dorsal View; 2: Basal View; 3: Nasal View; 4: Nuchal View. Note: BON 2053, from Level (7) of Phase II.

(ESBBC 1986:1-2). For this reason, in China usually *Bos taurus* remains are recovered during archaeological excavations and would probably be considered domesticated (e.g., Huang, Y. 1996; Qi 1988; Zhang, X. 1987). This may be the reason why X. Zhang (1985) identified gaur remains recovered from the Dadunzi site (*ca.* 3300 B.P.) in Yuanmou County as domesticated cattle. However, this cannot be taken for granted at Haimenkou, as *Bos gaurus* is a type of wild cattle still present in southern Yunnan Province today. It can still be found in scattered areas from India to Indochina and the Malay Peninsula (Nowak 1991:1430; Smith, A. and Xie 2008:472). To discuss whether the *Bos gaurus* remains found at Haimenkou were domesticated or wild, modern zoological results, archaeological records, and ethnological data will be employed in the following discussion.

The Gaur (Figure 4.9)

Gaurs inhabit forest hills and associated grassy clearings up to elevations of 1,800 metres (Nowak 1991:1429). Currently this animal has declined drastically due to hunting and habitat change by humans and exposure to the diseases of domestic cattle (Nowak 1991:1430). Only individual, scattered herds live in regions of southeast Asia today, stretching from Nepal and India to Indochina and the Malay Peninsula (Nowak 1991:1425). They are the most massive type of wild cattle in the world (Grzimek 1990:386). The gaur's typical head and body length is between 250–330 centimetres, the tail length is between 70–105 centimetres, the shoulder height is between 165–220 centimetres, the weight ranges from 650 to 1,500 kilograms, and the horn length is between 60–115 centimetres (Grzimek 1990:386). The gaur usually has dark reddish brown to almost blackish brown colouration, as well as marked white stockings (Nowak 1991:1429). In general, gaur males are approximately one-fourth larger and heavier than females and adult males have a large hump over the shoulder (Nowak 1991:1429). Males also have very heavy horns that sweep outward and then curve upward, growing from massive pale coloured crown humps, whereas females show more delicate horns (Nowak 1991:1429-1430).

Figure 4.9 Male Gaur (After Lydekker 1888–1890:176)

A normal gaur herd contains two to 40 individuals, and there is generally only one single mature bull in the herd, with some males living alone or in bachelor groups (Nowak 1991:1429). Females attain sexual maturity in their second or third year of life and then reproduction occurs throughout the year. However, in central India copulation seems to be centred from December to June, leading to calving seasons peaking in December and January (Nowak 1991:1430). The gestation period is 270–280 days and normally they produce a single offspring that is nursed for months (Nowak 1991:1430).

In China, the gaur was first scientifically studied in 1957 and then reported in 1958, although local hunters in southern Yunnan were familiar with this animal and it may have been mentioned in ancient Chinese records (Shou and Cai 1958). Gaurs in present China are only distributed in Xishuangbanna *Dai* Autonomous Prefecture, southern Yunnan, where the *Dai* People live (Gao, et al. 1962; Li, Z. 1963; Yang, Dehua, et al. 1988). According to a Chinese zoologist Z. Li (1963:18), male gaurs in Yunnan attain sexual maturity at four or five years of age, while females attain sexual maturity in their third or fourth years of life. Unlike central India, in Yunnan the mating of gaurs takes place from March to May, with a gestation period of 260–280 days and the calving season is from December to March (Li, Z. 1963:18).

The domestication of gaurs seems possible. Z. Li and colleagues (Li, Z. 1963; Li, Z., et al. 1980) presented the process and outcomes of experiments on the domestication of six adult gaurs and two calves caught in Xishuangbanna, Yunnan. In this experimental domestication process, Z. Li (1963:18) observed that all gaur individuals showed the potential of domestication, even adult males showed the potential to be dometicated. After only one month of domestication, the fierce adult male gaurs had become tamer and could accept physical contact with the feeder. After nine months, these adult males had allowed physiological measurements such as the checking of temperature and heartbeat. For gaur calves, this process was described as much easier since they were tamer and could get along with domestic cattle better than adult gaurs. This modern experiment may indicate the possibility of the domestication process occurring in ancient times. Other results of research suggesting a link between the gaur and a type of semi-domestic cattle, the gayal (*Bos frontalis*), are described below.

The Gayal (Figure 4.10)

In contemporary China, Myanmar, and eastern India, a rare semi-domestic *Bos* species occurs. The gayal (*Bos frontalis*), is thought to be a domesticated form of the gaur (Grzimek 1990:389; Nowak 1991:1430). Outside China, gayals are found in the hill regions of eastern India, Bangladesh, and adjacent Burma (Ma, G., et al. 2007; Simoons 1984; Zeng, et al. 1979). In China, gayals only

Figure 4.10 Female Gayals (After Lydekker 1888–1890:180)

inhabit hilly regions of the Dulong and Nu River Valleys in northwestern Yunnan (see Figure 2.3), and the Menyu and Luoyu regions in the Tibet Autonomous Region (Li, S., et al. 2008:66; Liu, Y., et al. 2010:59; Ma, G., et al. 2007:413; Mao, et al. 2005:258; Zeng, et al. 1979:662). Both 'gayal' and 'mithan' are common names for *Bos frontalis*. 'Mithan' is more commonly used in English (Simoons 1984:34), while 'gayal' is widely adopted in Chinese academic literature (e.g., He, et al. 2009; Li, S., et al. 2008; Li, S., et al. 2008; Ma, G., et al. 2007; Mao, et al. 2005; Wang, L., et al. 2009). Following the tradition of Chinese academia, the term 'gayal' is used to refer to this *Bos* species throughout this book. In China the gayal was not officially studied and reported until 1978 (Zeng, et al. 1979:662), although it was mentioned in Chinese documents as a semi-domestic animal herded by the Dulong people for at least the last 100 years (He, et al. 2009:226).

The origin of the gayal has not been agreed upon by scholars. There are three hypotheses: firstly, the gayal is a domestic form of the gaur (e.g., Li, S., et al. 2008; Simoons 1984; Smith, A. and Xie 2008); secondly, it is a hybrid resulting from the crossing of the bull gaur and zebu cow (e.g., Lan, et al. 1993; Winter, et al. 1984); and thirdly, it is descended from a wild Indian bovine that is now extinct (e.g., Ma, G., et al. 2007; Shan, et al. 1980; Wang, L., et al. 2009). At the present time, it seems that the first one is most advocated.

The gayal resembles the gaur in many ways, such as the colouration (especially the white stockings) and strong body build, although it is much smaller, chubbier, and characterised by a shortened skull and a normally well-developed double dewlap on the chin and neck (Grzimek 1990:389; Simoons 1984:35). Compared with the gaur, the gayal's horns are massive at the base, less curved and extending outward with the tips never pointing inward. Also, the gayal varies in colouration and some individuals are white or spotted (Grzimek 1990:389; He, et al. 2009:227). The decrease in size and the variation in colouration are regarded as typical features of domestic forms, along with the more open shape of the horns (Grzimek 1990:389-390). In addition, the gaur and gayal also share similar life behaviours in numerous ways, such as group structure, food, habitats, and means of communication (Chen, Y., et al. 1978; Grzimek 1990:390; He, et al. 2009; Li, Z. 1963; Simoons 1984:35; Yang, Dehua, et al. 1988; Zeng, et al. 1979).

There is also a line of genetic evidence supporting the first hypothesis of the origin of the gayal. Over several decades, male gaurs have been reported that could breed freely with female gayals (Gray 1972; Simoons 1984:35). There seems to be no sterility barrier between the two, although a sterility barrier exists between common cattle and the gayal (Gray 1972; Simoons 1984:35). Recent studies provide more proof. For example, based on the phylogenetic tree constructed by the analysis of the cytochrome b gene of the entire sequences of 11 gayals from Yunnan, Li and colleagues (Li, S., et al. 2008; Li, S., et al. 2008) suggested that the gayal might be the domestic form of the gaur. However, this study also showed that a great proportion of the gayal bloodline included other bovine species.

The Dulong people, who inhabit the Dulong River Valley in northwestern Yunnan, are the only ethnic minority group who herd gayal in China today (Zeng, et al. 1979:662). Gayals play significant roles in the normal life of the Dulong people. They are reared for meat, for sacrifice, and for prestige but not as draught animals or for labour (ECBRLPYP 1987:84-85; Li, K. 1980:268-269). Similar usage also occurs in the regions where humans raise gayals outside China (Simoons 1984:37-38). According to Simoons (1984:37), there were native traditions that the gayal was attracted to humans by its craving for salt. At present in China, the Dulong people maintain this relationship of semi-domestication with the gayal by feeding them salt (He, et al. 2009:229; Li, G. and Lu 1987:163; Zeng, et al. 1979:663). The craving for salt is also a habit of gaurs, therefore Simoons (1984:37) has assumed the process of domestication as follows: firstly, the gaur came into contact with humans around the agricultural clearings; secondly, humans recognised the gaur's craving for salt and; thirdly, the gaur was drawn into more permanent association by humans who regularly provided it with salt. A similar domestication process of wild cattle in ancient times is recorded through folk songs of the modern *Yi* people (Liu, Z. 1997:216), an ethnic minority group who live in scattered regions in Yunnan.

Archaeological Records

Both the gaur and gayal are present in the archaeological records of Yunnan Province. As reviewed in chapter one, there is a lack of detailed faunal analyses in Yunnan. Although cattle remains were unearthed from many of the sites in Yunnan, most of them were only briefly described and mentioned as 'cattle' or '*Bos* sp.' in archaeological reports. Only at two sites were the cattle remains identified to species. One of these two sites is Tangzigou (*ca.* 7000 B.P.), a late Palaeolithic site in Baoshan city (Zhang, X., et al. 1992; Zong and Huang 1985). Relatively scarce cattle remains found at this site were identified as *Bos gaurus* based on the shape and size of horn cores (Zhang, X., et al. 1992; Zong and Huang 1985). The other site is the Dadunzi site (*ca.* 3300 B.P.), Yuanmou County (YPM 1977; Zhang, X. 1985), which was mentioned above. The cattle remains recovered at Dadunzi were identified as *Bos gaurus*, but the common name used to refer to this species is 'gayal'(*daeniu*) rather than 'gaur' (*yeniu*) (Zhang, X. 1985:31). X. Zhang (1985:32) treated this *Bos* species, which made up the second most abundant species just after pigs, as a domestic animal taxa. X. Zhang (1985) may have meant to identify this *Bos* species as *Bos frontalis* rather than *Bos gaurus*, or he may have considered the gayal as a domesticated form of the gaur. However, according to the description of the horn core shape, the *Bos* species from Dadunzi is more like the gaur than the gayal, and thus is probably the same *Bos* species revealed at Haimenkou. X. Zhang (1985) did not explain why the *Bos gaurus* remains were identified as domesticated, but it may be due to the traditional Chinese opinion of cattle as always domesticated, as discussed above. Therefore, the identification of domestic gaurs at Dadunzi is relatively problematic. There is no strong evidence to support the existence of gaur domestication in Yunnan during the period of human occupation at Haimenkou in prehistory (*ca.* 5000–2400 B.P.).

The bronze cattle figures and cattle motif decorations found at several sites dating from the late Bronze Age to the Iron Age may show the domestication of the gayal (Figure 4.11) and another *Bos* species in Yunnan, the zebu (*Bos indicus*) (Figure 4.12). These sites include Dabona (*ca.* 2450 B.P.), Xiangyun County (Li, J. 1965; LIACAS 1977:203; YPAT 1964), Shizhaishan (*ca.* 2500–2100 B.P.), Jinning County (Jiang, Z. 1999; Zhang, Zengqi 1998), and Lijiashan (*ca.* 2100 B.P.), Jianchuan County (YPICRA 2007) (see locations in Figure 2.28). These artefactual and artistic data evidently reflect the domestication of the gayal and zebu. On the other hand, despite the gaur remains recovered at archaeological sites in Yunnan, there are no artefactual or artistic pieces of data recovered from the Neolithic Period onward to suggest the existence of gaurs in prehistoric Yunnan.

In summary, the modern zoological results, archaeological records, and ethnological data may indicate that it is possible that gaurs recovered at Haimenkou were domesticated or were undergoing a domestication process. However, this assumption requires further research due to the limited sample size and high fragmentation of gaur specimens (see details of size in Appendix 2). Thus, the gaur can only be conservatively recognised as wild in this study. A description of the amount and relative proportions of the gaur will be presented below.

Amount and Relative Proportions

The gaur accounts for 21.6 percent of the total of identified bone specimens based on NSIP, 7.1 percent based on MNI, and 43.4 percent based on bone weight. Gaur bones make up the second most abundant mammal species, after pigs, based on NISP during Phases I–III, whereas the gaur is the most abundant species based on weight in Phases I–III. The abundance of gaur remains may indicate their important status in the subsistence economy of the Haimenkou people in ancient times.

The numbers of gaur specimens show similar trends to that of pig remains, increasing in number from Phases I to III based on NISP, MNI, and weight (Figures 4.13a–c). The abundance of gaur grew in Phases II and III, suggesting an increase in hunting of gaurs from Phase I to III, but especially between Phases I and II. This phenomenon may indicate an intensification of hunting during the transition from the late Neolithic Period to the early Bronze Age. The MNI is much lower compared with that of pigs. Nevertheless, since a gaur can weigh between 650–1,500 kilograms (Grzimek 1990:386), its meat could contribute

Figure 4.11 Two Men Binding Gayal (Metal Buckle from Shizhaishan) (ca. 2500–2100 B.P.) (After Zhang, Zengqi 1998: plate 97; photo by Yi Xing, used with his permission)

Figure 4.12 Cavalier and Four Zebus (Cover of Storage Vessel for Cowry Shells from Shizhaishan) (ca. 2500–2100 B.P.) (After Zhang, Zengqi 1998: plate 31; photo by Yi Xing, used with his permission)

to the human diet dramatically more than a pig, which weigh less than one-tenth of a gaur (ESPBC 1986:172). The status of the gaur in the human subsistence economy may have been relatively high.

From the trends of relative proportions based on NISP, MNI, and weight (Figure 4.13a–c), it appears that the relative importance of gaurs decreased from Phases I to II and then slightly increased in Phase III. This may have been due to the sustained growth in relative importance of pigs from Phase I to III, especially in Phase II. In other words, pigs may have become more and more important in the subsistence economy of the Haimenkou people, leading to a slight decline in the relative importance of gaurs as hunted game. This scene may indicate that people were becoming more efficient in husbanding pigs, and domestic animals were occupying a more crucial status in subsistence economy at the prehistoric Haimenkou during the transition between the late Neolithic Period and the early Bronze Age.

Summary

The cattle remains revealed at Haimenkou were identified as gaur (*Bos gaurus*) due to their extremely large body size as well as distinct morphological features on horn cores. The presence of gaurs indicates that in prehistory this large

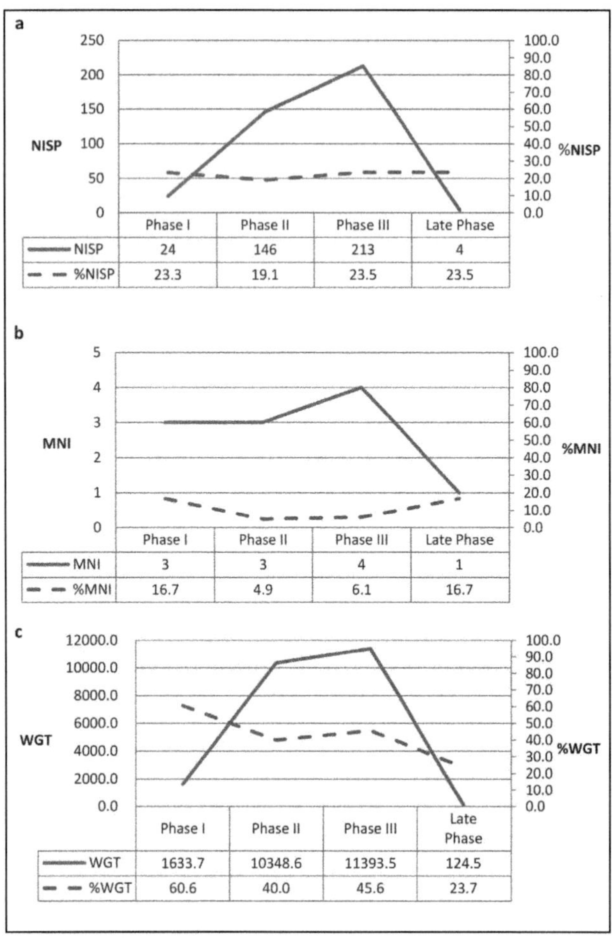

Figure 4.13a–c Numbers and Relative Proportions of Gaur Based on NISP, MNI, and Weight

Bos species, which currently can only be found in southern Yunnan, used to be distributed in northwestern Yunnan as far as Jianchuan County. The large number of gaur specimens discovered suggests this species had an important status in the subsistence economy throughout the entire human occupation at Haimenkou. Based on circumstantial evidence, the possibility of gaur domestication at Haimenkou in prehistory cannot be absolutely excluded. However, due to the lack of data to support a body size reduction, it is not possible to demonstrate the process of domestication in this study. Consequently, gaurs recovered at Haimenkou were recognised as wild animals, and more evidence is required from future research.

Cervidae

Based on body size, the cervids revealed at Haimenkou were classified into three categories: large cervid, medium cervid, and small cervid. In zooarchaeological studies in China, usually large cervids refer to the cervid species such as elk, red deer, and sambar. Medium cervids refer to the cervid species such as sika deer and hog deer. Other cervid species including a variety of muntjacs and musk deer are assigned into the small cervid class. However, in this study the size of the red muntjac (*Muntiacus muntjak*) is considered to be a medium cervid due to the scarcity of cervid remains similar to sika deer. All the cervids that are smaller than the red muntjac, such as the Reeve's muntjac (*Muntiacus reevesi*) and the forest musk deer (*Moschus berezovskii*), are classified as small cervids; while those larger than the red muntjac, such as the sambar (*Rusa unicolor*) and the hog deer (*Axis porcinus*), are classified as large cervids. According to this classification, all cervid remains that could not be identified to certain species were defined as large, medium, or small cervids based on their sizes.

Large cervid remains account for more than half of the total identified cervid specimens based on both NISP and weight, and make up nearly half of the MNI. Small cervids contribute less than 15.6 percent based on MNI, less than five percent based on NISP and even less based on weight. It is clear that the large cervid is dominant in the cervid category. As with the domestic animals and gaurs, cervids are also dramatically abundant in Phases II and III at Haimenkou, indicating more intensive hunting activities as well as a considerable improvement in hunting techniques. This is also suggested by the presence of the remains of other wild animal species.

Sambar (*Rusa unicolor*)

The largest wild animal recovered at Haimenkou is the sambar, which is a large deer whose head and body length is between 180–220 centimetres (Smith, A. and Xie 2008:466). The height at the shoulder is between 140–160 centimetres, and its weight ranges from 185 to 260 kilograms (Smith, A. and Xie 2008:466). It is generally dark brown in colour and males usually have lyre-like and three-tine antlers (Smith, A. and Xie 2008:466; Walker 1968:1375).

The sambar inhabits various environments including tropical forests, scrubs, hills, and secondary growth marshes, although it seems to prefer wooded areas (Smith, A. and Xie 2008:467; Walker 1968:1375). It may be found on forest mountain slopes up to 3,700 metres (Walker 1968:1375). More of a browser than a grazer, the sambar generally feeds in areas of secondary growth on leaves, buds, grass, berries, and fallen fruit (Walker 1968:1375). At present its distribution in China is throughout the tropical and subtropical zones (Smith, A. and Xie 2008:467) such as Yunnan Province, in the mountainous regions of southern China.

Based on archaeological records, the sambar was usually found at sites scattered in the southern region of China. The Xiawanggang site (*ca.* 7000–4000 B.P.) (Jia and Zhang 1989) is an exception to this, as this site is located in the Yellow River Valley in Henan Province. In the early Neolithic Period, sambar remains were unearthed from the Xianrendong Cave (*ca.* 10,000 B.P.) in Jiangxi Province (CPAMJP 1963; Huang, W. and Ji 1963; Li, Y. 1963), and the Zengpiyan site (*ca.* 8000 B.P.) in Guangxi Province (Li, Y. and Han 1978). The sambar was also found at Middle Neolithic sites such as Hemudu (*ca.* 7000–5000 B.P.) in Yuyao County, Zhejiang Province (Wei, et al. 1989), and at sites of the late Neolithic Period, such as Tanshishan (*ca.* 3100 B.P.) in Minhou County, Fujian Province (FPM 1976; Qi 1977).

In Yunnan Province, the sambar was present as an important game animal from the late Palaeolithic to the late Neolithic Periods. Its remains were found at Tangzigou (*ca.* 7000 B.P.) (Zhang, X., et al. 1992; Zong and Huang 1985), Huangjiaying (*ca.* 5000 B.P.) (Yang, Z., et al. 1985), and Dadunzi (*ca.* 3300 B.P.) (Zhang, X. 1985). In this study it also appears to have been present at Haimenkou in the Bronze Age.

The sambar contributes a relatively large proportion of bones to total cervid remains (see Table 4.3). Their representation is probably higher given that most of the specimens assigned to the large cervid category may be sambars, as there were no antlers from similar-sized cervids found at Haimenkou. If this is the case, then the sambar may account for around half of the total cervid specimens based on NISP, MNI, and weight. This indicates a marked dominance of this taxon in the faunal assemblage.

Hog Deer (*Axis porcinus*)

The hog deer is usually recognised as a medium-sized deer ranging from 105 to 115 centimetres in head and body length (Smith, A. and Xie 2008:460). It is between 60–72 centimetres in height at the shoulder and weights between 36–50 kilograms (Smith, A. and Xie 2008:460). Nevertheless, it is assigned to the large cervid category in this book since it is larger than the red muntjac. Its appearance is characterised by short legs as well as trifurcate antlers with obtusely angled front tines in males (Nowak 1991:1371-1373).

The hog deer prefers low lying rather than open grassy habitats, especially along river banks, on floodplains, and in swamps (Smith, A. and Xie 2008:460). It is much more of a grazer than the sambar, and small groups of hog deer gather during the mating season (Sheng 1992c; Smith, A. and Xie 2008:460). The hog deer was first scientifically discovered in China in the early 1960s, as it was distributed only in the lowest valleys of Xishuangbanna in southern Yunnan (Sheng 1992c). Unfortunately, it is now extinct in the wild in China, as it is reported that the hog deer has not been seen since the 1960s (Sheng 1992c).

The hog deer barely appears in the archaeological records of China except for Xiawanggang (Jia and Zhang 1989). It is also apparently absent from sites in Yunnan that date to the period before Haimenkou. It is possible that archaeologists did not recognise it when its remains were encountered, as it has been very rare during the last century, and thus it is hard to collect its skeleton for comparison. In this study, only two antler specimens were identified as hog deer. This was based on a comparison with the faunal skeleton collection of the Reference Collection Department of the KIZCAS. Based on this scarce find, it seems that hog deer contributed little to the subsistence economy of humans at Haimenkou. It is possible that some of its limb bones were present in the assemblage but were not recognised as distinct so were assigned to the large cervid category.

Large Cervids

As mentioned earlier, most of the specimens that were assigned to the large cervid category are likely to be the sambar. However, other species of large cervids may also exist in the faunal assemblage. There are two specimens from ancient Haimenkou (BON1903 and BON636), which resemble the sika deer (*Cervus nippon*) as both shape and size comparison with the modern reference collection suggests. This resemblance may indicate the existence of sika deer however it is not certain due to the lack of corresponding and diagnostic antlers.

Red Muntjac (*Muntiacus muntjak*)

The red muntjac is the largest species of the genus *Muntiacus*, with a head and body length between 98–120 centimetres. Their height at the shoulder is between 50–72 centimetres and their weight ranges from 17 to 40 kilograms (Sheng 1992a:164; Smith, A. and Xie 2008:464). The medium-sized reddish-coloured muntjac is characterised by simple antlers that emerge from the long and narrow pedicle in males, while females only have small bony pedicles covered with tufts of hair (Smith, A. and Xie 2008:464). Red muntjacs inhabit mountain forests where they live singly or sometimes in small groups of two to four individuals (Smith, A. and Xie 2008:465). Nowadays they are distributed across southern China, including Guangdong, Guangxi, Hainan, Yunnan, southwestern Guizhou, Fujian, Sichuan, southern Jiangxi, and the south slope of the Himalayas and Tibet (Sheng 1992a:166). Their main distribution is in Yunnan (Sheng 1992a:166).

Although its modern distribution is limited to southern China, the red muntjac used to range widely during prehistory. According to archaeological records, the distribution of the red muntjac can be found as far north as the Cishan site (*ca.* 8000 B.P.) in Hebei Province, North China (Zhou 1981), and as far east as the Hemudu site (*ca.* 7000–5000 B.P.) in Zhejiang Province in the Yangtze River Delta (Wei, et al. 1989). There are also numerous reports of remains identified as *Muntiacus* sp. (e.g., Jia and Zhang 1989) or small-sized deer, which may be the red muntjacs suggesting a wider distribution covering half of the Chinese mainland. In Yunnan, red muntjacs can be found throughout the prehistoric sites such as Tangzigou (*ca.* 7000 B.P.) of the late Palaeolithic Period (Zhang, X., et al. 1992; Zong and Huang 1985). They are also found and sites ranging from the middle Neolithic Period to the Bronze Age, such as at Shishanzui (*ca.* 5000 B.P.) (Yang, Z., et al. 1985) and Dadunzi (*ca.* 3300 B.P.) (LIACAS 1974:337; YPM 1977; Zhang, X. 1985).

Antlers and mandibles are the main evidence for the identification of the red muntjac in this study, as limb bones were generally assigned to the category of medium cervid. This was because of the variety of species of the genus *Muntiacus* found in Yunnan and the fact that their limb bones share many morphological similarities. The NISP and weight of red muntjac only contributed 6.7 percent and 9.8 percent of the total identified cervid specimens respectively, although it accounts for one fifth of the MNI. As discussed in chapter three, this obvious discrepancy may be attributed to the absence of proper sieving methods, which may have led to inadequate faunal sample collection.

Medium Cervids

The unidentified medium cervid remains are probably those of the red muntjac. In addition, the vertebrae identified as medium cervid may be assigned to sheep/goat, due to their similar sizes and the fragmentation of the specimens. This problem should not influence the general proportions in this study, although it may have inflated the sample size of medium cervids.

Reeve's Muntjac (*Muntiacus reevesi*)

The Reeve's muntjac is the smallest species of the genus *Muntiacus.* It has a head and body length between 64–90 centimetres, the height at the shoulder is between 40–49 centimetres, and its weight is between 11–16 kilograms (Smith, A. and Xie 2008:465). In appearance it is a small reddish-chestnut muntjac with blackish brown limbs. Males have upper canines and small antlers (*ca.* 7 cm) characterised by a tiny branch (*ca.* 1–2 cm) and a short pedicle (*ca.* 4 cm) (Sheng 1992b:126). The Reeve's muntjac is found widely in central, southern, and southeastern China today (Sheng 1992b:126; Smith, A. and Xie 2008:465). It is present in the archaeological records at sites such as Zengpiyan (*ca.* 8000 B.P.) in Guangxi Province (Li, Y. and Han 1978), and Hemudu (*ca.* 7000–5000 B.P.) in Zhejiang Province (Wei, et al. 1989). The animal remains identified as *Muntiacus* sp. or small-sized deer which were found at sites scattered throughout more than half of China may have been Reeve's muntjacs as discussed above. In Yunnan, Reeve's muntjacs have been encountered at sites such as Tangzigou (*ca.* 7000 B.P.) of the late Palaeolithic Period (Zhang, X., et al. 1992; Zong and Huang 1985).

At Haimenkou, the Reeve's muntjac was identified by the presence of one antler and one mandible in Phases II and III. Other limb bones that possibly belonged to this muntjac could only be identified as small cervids since there were other small deer species in this faunal assemblage.

Moschidae

Forest Musk Deer (*Moschus berezovskii*)

Musk deer (*Moschus* sp.) are small deer without antlers and both genders have prominent upper canines (Smith, A. and Xie 2008:454). The forest musk deer (*Moschus berezovskii*) is a species of the genus *Moschus* whose head and body length is between 63–80 centimetres, height at the shoulder is less than 50 centimetres, and weight is only 6–9 kilograms. Today the forest musk deer is widely distributed in central and southern China (Smith, A. and Xie 2008:455). They inhabit coniferous or broadleaf forests, or mixed forests at high elevations of 2,000–3,800 metres, and feed on leaves, grasses and shoots (Smith, A. and Xie 2008:455).

The genus *Moschus* does not have antlers, which is an important and distinct element for cervid species identification. Moreover, there are at least six species of the genus *Moschus* distributed in China, and they have similar skeletal characteristics. Consequently, it is difficult to identify musk deer remains with certainty. This has led to most musk deer remains being assigned to the genus *Moschus*[3] , such as at Jiangzhai (*ca.* 6700 B.P.) (Qi 1988). In Yunnan, the *Moschus* sp. was only present at the Dadunzi site (*ca.* 3300 B.P.) (Zhang, X. 1985). At Haimenkou, only one canine found in Phase III was considered to belong to a forest musk deer, based on comparisons with the faunal skeleton collection of the Reference Collection Department of the KIZCAS, and the modern geographic distributions of different *Moschus* species. The remains that were assigned to the small cervid category may also contain bone fragments of the forest musk deer.

Small Cervids

The small cervid category may contain more species of

[3] All the musk deer found from the archaeological sites of China were identified as '*she*', which in Chinese means 'musk deer', and refers to the genus *Moschus*. However, the scientific name used with '*she*' was '*Moschus moschiferus*', which is a species of the genus *Moschus* that is only found in northeastern and northwestern China today. This confusion might result from the confusion in the classification of *Moschus* species. Hence in this study all the '*she*' found in previous archaeological reports are described as '*Moschus* sp.'.

small-sized deer other than the Reeve's muntjac and forest musk deer. The remains of small cervids appeared during Phases II and III. Even when the three specimens of the Reeve's muntjac and forest musk deer are added, small cervids account for less than five percent of the total cervid specimens based on NISP, 15.6 percent based on MNI, and just less than one percent based on bone weight.

Suidae

Wild Boar (*Sus scrofa*)

Wild boars are still widespread across China, except in the driest desert and high plateau areas. They are found in all wild habitats from forest to scrub, grasslands, and swamps areas (Smith, A. and Xie 2008:452).

Wild boar remains are commonly discovered at archaeological sites in China. They are distinguished from the domestic pig in several ways. The most prominent of these is a larger body size, especially in later prehistory. In this study, the identification of wild boar is based on the size of the specimens. These are consistently larger than those of the domestic pig recovered throughout the four phases, and even larger than the modern female wild boar of the comparative collection (see Appendix 2).

Only two specimens of the wild boar were found in the Haimenkou faunal assemblage, including one almost complete left radius and one left ulna which was broken into two pieces before excavation. These two elements can be re-fitted together and thus probably belonged to one individual. Their low representation may indicate the high level of the development of pig domestication.

4.3.2.2 Primates

Cercopithecidae

Macaque (*Macaca* sp.)

Macaques are highly intelligent and versatile monkeys that are currently distributed mostly in central and southern China (Smith, A. and Xie 2008:161). They have occasionally been found at archaeological sites in China (e.g., Qi 1988; Wei, et al. 1989; Zhou 1981), and most macaque remains were identified as Rhesus macaque (*Macaca mulatta*) although at some sites just a few broken specimens appeared (e.g., Huang, W. and Ji 1963). In the reports concerning the sites in Yunnan, macaque remains were identified as *Macaca* sp. (e.g., Zhang, X. 1985) or other species of the genus *Macaca*, such as the stump-tailed macaque (*Macaca arctoides*) (Zhang, X., et al. 1992). Indistinct remains of macaque are often identified as *Macaca* sp. because the bones of different species, especially the limb bones, within this genus resemble each other in many ways. Due to the insufficiency of macaque specimens in the faunal assemblage from Haimenkou (only limb bones, NISP = 3), the remains of macaque could only be recognised as *Macaca* sp. in this book.

4.3.2.3 Lagomorpha

Leporidae

Hare (*Lepus* sp.)

At least ten species of the genus *Lepus* (hare) can be found in China, with their distribution almost covering the entire mainland (Smith, A. and Xie 2008:287-292). Therefore, it is not surprising that hare remains are often recovered from archaeological sites in China (e.g., Li, Y. 1963; Li, Y. and Han 1959; Qi 1988). Usually it is difficult to identify them to a particular species. Thus hare remains found at sites are often assigned to *Lepus* sp., as has been done at other sites in Yunnan (e.g., Zhang, X. 1985).

Seven specimens of hares were revealed at Haimenkou. They were identified as *Lepus* sp., although it is possible that they are in fact the Yunnan hare (*Lepus comus*) based on analysis of modern geographic hare distribution. The hare only contributes 0.4 percent of the total identified specimens in the faunal assemblage based on NISP, 1.3 percent based on MNI, and less than 0.05 percent based on weight. However, if gaur and cervids are removed from the analysis, in terms of the proportion of other smaller wild mammal species only, the hare accounts for 15 percent of remaining wild mammal taxa based on NISP. This indicates its importance in Haimenkou subsistence, and that hare may have been a common game animal at Haimenkou.

4.3.2.4 Rondentia

Spalacidae

Bamboo Rat (*Rhizomys* sp.)

The genus *Rhizomys* comprises three large stocky burrowing rodents, and all these species can be found in China (Smith, A. and Xie 2008:213). At present, they are widespread in southern China (Smith, A. and Xie 2008:213-214), and are often found at prehistoric sites (e.g., Li, Y. and Han 1978; Zhou 1994) scattered in most Chinese territories, including Yunnan (e.g., Zhang, X. 1985; Zhang, X., et al. 1992). Usually the bamboo rat found at sites is identified as the Chinese bamboo rat (*Rhizomys sinensis*). Nevertheless, in this study the bamboo rat remains could only be identified as *Rhizomys* sp., because only one upper incisor was present meaning that no accurate identification of the tooth could be made.

Muridae

Rat (*Rattus* sp.)

A few tiny specimens found in the faunal assemblage from Phases II and III of Haimenkou, including a tooth and five limb bones, are thought to belong to rats. Due to the excessive number of species within the genus *Rattus* and other similar genus of the family Muridae, and the wide distribution of rats in modern China, it is nearly impossible to identify these tiny and fragile bones to a species level.

Hystricidae

Porcupine (*Hystrix brachyuran*)

The porcupine is widely distributed across southern China and inhabits forests and open fields where large burrows are dug into banks and under rocks (Smith, A. and Xie 2008:274). Its diet consists of roots, tubers, bark, and so on (Smith, A. and Xie 2008:274). This species is occasionally encountered at prehistoric sites in China (e.g., Li, Y. and Han 1978; Wei, et al. 1989), and it also appears at sites in Yunnan (e.g., Yang, Z., et al. 1985; Zhang, X. 1985). At Haimenkou there is just one ulna from Phase II that was identified as coming from a porcupine. The identification was made based on its similarity to the porcupine skeleton from the faunal skeleton collection of the Reference Collection Department of the KIZCAS.

As these three kinds of rodents, bamboo rat, rat, and porcupine, are animals that burrow, they may have invaded the Haimenkou site after occupation by prehistoric people had ceased. Possible disturbance of the strata by burrowing animals at the site is currently unclear. They may have made no contribution to the subsistence economy of the Haimenkou people.

4.3.2.5 Carnivora

Ursidae

Bear (*Ursus* sp.)

The genus *Ursus* contains the largest terrestrial carnivores in the world. In China, they are massive in size and brown, dark brown, or black in colouration (Smith, A. and Xie 2008:425). There are two species of the genus *Ursus* that occur in China, the brown bear (*Ursus arctos*) and the Asian black bear (*Ursus thibetanus*) (Smith, A. and Xie 2008:425-427). The brown bear can weigh more than 200 kilograms. It used to be one of the most widely distributed land mammals on earth, while today its distribution in China is reduced to the northeast, northwest, and southwest regions (Lu, C. 1962; Smith, A. and Xie 2008:425). It lives in a wide range of habitats including dense forest, subalpine mountain areas, and tundra (Smith, A. and Xie 2008:426). The Asian black bear is smaller than the brown bear. There is a distinct white patch on its chest that is sometimes in the shape of a 'V' (Smith, A. and Xie 2008:426). The Asian black bear is widespread in central, southern, and northeastern China, inhabiting oak, broadleaf, and mixed forests, and preferring forest hills and mountains (Smith,. and Xie 2008:427). It may be encountered at altitudes over 3,000 metres in summer (Smith, A. and Xie 2008:427).

A small number of remains from both bear species have been encountered in the faunal assemblages from prehistoric sites in China (e.g., Jia and Zhang 1989; Qi 1977; 1988; Wei, et al. 1989). Both of the species can still be found in Yunnan today (KIZCAS 1999:37), and their remains have been reported in archaeological records (e.g., YPM 1981; Zhang, X. 1985; Zhang, X., et al. 1992; Zong and Huang 1985).

There were seven specimens of limb bones of the genus *Ursus* found at Haimenkou in Phases I–III, including one radius and one ulna that could be re-fitted. Due to the lack of comparative skeletal collections, these specimens could only be identified as *Ursus* sp.

Canidae

Red Fox (*Vulpes vulpes*)

The red fox is the largest animal in the genus Vulpes. It is usually reddish brown in colouration, with long and slender black legs (Smith, A. and Xie 2008:421). At present the red fox is distributed throughout almost all of China, and adapted to all kinds of habitats including deserts, forests, and major metropolitan areas (Smith, A. and Xie 2008:422). The red fox is a species occasionally found at archaeological sites (e.g., Lu, H. and Zhou 1990), but it has so far been absent in the archaeological records of Yunnan.

There was only one mandible found at Haimenkou from Phase II that was identified as a red fox, based on the comparison with its counterpart from the faunal skeleton collection of the Reference Collection Department of the KIZCAS. There were a few specimens assigned to the category of small carnivore, as they lack important features for identification. It is possible that some of these remains identified as small carnivore actually belonged to the red fox.

Viverridae

Masked Palm Civet (*Paguma larvata*)

As could be inferred from its common name, the masked palm civet (*Paguma larvata*) generally has a 'mask' in the facial region. This 'mask' consists of a median white strip from the top of the head to the nose, a white mark below each eye, and a white mark above each eye extending to the base of the ear and below (Nowak 1991:1154). It is widespread across southeastern China, lives in a variety of forest habitats from primary growth evergreen to second-growth deciduous forest, and frequents agricultural areas (Smith, A. and Xie 2008:409).

Like the red fox, the masked palm civet is occasionally encountered at prehistoric sites in southern China (e.g., CPAMJP 1963; Wei, et al. 1989). It was also found at the Cishan site in North China, which may indicate warmer weather there around 8,000 years ago (Zhou 1981). In Yunnan, the masked palm civet has been recovered at the late Palaeolithic sites (e.g., Zhang, X., et al. 1992), and was also present in the middle Neolithic faunal assemblages (e.g., Yang, Z., et al. 1985).

Three humeri of the masked palm civet from Phases II and III, representing two individuals, were indentified at Haimenkou. It was the first time that the masked palm civet had been found at a site dating from the late Neolithic Period to the Bronze Age in Yunnan.

Small Felids

Three specimens, comprising one radius from Phase III, and two tibiae from Phases II and III respectively, were only identified as small felids. Based on the size and characteristics of these bones, they may belong to species such as the marbled cat (*Pardofelis marmorata*), the leopard cat (*Prionailurus bengalensis*), or some other species belonging to the family Felidae.

Small Carnivores

Eight specimens were assigned to the 'small carnivore' category, consisting of three mandibles, one canine, two humeri, and two femurs. These specimens from Phases I–III lack diagnostic features to allow more precise identification.

4.3.3 Birds

There are 802 species of bird distributed in Yunnan. These account for more than 60 percent of birds in China today and about 9 percent of birds worldwide (KIZCAS 1999:61). The bird remains from Haimenkou were identified by Mr. Shuo Wang from the Key Laboratory of Evolutionary Systematics of Vertebrates, IVPPCAS.

Twenty-one specimens of identified bird were recovered at Haimenkou, contributing 2.7 percent of the total identified wild animals based on NISP, 12.1 percent based on MNI, and less than 0.2 percent based on weight. Due to the fragmentary nature of these specimens and the lack of a comparative collection, most of the specimens were only identifiable to genus or family level.

4.3.3.1 Anseriformes

Anatidae

Anas sp.

The genus *Anas* includes a variety of dabbling ducks such as wigeons, pintails, teals, mallards, and shovelers (Zhang, C., et al. 1997:10-15). Fourteen specimens were identified as *Anas* sp. in the faunal assemblage from Haimenkou, and these belonged to Phases I–III. These include humeri, coracoids, and tibiotarsi. The remains of the genus *Anas* dominate the bird remains, accounting for 66.7 percent of the total of identified bird bones based on NISP, more than half based on MNI, and a quarter based on weight.

Cygnus sp.

Cygnus is a genus of swans that includes the mute swan, whooper swan, whistling swan, and others (Zhang, C., et al. 1997:8). There are only two ulnae identified as *Cygnus* sp. that were recovered from Haimenkou. Since there is only one species, the whooper swan (*Cygnus cygnus*), currently present in the region of Yunnan (KIZCAS 1999:70), the remains probably belong to this species.

In addition, there are two humeri that could be only assigned to the family Anatidae due to the lack of a comparative collection.

4.3.3.2 Galliformes

Phasianidae

Gallus gallus

The *Gallus gallus* is a species under the family Phasianidae. Its remains have been unearthed from several sites in China. Some were identified as wild (i.e., the red junglefowl) (e.g., Chen, Q. 1993), while others were recognised as domestic chicken (e.g., Zhou 1981; 1994). Zhou and his colleagues (West and Zhou 1988; Zhou 1981) suggested that the first domestic chicken in China was recovered at Cishan in North China dating to around 8000 B.P., but this is debatable (e.g., Yuan 2001).

Only one specimen, a femur, of *Gallus gallus* was recovered at Haimenkou. The *Gallus gallus* specimen was described as possibly domesticated at the Dadunzi site (*ca.* 3300 B.P.) in Yunnan, in a similar time span to Haimenkou, because of the presence of a chicken-form pottery pot (Figure 4.14) (YPM 1977; Zhang, X. 1987). However, so far there are no reliable criteria to identify

Figure 4.14 Chicken-Form Pottery Pot from Dadunzi (ca. 3300 B.P.) (After YPM 1977: plate 10-4; reproduced with permission of Archaeology Press)

domestic chicken from the junglefowl based on the femur (Deng, et al. 2013). Thus this specimen is considered to be from a wild animal in this study temporarily.

In addition, there is one tibiotarsus from bird remains that could only be identified to the family Phasianidae.

4.3.3.3 Gruiformes

Gruidae

One femur of the bird remains could only be identified to the family Gruidae, which contains cranes, such as common cranes, red-crowned cranes, as well as others (ECFSAS 1963:230-235).

4.3.4 Other Taxa

4.3.4.1 Fish

Haimenkou is a typical waterside site, with net sinkers frequently present in the artefact assemblage (see chapter two). This indicates the significant contribution of fish to the daily diet of the Haimenkou people. Even several decades ago, people who lived around Jianhu Lake were still part-time fishermen (POYPM 1958:11). However, tiny bones of fish were rarely found in the faunal assemblage from the Haimenkou site. This may indicate the inadequate collection of animal remains, as discussed in chapter three. Only seven specimens of fish were collected at Haimenkou, including six dorsal fins and one rib. These specimens could not be identified even to family level.

4.3.4.2 Molluscs

As a site located beside a lake, Haimenkou should have had a large quantity of freshwater molluscs. Shells of molluscs are expected to survive much better than the bones and teeth of mammals (Reitz and Wing 2008:203). Therefore, it is reasonable to assume that large quantities of mollusc shells were present in the faunal assemblage from Haimenkou. However, the results of the 2008 excavation season point to the contrary. Only two pieces of shell representing one individual were recovered at Haimenkou from Phase II. This was identified as *Unio douglasiae*, a freshwater mussel that is widespread across China. As with fish remains, this under-representation may be attributed to the lack of the systematic sieving during excavation.

4.3.5 Comparison with Faunal Assemblage from Dadunzi

Only a few Neolithic sites in Yunnan have animal remains that were systematically collected, and even fewer zooarchaeological reports have been published. Of those sites which have published zooarchaeological data, the Dadunzi site (*ca.* 3300 B.P.) (LIACAS 1974:337; YPM 1977) is the only one contemporaneous with Haimenkou. Therefore, comparison between the faunal assemblages from Dadunzi and Haimenkou may provide more information on people's subsistence patterns.

The Dadunzi site is located in Yuanmou County, Chuxiong *Yi* Minority Autonomous Prefecture in northern Yunnan Province (see location in Figure 2.23). The site covers about 5,000 square metres, of which an area of 496 square metres has been excavated through three seasons, between February 1972 and January 1973 (YPM 1977; Zhang, X. 1987).

The remains of 15 dwellings, seven hearths, four storage pits, 37 tombs, and a large quantity of implements, objects of daily use, and animal remains were excavated at Dadunzi. The finds were assigned to two stages of development on the basis of stratigraphic evidence, of which the earlier stage is dated to 1260±90 B.C. (LIACAS 1974:337).

There were more than 1,300 animal bone fragments recovered at Dadunzi, of which more than 300 were identified (not including molluscs) (Zhang, X. 1985:29). The Dadunzi report only provides the NISP of each species (Zhang, X. 1985), and comparisons are made here with the Haimenkou site in Table 4.8. As discussed in Section 4.3.2.1, X. Zhang (1985) did not provide strong evidence to support his identification of domestic gaur. Therefore, in the comparison below the gaur from Dadunzi is also treated as a wild animal. To allow comparison, the domestic animals from Haimenkou and the domestic and possibly domestic animals from Dadunzi are discussed in one category as 'domestic and possibly domestic animals', the forest musk deer from Haimenkou and the Siberian musk deer from Dadunzi are combined into one category as 'musk deer' (*Moschus* sp.), and the rats from both sites are combined into one category as 'rat' (family Muridae). The remains of fish and molluscs are not included in this comparison due to the inadequate collection of tiny bones at Haimenkou as well as the insufficient description of fish and molluscs from the Dadunzi site.

From Table 4.8, it is clear these two sites have common major species. Pigs and dogs were identified as domestic animals at both sites, and sheep & goats were considered domesticated at Haimenkou and possibly domesticated at Dadunzi due to the dearth of specimens. These animals account for the majority of the faunal assemblages in both sites. Based on NISP, the domestic and possibly domestic animals make up approximately 57 percent of the total indentified specimens of mammals at Haimenkou, and about 43 percent at Dadunzi. The pig is apparently the most important domestic animal at the two sites.

The relative proportions of wild animals are similar at both sites. The gaur dominates wild mammals, accounting for 56 percent of the wild animals at Dadunzi and 50 percent at Haimenkou. Moreover, at both sites large cervid remains are the second most common wild animal, followed by medium cervids and small cervids.

Table 4.8 Comparison between Dadunzi and Haimenkou Specimens by Range of Taxa and Relative Abundance based on NISP

Domestic/Wild	Taxon	Common Name	DDZ	HMK
Domestic and Possibly Domestic	*Sus domesticus*	Pig	108	931
	Canis familiaris	Dog	22	25
	Ovis & Capra	Sheep & Goat	1	65
Wild	*Bos gaurus*	Gaur	95	387
	Cervus unicolor	Sambar	31	23
	Axis porcinus	Hog Deer	—	2
	Large Cervid	Large Deer	—	147
	Muntiacus muntjak	Red Muntjac	>10	21
	Medium Cervid	Medium Deer	—	108
	Muntiacus reevesi	Reeve's Muntjac	—	2
	Moschus sp.	Musk Deer	2	1
	Small Cervid	Small Deer	—	11
	Sus scrofa	Wild Boar	—	3
	Macaca sp.	Macaque	1	3
	Vespertilionidae	Bat	1	—
	Ursus sp.	Bear	7	7
	Vulpes vulpes	Red Fox	—	1
	Prionailurus bengalensis	Leopard Cat	—	1
	Paguma larvata	Masked Palm Civet	—	3
	Small Felid	Small Felid	—	3
	Small Carnivore	Small Carnivore	—	8
	Sciurus sp.	Squirrel	>1	—
	Lepus sp.	Hare	1	7
	Rhizomys sp.	Bamboo Rat	6	1
	Hystrix brachyura	Porcupine	3	1
	Muridae	Rat	6	7
	Bird	Bird	5	21
Total			≈300	1789

Key: DDZ = Dadunzi; HMK = Haimenkou; — = no data available in this category; > = more than; ≈ = about.
Note: Data of Dadunzi is from Zhang (1985).

Based on the phases or stages of development, both of the sites show similar trends of amount, proportion, and diversity of animals. As displayed by Table 4.9[4], the amount of animal remains from both sites increased from the earlier to the later periods, with an overall dominance of domestic and possible domestic animals and an increased diversity of wild animal taxa. As Haimenkou has four phases and three of them fall within the scope of this study (Phases I–III), whereas Dadunzi only presents two stages of development, the analysis of Haimenkou is more complex than that of Dadunzi. However, the trend of increasing numbers of animal bones at both sites is similar.

[4] The data of the Dadunzi site is limited to five grids, so the sum of NISPs of Dadunzi in table 4.9 is less than in Table 4.8.

Table 4.9 Comparison between Dadunzi and Haimenkou by Period

SIT	Period	NISP-D	NISP-W	Diversity-W
DDZ	Earlier Stage	12*	3*	>4
	Later Stage	77*	11*	>6
HMK	Phase I	57	46	>6
	Phase II	454	310	>15
	Phase III	500	405	>14

Key: SIT = site; NISP-D = NISP of domestic and possibly domestic animals;
NISP-W = NISP of wild animals; Diversity-W = diversity of wild animals.
For other abbreviations, see Table 4.8.
Note: *The numbers in the category NISP-D contains gaur bone fragments at Dadunzi, but detailed numbers were not given in the Dadunzi report (Zhang, X. 1985). Therefore, in this table the numbers in NISP-D and NISP-W of Dadunzi is not precise, and the actual situation seems similar to the situation at Haimenkou.

4.4 Summary

Based on the analysis of range and relative importance of identified taxa, it is evident that domestic animals contribute a significant proportion of the faunal assemblage from the Haimenkou site. This trend increases throughout Phases I to III. Of the domestic animals, pigs are most common and play an important role in the subsistence economy of the Haimenkou people. The remains of the small-sized dog at Haimenkou demonstrate its presence in Yunnan from the late Neolithic Period to the Bronze Age, and suggest some connection between Haimenkou and southwestern and eastern Yunnan in prehistory. The remains of domestic sheep may indicate animal introductions and cultural influences from northwestern China, Gansu and Qinghai Provinces, from the beginning of Phase II onwards.

The gaur is the most common wild animal, possibly showing its significant status as a game animal. Large cervids are also important in the subsistence economy of the Haimenkou people. At the same time, the presence of other wild mammal remains suggests a mixed spectrum of prehistoric human subsistence at Haimenkou. In addition, the low numbers of fish remains and other aquatic animals may reflect inadequate collection of specimens, due to a lack of sieving during the excavation.

The abundance of all domestic and wild animals is concentrated in Phases II and III, after a dramatic increase in faunal remains between Phases I and II. This pattern may suggest greater progress in both animal husbandry and hunting during the period from the early to the middle Bronze Age. It may also indicate the marked human population growth at the site over the same period, corresponding with the considerable development that took place at the transition from the late Neolithic Period to the early Bronze Age.

In the next chapter, the analysis of skeletal part representation and bone modification of the faunal assemblage recovered from Haimenkou will be discussed.

5

Skeletal Part Representation and Bone Modification

5.1 Introduction

It seldom happens that the skeletal part representation of a faunal assemblage recovered from an archaeological site matches the expected anatomical scenario. If as simply assumed by Theodore White (1952; 1953; 1954; 1955) that each bone had an equal chance to survive post-mortem disturbance, patterns of skeletal part representation revealed at sites could be interpreted just from the aspects of human behaviours. These human activities include the choice of which body parts of prey were carried back to the residential site and which parts were dumped at the kill site (e.g., Perkins, Dexter Jr. and Daly 1968). These human choices have been labelled 'selective transport' by Klein (1989:363), who pointed out that the discrepancies between the anatomical expectation and the real skeletal part representation present at different sites could not only reflect the selective transport by humans (preferential import or removal of some parts), but also selective destruction including a variety of pre- and post-depositional destructive processes. Therefore, both human behaviours and natural taphonomic processes play important roles in affecting representation of animal skeletal parts.

In addition to frequencies of specimens from different parts of animal skeletons, the modification of animal bones, including butchering marks, traces of burning, and animal gnaw marks also constitute basic data for the study of the skeletal part representation (e.g., Behrensmeyer 1978; Fisher 1995; Marshall L. 1989; Shipman and Rose 1983; White, Theodore 1956).

Based on multiple lines of evidence, this chapter will further discuss the disparities between the anatomical expectation and the skeletal part representation occurring among various taxa and different periods at Haimenkou. Human behaviours, such as animal husbandry, hunting pattern, butchering custom, transport choice, and food preparation, as well as nonhuman forces acting on the faunal assemblage, will be discussed in this chapter.

5.2 Skeletal Part Representation

5.2.1 Methods

The study of skeletal part representation in this book begins by classifying specimens into different skeletal portions. These are usually based on skeletal elements, anatomical regions, or butchering units (Reitz and Wing 2008:216). The sum of skeletal part representation for each taxon is displayed in Tables 5.1 to 5.6, based on MNE (minimum numbers of elements), MAU (minimum animal unit), and %MAU.

The term MNE was defined by Binford (1984:50-51) as the 'minimum numbers of elements', referring to the minimum number of different specimens identifiable to a given anatomical part used in classification. The 'elements' used in this definition actually do not represent complete elements, but parts of large elements such as the proximal tibia and the distal femur or small elements such as the astragalus. The estimate of the MNE for a certain skeletal part is based on the assumption that a complete or anatomically recognisable distal tibia had originally been present. For example, the way to count the MNE for proximal tibiae in a faunal assemblage is as follows. If there are three left and four right proximal tibia fragments in a faunal assemblage, and two of the left ones could come from the same individual based on visual matching, then the MNE for proximal tibiae would be (3 – 1) + 4 = 6. MNE was also defined by Lyman (1994b:290) as the 'minimum number of complete skeletal elements necessary to account for all observed specimens' focusing on the complete element rather than fragments. Lyman's method will not be applied in this study.

MAU (minimum animal unit) employed in this book was also developed by Binford (1984:50-51) as a conversion of MNE for comparative purpose. The anatomy of living animals is taken as the standard for calculating MAUs. For instance, if in the body of a living animal there are two distal tibiae and the MNE of this species is 10, then the MAU is derived from dividing the MNE for distal tibia by two (10 ÷ 2 = 5). After deriving MAUs based on different skeletal parts, the largest MAU is employed as the denominator and each MAU is used as the numerator. Then, the resulting decimal is multiplied by 100, leading to the percentage of MAU (%MAU) as the measure for comparison.

For further discussion, FUI (food utility index) and bone density will be used to provide auxiliary information in this book. FUI, which simply scales variation in the amount of meat, marrow, and bone grease associated with different caribou body parts (Metcalfe and Jones 1988), is a simpler version of the MGUI (modified general utility index) proposed by Lewis Binford (1978). FUI is employed to explain the human choice of transport or discarding of different skeletal parts. For example, if more skeletal parts (i.e., high %MAU values) with high FUI values, such as proximal femurs, of an animal were found at site A, whereas less skeletal parts with low FUI values, such as phalanges, of the animal were present at site B, it may signify that this animal was slaughtered at site B rather than site A, and then the skeletal parts with high FUI values were transported to site A while the skeletal parts with low

FUI values were discarded at site B. Therefore, it may be inferred that this animal may have been a game animal, site A was a residential settlement, and site B was a kill site. From another perspective, it is also possible there were commercial activities between sites A and B, and the skeletal part representation of this animal was a result of commercial activities.

According to Binford (1981:217), it is 'apparent that the survival potential of a given bone subjected to destructive agents is a function of its strength or simple ability to resist destructive agents...resistance to destruction was monitored by measures of bone density'. Therefore, to determine which bones are more likely to survive, bone density measures are employed in this book. The volume density (VD) which was recommend by Lyman (1984:282-283) is used here, also the values of volume density are taken from Lyman (1984; 1985).

In the next section, MNE, MAU, and %MAU will be calculated for the main taxa of mammals in the faunal assemblage from Haimenkou, and %MAU will be plotted against FUI and bone density respectively when available. It will then be possible to conduct comparisons among different animal taxa and phases based on these data.

5.2.2 Results and Analysis

Tables 5.1 to 5.6, with derived Figures 5.1 to 5.11, display the skeletal part representation of the different taxon of mammal remains at Haimenkou.

A total of 29 skeletal parts are summarised in these tables, which could be divided into four categories for the purpose of comparison: the head includes the cranium and mandible; vertebral column contains the atlas, axis, and sacrum; the upper limbs consist of the scapula, humerus, radius, ulna, pelvis, femur, and tibia; and the lower limbs comprise carpal, tarsal, metacarpal, metatarsal, and phalanges. The vertebrae other than the atlas, axis, and sacrum are not included in these tables, due to the difficulty of identifying these bones. In these tables, each long bone is represented by proximal and distal portions, which means even an intact long bone, with both ends existing, is still described by two skeletal parts. Bone shafts are not present in these tables, except when there is an absence of either proximal or distal end of the long bone. For the cranium, the number of the most frequent element was counted, such as the temporal bone. This is also done for the carpals: the most frequently encountered carpal bones are counted and used to represent the MNE for the carpal, for example, the ulnar carpal.

Due to the lack of taxonomic abundance, the analysis of skeletal representation is less meaningful for some animal taxa, such as bear, fox, and hare. Therefore, they will not be included in this chapter. For the sake of comparison, animal categories in this chapter include gaur, pig, sheep & goat, large cervids, medium cervids, and small cervids. The comparisons are based on the total skeletal part representation of each animal category from the whole human occupation at Haimenkou in prehistory (Phases I–III). The Late Phase is present as supplementary data. The results will be shown as follows.

5.2.2.1 Gaur

As summarised by Tables 5.1a–c, the skeletal part representation of gaurs in each phase is clear. Similar distribution of skeletal parts could be observed from Figures 5.1a–b that were derived from Tables 5.1a–b and display the skeletal part representation from Phases I to III.

As can be seen in Figures 5.1a–b, in Phase I there are no head parts present, and the kinds of skeletal parts are more complete in the later phase. Phase III exhibits the most complete kinds of skeletal parts aside from the proximal radius and distal metacarpal. Apart from this, no significant changes occur from Phases I to III, especially between Phases II and III. Although no fragments representing head parts appear in Figure 5.1a, there were several loose teeth of gaur found in Phase I. Also, according to the small sample size and disappearance of other skeletal parts in Phase I, the absence of gaur head parts in this phase does not necessarily indicate they were not returned to the site. It may indicate some sampling or recovery bias. Moreover, as figures for Phase II and III suggest, almost every body part of gaur is represented. Therefore, it is more likely that gaurs were slaughtered at the site.

To show the complete presence of the skeletal representation of the gaur, Figure 5.1c displays the total data of gaur remains from Haimenkou throughout the time span of human occupation (the data of the Late Phase is also added). Obviously the %MAUs for the calcaneus and pelvis are highest, whereas the patella and third phalanx are poorly represented and other skeletal parts are encountered more evenly. Almost every long bone element has been unevenly presented at two ends, especially in the humerus and radius, which may be caused by the differential densities of two ends. Because of the large body size, even the small elements of gaur, such as the calcaneus and a variety of phalanges, were relatively easy to collect and thus the %MAUs was reasonably high, although sieving was not used in sample collection. To more clearly elucidate this scene, %MAU is plotted against FUI and VD (volume density) respectively in Figures 5.2 and 5.3. FUI values are from Metcalfe and Jones (1988: table 2), and VD values are taken from Lyman (1984: tables 6 and 10; 1994c: table 7.6, column bison).

Because of the similarities of body size and skeleton structure shared by the *Bos* sp. and large cervids, the FUI values for the caribou, a large-sized deer, are employed here as no FUI data for *Bos* has been published. According to Figure 5.2, a positive but not significant (R^2=0.099) relationship between FUI values and %MAU can be observed, which indicates no strong relationship between economic utility and representation of the skeletal parts of gaurs.

Table 5.1a Skeletal Part Representation of Gaur from Haimenkou (Phases I and II)

Skeletal Part	Phase I			Phase II		
	MNE	MAU	%MAU	MNE	MAU	%MAU
Cranium	0	0.00	0.0	1	0.50	14.3
Mandible	0	0.00	0.0	2	1.00	28.6
Atlas	0	0.00	0.0	0	0.00	0.0
Axis	0	0.00	0.0	0	0.00	0.0
Scapula	1	0.50	50.0	2	1.00	28.6
Humerus, proximal	0	0.00	0.0	1	0.50	14.3
Humerus, distal	1	0.50	50.0	3	1.50	42.9
Radius, proximal	0	0.00	0.0	1	0.50	14.3
Radius, distal	1	0.50	50.0	3	1.50	42.9
Ulna, proximal	0	0.00	0.0	1	0.50	14.3
Pelvis	1	0.50	50.0	7	3.50	100.0
Sacrum	0	0.00	0.0	1	1.00	28.6
Femur, proximal	0	0.00	0.0	2	1.00	28.6
Femur, distal	2	1.00	100.0	2	1.00	28.6
Patella	0	0.00	0.0	0	0.00	0.0
Tibia, proximal	0	0.00	0.0	2	1.00	28.6
Tibia, distal	1	0.50	50.0	2	1.00	28.6
Carpal	1	0.50	50.0	3	1.50	42.9
Metacarpal, proximal	1	0.50	50.0	2	1.00	28.6
Metacarpal, distal	0	0.00	0.0	0	0.00	0.0
Calcaneus	0	0.00	0.0	6	3.00	85.7
Astragalus	1	0.50	50.0	3	1.50	42.9
Tarsal, C+4	0	0.00	0.0	1	0.50	14.3
Metatarsal, proximal	0	0.00	0.0	2	1.00	28.6
Metatarsal, distal	0	0.00	0.0	4	2.00	57.1
Metapodial, distal	0	0.00	0.0	3	2.00	57.1
First phalanx	1	0.13	12.5	5	0.63	17.9
Second phalanx	0	0.00	0.0	8	1.00	28.6
Third phalanx	0	0.00	0.0	1	0.13	3.6

Key: Tarsal, C+4 = the central and fourth tarsal.
Note: Cranium and carpal are counted by most frequent elemental part.

The VD values in Figure 5.3 are taken from the values for bison (Lyman 1984: tables 6 and 10; 1994c: table 7.6, column bison). There is still no significant relationship appearing between VD values and %MAU of skeletal parts of gaurs ($R^2 = 0.0608$), although the tendency is positive. Nevertheless, this positive tendency may be interpreted partly by the differential volume densities for various skeletal parts, since the most frequently present skeletal part, the calcaneus, exhibits the highest density among all the elemental parts of a gaur skeleton. Gaur astragalus and pelvis bones also have a relatively high density.

The degree of fragmentation for gaur specimens is relatively high as almost every long bone is broken. This

Table 5.1b Skeletal Part Representation of Gaur from Haimenkou (Phase III and the Late Phase)

Skeletal Part	Phase III			Late Phase		
	MNE	MAU	%MAU	MNE	MAU	%MAU
Cranium	5	2.50	62.5	0	0.00	0.0
Mandible	2	1.00	25.0	0	0.00	0.0
Atlas	3	3.00	75.0	0	0.00	0.0
Axis	2	2.00	50.0	0	0.00	0.0
Scapula	2	1.00	25.0	1	0.50	100.0
Humerus, proximal	1	0.50	12.5	0	0.00	0.0
Humerus, distal	2	1.00	25.0	0	0.00	0.0
Radius, proximal	0	0.00	0.0	0	0.00	0.0
Radius, distal	5	2.50	62.5	0	0.00	0.0
Ulna, proximal	2	1.00	25.0	0	0.00	0.0
Pelvis	5	2.50	62.5	0	0.00	0.0
Sacrum	2	2.00	50.0	0	0.00	0.0
Femur, proximal	6	3.00	75.0	0	0.00	0.0
Femur, distal	5	2.50	62.5	0	0.00	0.0
Patella	2	1.00	25.0	0	0.00	0.0
Tibia, proximal	2	1.00	25.0	0	0.00	0.0
Tibia, distal	3	1.50	37.5	0	0.00	0.0
Carpal	2	1.00	25.0	0	0.00	0.0
Metacarpal, proximal	2	1.00	25.0	0	0.00	0.0
Metacarpal, distal	0	0.00	0.0	0	0.00	0.0
Calcaneus	8	4.00	100.0	0	0.00	0.0
Astragalus	6	3.00	75.0	0	0.00	0.0
Tarsal, C+4	3	1.50	37.5	0	0.00	0.0
Metatarsal, proximal	6	3.00	75.0	0	0.00	0.0
Metatarsal, distal	1	0.50	12.5	0	0.00	0.0
Metapodial, distal	4	2.50	62.5	0	0.00	0.0
First phalanx	17	2.13	53.1	0	0.00	0.0
Second phalanx	12	1.50	37.5	1	0.13	25.0
Third phalanx	8	1.00	25.0	0	0.00	0.0

Key: See Table 5.1a.
Note: See Table 5.1a.

may be due to gaur bodies having been dismembered into smaller chunks than smaller sized animals, such as pigs. At present, the people of the Dulong minority who live in northwestern Yunnan Province, mentioned in chapter four, still maintain age-old customs of sacrificing gayals, the semi-domestic cattle which may be the offspring of gaurs, when the New Year is coming. After the ceremony, the gayal's body is chopped into numerous chunks and distributed to the whole village (Li, G. and Lu 1987:175; Wang, W. and Long 2009:296-297). A similar custom in prehistory may be the reason for the high fragmentation of the large elements of gaurs.

In sum, the skeletal part representation of the gaur does not

Table 5.1c Skeletal Part Representation of Gaur from Haimenkou (All Phases)

Skeletal Part	Total		
	MNE	MAU	%MAU
Cranium	6	3.00	42.9
Mandible	4	2.00	28.6
Atlas	3	3.00	42.9
Axis	2	2.00	28.6
Scapula	6	3.00	42.9
Humerus, proximal	2	1.00	14.3
Humerus, distal	6	3.00	42.9
Radius, proximal	1	0.50	7.1
Radius, distal	9	4.50	64.3
Ulna, proximal	3	1.50	21.4
Pelvis	13	6.50	92.9
Sacrum	3	3.00	42.9
Femur, proximal	8	4.00	57.1
Femur, distal	9	4.50	64.3
Patella	2	1.00	14.3
Tibia, proximal	4	2.00	28.6
Tibia, distal	6	3.00	42.9
Carpal	4	2.00	28.6
Metacarpal, proximal	5	2.50	35.7
Metacarpal, distal	0	0.00	0.0
Calcaneus	14	7.00	100.0
Astragalus	10	5.00	71.4
Tarsal, C+4	4	2.00	28.6
Metatarsal, proximal	8	4.00	57.1
Metatarsal, distal	5	2.50	35.7
Metapodial, distal	7	3.50	50.0
First phalanx	23	2.88	41.1
Second phalanx	21	2.63	37.5
Third phalanx	9	1.13	16.1

Key: See Table 5.1a.
Note: See Table 5.1a.

indicate that gaurs revealed at Haimenkou were slaughtered elsewhere and then transported to Haimenkou. It is more possible that gaurs were slaughtered and consumed at Haimenkou itself, according to the relatively high frequencies for skeletal parts with low FUI values, such as

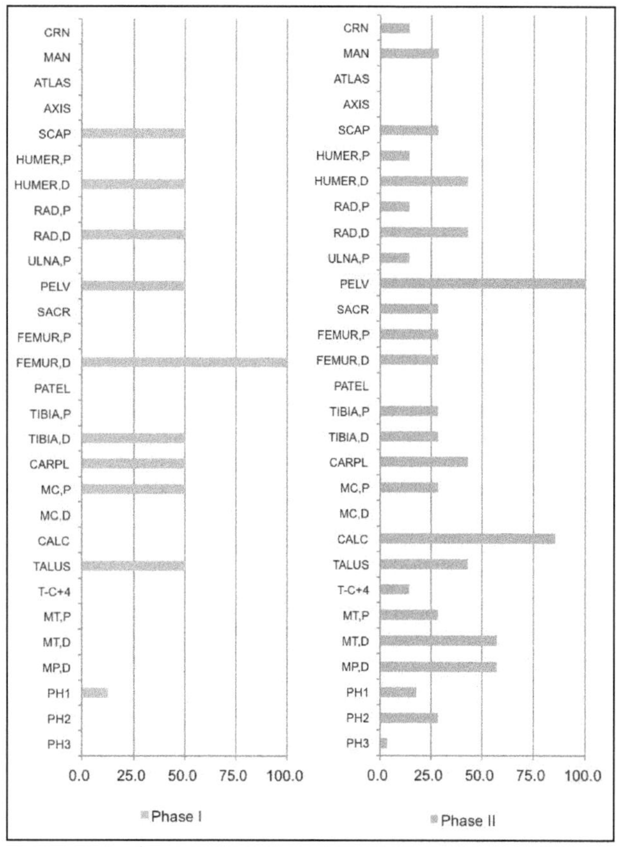

Figure 5.1a %MAU for Gaur from Haimenkou (Phases I and II)
Key: P = proximal, D = distal.
Note: See abbreviation for each element in Appendix 1: ELM.

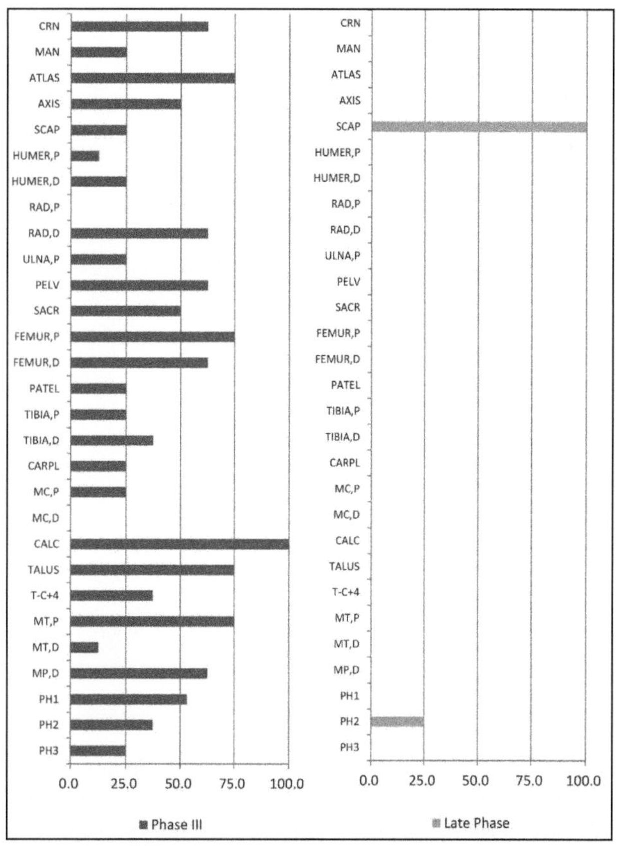

Figure 5.1b %MAU for Gaur from Haimenkou (Phase III and the Late Phase)
Key: See Figure 5.1a. Note: See Figure 5.1a.

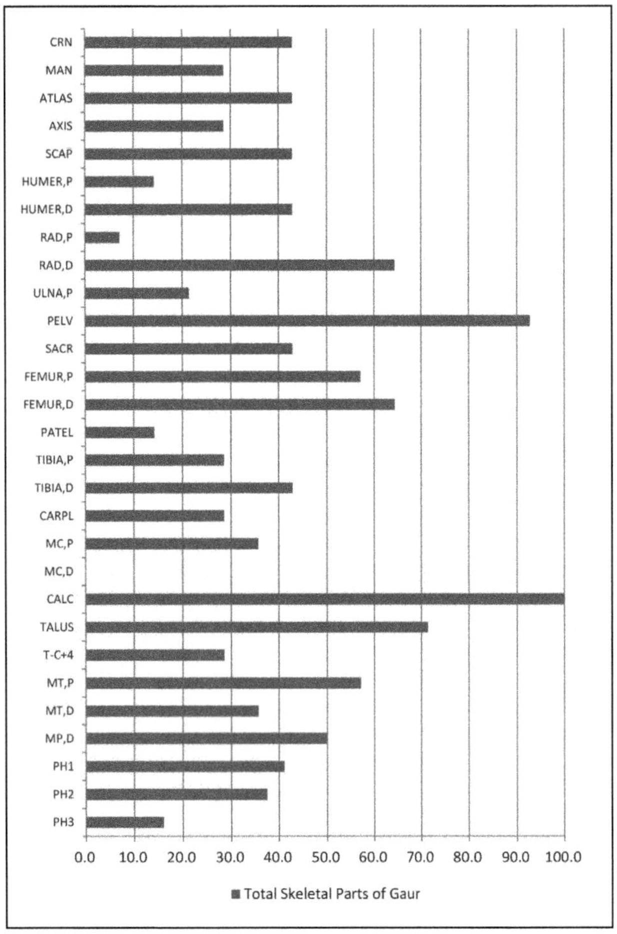

Figure 5.1c %MAU for Gaur from Haimenkou (Total Skeletal Parts)

Key: See Figure 5.1a. Note: See Figure 5.1a.

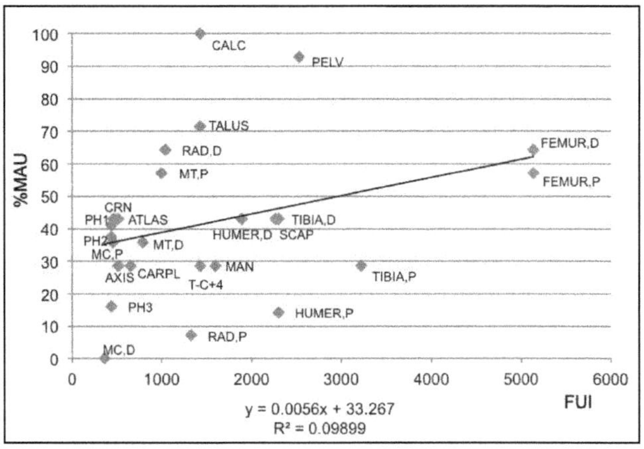

Figure 5.2 Scatterplot of the %MAU of Gaur Skeletal Parts from Haimenkou against FUI Values for Caribou

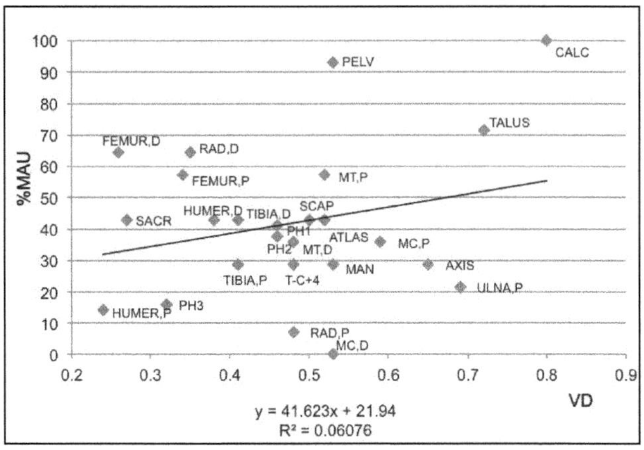

Figure 5.3 Scatterplot of %MAU of Gaur Skeletal Parts from Haimenkou against VD Values for Bison

the calcaneus, talus, and phalanges, and the high fragmentation of the large bones of gaurs.

5.2.2.2 Pig

The skeletal part representation of the pig is summarised in Tables 5.2a–c and also shown in Figures 5.4a–c. Unlike the gaur, there is no carpal or tarsal (except for the calcaneus and astragalus which have bigger sizes) present in the tables showing the skeletal part representation of the pig. There are also few appearances of metapodials (both proximal and distal ends), and the appearance of the patella and phalanges is even less frequent. The pig is considered a domestic animal at Haimenkou, thus it may not have undergone the butchering process outside this residential site, and all its body parts should have been present at the site. Therefore, the absence of these small elements of pigs may indicate inadequate or biased sample collection. The pig elements are much smaller than those of gaurs, for instance, a first phalanx of a gaur is larger than the biggest metapodial in a pig skeleton. Consequently, it is reasonable to attribute the absence of the small elements of pig to inadequate sample collection.

There are no published data available for pigs either by FUI or VD, thus no scatterplot figure for pig could be made for this book. To further elucidate the skeletal part representation of pig, the comparisons with other animal taxa will be conducted in Section 5.2.2.9.

5.2.2.3 Sheep & Goat

As discussed in chapter four, the sheep or goat remains from Haimenkou probably all belong to sheep. Since these two species have similar body sizes and skeletal structures, the bones that were identified as sheep or sheep/goat are combined in this section to built the skeletal part representation for sheep & goats.

The skeletal part representation of the sheep & goat is presented in Tables 5.3a–c, along with Figures 5.5a–b derived from these tables (there are no sheep & goat remains appearing in the Late Phase, so no figure is provided for this period). It is interesting that there are hardly any sheep & goat specimens appearing in Phase I, as this is in apparent contrast to the situation in Phases II and III. As discussed in chapter four, the sole specimen of sheep & goats found in Level (8) of Phase I may actually belong to Phase II, and sheep & goats may have been introduced into Haimenkou from northwestern China along with wheat from the beginning of Phase II (*ca.* 3700–3200 B.P.). Also, unlike the other

Table 5.2a Skeletal Part Representation of Pig from Haimenkou (Phases I and II)

Skeletal Part	Phase I			Phase II		
	MNE	MAU	%MAU	MNE	MAU	%MAU
Cranium	2	1.00	40.0	21	10.50	63.6
Mandible	1	0.50	20.0	33	16.50	100.0
Atlas	1	1.00	40.0	8	8.00	48.5
Axis	0	0.00	0.0	2	2.00	12.1
Scapula	2	1.00	40.0	18	9.00	54.5
Humerus, proximal	1	0.50	20.0	11	5.50	33.3
Humerus, distal	5	2.50	100.0	22	11.00	66.7
Radius, proximal	0	0.00	0.0	13	6.50	39.4
Radius, distal	0	0.00	0.0	9	4.50	27.3
Ulna	0	0.00	0.0	17	8.50	51.5
Pelvis	1	0.50	20.0	19	9.50	57.6
Sacrum	0	0.00	0.0	2	2.00	12.1
Femur, proximal	2	1.00	40.0	12	6.00	36.4
Femur, distal	1	0.50	20.0	14	7.00	42.4
Patella	0	0.00	0.0	1	0.50	3.0
Tibia, proximal	1	0.50	20.0	11	5.50	33.3
Tibia, distal	3	1.50	60.0	13	6.50	39.4
Carpal	0	0.00	0.0	0	0.00	0.0
Metacarpal, proximal	0	0.00	0.0	5	2.50	15.2
Metacarpal, distal	0	0.00	0.0	4	2.00	12.1
Calcaneus	0	0.00	0.0	3	1.50	9.1
Astragalus	0	0.00	0.0	5	2.50	15.2
Other Tarsal	0	0.00	0.0	0	0.00	0.0
Metatarsal, proximal	0	0.00	0.0	1	0.50	3.0
Metatarsal, distal	0	0.00	0.0	1	0.50	3.0
Metapodial	1	0.50	20.0	0	0.00	0.0
First phalanx	0	0.00	0.0	1	0.13	0.8
Second phalanx	1	0.13	5.0	0	0.00	0.0
Third phalanx	0	0.00	0.0	0	0.00	0.0

Key: See Table 5.1a.
Note: See Table 5.1a.

domestic or possibly domestic animals at Haimenkou, the relative abundance of sheep & goats decreased from Phases II to III by whichever measurement is used (see chapter four). This trend is also evident in skeletal part representation, and is obvious from the greater numbers of types of skeletal parts as well as the higher frequencies of main skeletal parts of sheep & goats in Phase II when compared to in Phase III. Therefore, Figures 5.5a–b supports the results from chapter four which suggest that sheep & goats were not local animals and their abundance declined over time. As introduced animals which may have been traded from a distant location (northwestern China), sheep & goats may have arrived at Haimenkou as a small group. They may not have adjusted well to the husbandry

Table 5.2b Skeletal Part Representation of Pig from Haimenkou (Phase III and the Late Phase)

Skeletal Part	Phase III			Late Phase		
	MNE	MAU	%MAU	MNE	MAU	%MAU
Cranium	25	12.50	59.5	1	0.50	50.0
Mandible	42	21.00	100.0	0	0.00	0.0
Atlas	13	13.00	61.9	0	0.00	0.0
Axis	3	3.00	14.3	0	0.00	0.0
Scapula	21	10.50	50.0	1	0.50	50.0
Humerus, proximal	2	1.00	4.8	0	0.00	0.0
Humerus, distal	15	7.50	35.7	1	0.50	50.0
Radius, proximal	13	6.50	31.0	2	1.00	100.0
Radius, distal	4	2.00	9.5	0	0.00	0.0
Ulna	29	14.50	69.0	0	0.00	0.0
Pelvis	24	12.00	57.1	2	1.00	100.0
Sacrum	0	0.00	0.0	0	0.00	0.0
Femur, proximal	11	5.50	26.2	0	0.00	0.0
Femur, distal	16	8.00	38.1	0	0.00	0.0
Patella	1	0.50	2.4	0	0.00	0.0
Tibia, proximal	12	6.00	28.6	0	0.00	0.0
Tibia, distal	12	6.00	28.6	1	0.50	50.0
Carpal	0	0.00	0.0	0	0.00	0.0
Metacarpal, proximal	2	1.00	4.8	0	0.00	0.0
Metacarpal, distal	1	0.50	2.4	0	0.00	0.0
Calcaneus	5	2.50	11.9	0	0.00	0.0
Astragalus	6	3.00	14.3	0	0.00	0.0
Other Tarsal	0	0.00	0.0	0	0.00	0.0
Metatarsal, proximal	2	1.00	4.8	0	0.00	0.0
Metatarsal, distal	2	1.00	4.8	0	0.00	0.0
Metapodial	2	1.00	4.8	0	0.00	0.0
First phalanx	2	0.25	1.2	0	0.00	0.0
Second phalanx	1	0.13	0.6	0	0.00	0.0
Third phalanx	0	0.00	0.0	0	0.00	0.0

Key: See Table 5.1a.
Note: See Table 5.1a.

pattern of the Haimenkou people, and thus their group size reduced through time.

As with the representation of skeletal parts of pig, small elements are also absent in sheep & goat, including the patella, carpal, tarsal, and phalanges. The absence of these may be attributed to the inadequate sample collection using the technique of hand collection during the excavation. This issue will be further discussed in Section 5.2.2.9.

There are no published data of FUI for sheep or goats, thus %MAU for sheep & goats is only plotted against VD

Table 5.2c Skeletal Part Representation of Pig from Haimenkou (All Phases)

Skeletal Part	Total		
	MNE	MAU	%MAU
Cranium	49	24.50	64.5
Mandible	76	38.00	100.0
Atlas	22	22.00	57.9
Axis	5	5.00	13.2
Scapula	42	21.00	55.3
Humerus, proximal	14	7.00	18.4
Humerus, distal	43	21.50	56.6
Radius, proximal	28	14.00	36.8
Radius, distal	13	6.50	17.1
Ulna	46	23.00	60.5
Pelvis	46	23.00	60.5
Sacrum	2	2.00	5.3
Femur, proximal	25	12.50	32.9
Femur, distal	31	15.50	40.8
Patella	2	1.00	2.6
Tibia, proximal	24	12.00	31.6
Tibia, distal	29	14.50	38.2
Carpal	0	0.00	0.0
Metacarpal, proximal	7	3.50	9.2
Metacarpal, distal	5	2.50	6.6
Calcaneus	8	4.00	10.5
Astragalus	11	5.50	14.5
Other Tarsal	0	0.00	0.0
Metatarsal, proximal	3	1.50	3.9
Metatarsal, distal	3	1.50	3.9
Metapodial	3	1.50	3.9
First phalanx	3	0.38	1.0
Second phalanx	2	0.25	0.7
Third phalanx	0	0.00	0.0

Key: See Table 5.1a.
Note: See Table 5.1a.

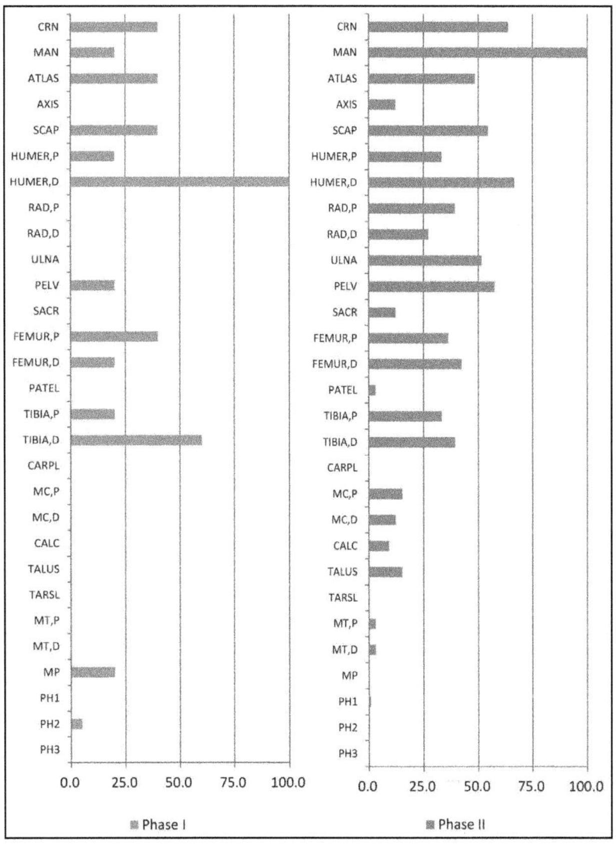

Figure 5.4a %MAU for Pig from Haimenkou (Phases I and II)
Key: See Figure 5.1a. Note: See Figure 5.1a.

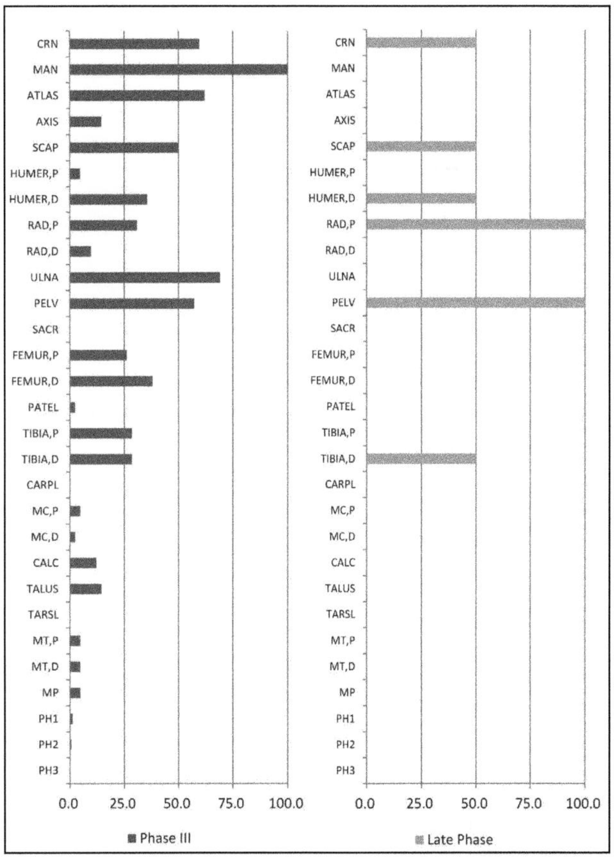

Figure 5.4b %MAU for the Pig from Haimenkou (Phase III and the Late Phase)
Key: See Figure 5.1a. Note: See Figure 5.1a.

(volume density) values for sheep in Figure 5.6, in which the VD values are taken from Lyman (1984:tables 6 and 10; 1994c:table 7.6, column sheep).

From Figure 5.6, no significant relationship is shown between VD values and %MAU of skeletal parts of

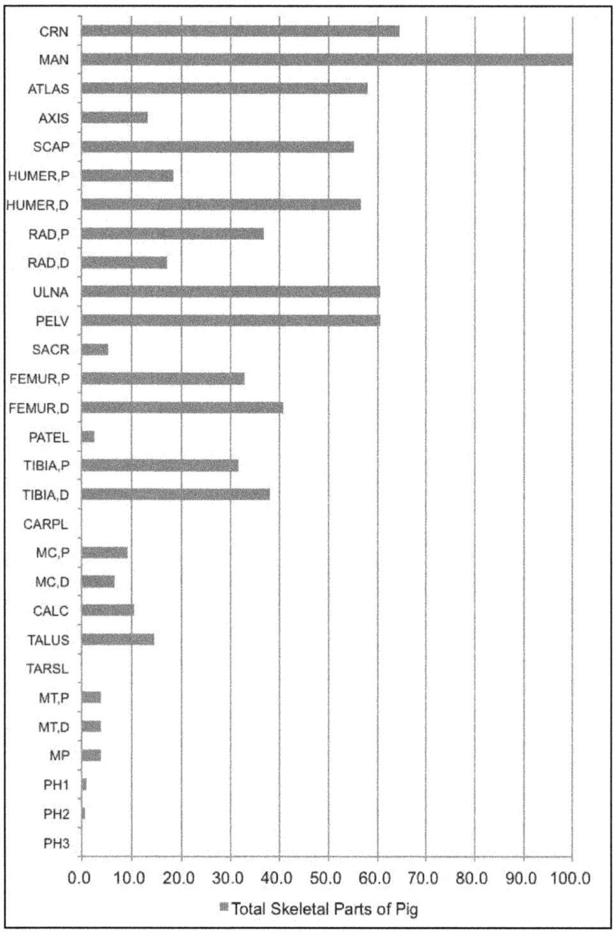

Figure 5.4c %MAU for Pig from Haimenkou (Total Skeletal Parts)
Key: See Figure 5.1a. Note: See Figure 5.1a.

sheep & goats (R^2 = 0.3852). Nevertheless, the tendency displayed is negative, which may imply that some relatively soft skeletal parts survived the post-mortem process rather than hard ones. However, as mentioned above, the relatively hard skeletal parts, such as the calcaneus, astragalus, metapodial, carpal, tarsal, and phalanges, are small elements that could have been missed in excavation due to inadequate sample collection. If these small but hard elements were excluded from Figure 5.6, no negative tendency would emerge. Moreover, the mandible is

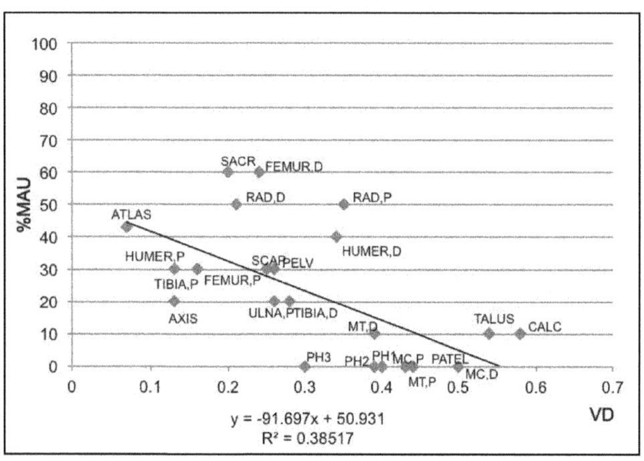

Figure 5.6 Scatterplot of %MAU of Sheep & Goat Skeletal Parts from Haimenkou against VD Values for Sheep

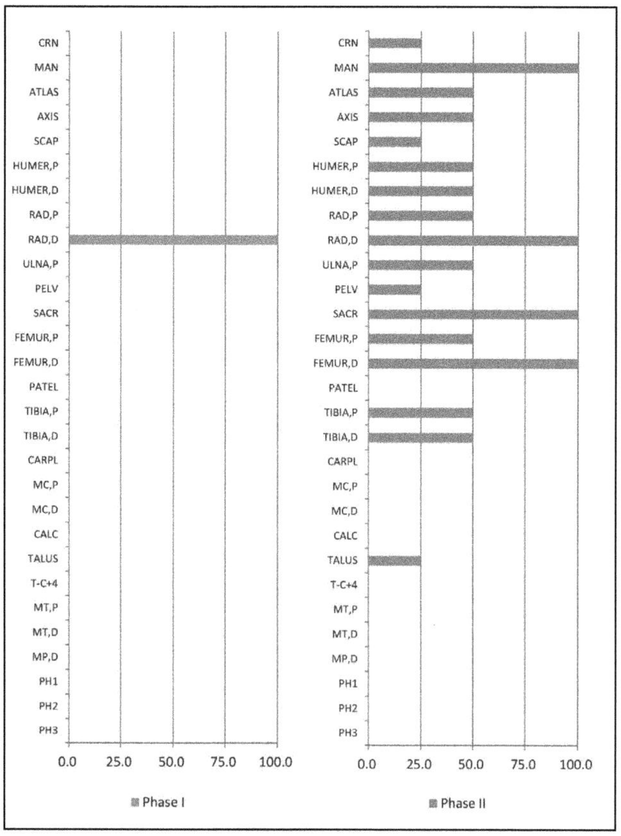

Figure 5.5a %MAU for Sheep & Goat from Haimenkou (Phases I and II)
Key: See Figure 5.1a. Note: See Figure 5.1a.

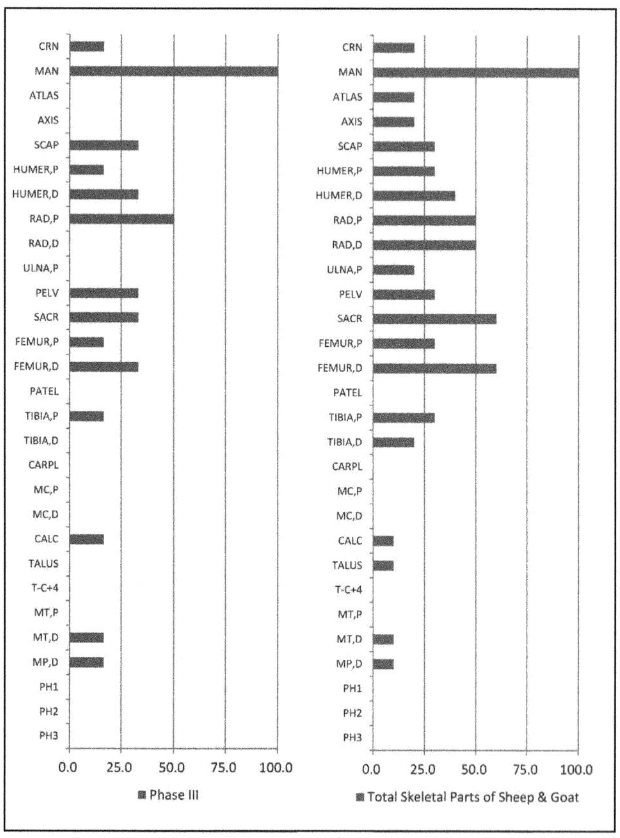

Figure 5.5b %MAU for Sheep & Goat from Haimenkou (Phase III and Total Skeletal Parts)
Key: See Figure 5.1a. Note: See Figure 5.1a.

Table 5.3a Skeletal Part Representation of Sheep & Goat from Haimenkou (Phases I and II)

Skeletal Part	Phase I			Phase II		
	MNE	MAU	%MAU	MNE	MAU	%MAU
Cranium	0	0.00	0.0	1	0.50	25.0
Mandible	0	0.00	0.0	4	2.00	100.0
Atlas	0	0.00	0.0	1	1.00	50.0
Axis	0	0.00	0.0	1	1.00	50.0
Scapula	0	0.00	0.0	1	0.50	25.0
Humerus, proximal	0	0.00	0.0	2	1.00	50.0
Humerus, distal	0	0.00	0.0	2	1.00	50.0
Radius, proximal	0	0.00	0.0	2	1.00	50.0
Radius, distal	1	0.50	100.0	4	2.00	100.0
Ulna, proximal	0	0.00	0.0	2	1.00	50.0
Pelvis	0	0.00	0.0	1	0.50	25.0
Sacrum	0	0.00	0.0	2	2.00	100.0
Femur, proximal	0	0.00	0.0	2	1.00	50.0
Femur, distal	0	0.00	0.0	4	2.00	100.0
Patella	0	0.00	0.0	0	0.00	0.0
Tibia, proximal	0	0.00	0.0	2	1.00	50.0
Tibia, distal	0	0.00	0.0	2	1.00	50.0
Carpal	0	0.00	0.0	0	0.00	0.0
Metacarpal, proximal	0	0.00	0.0	0	0.00	0.0
Metacarpal, distal	0	0.00	0.0	0	0.00	0.0
Calcaneus	0	0.00	0.0	0	0.00	0.0
Astragalus	0	0.00	0.0	1	0.50	25.0
Tarsal, C+4	0	0.00	0.0	0	0.00	0.0
Metatarsal, proximal	0	0.00	0.0	0	0.00	0.0
Metatarsal, distal	0	0.00	0.0	0	0.00	0.0
Metapodial, distal	0	0.00	0.0	0	0.00	0.0
First phalanx	0	0.00	0.0	0	0.00	0.0
Second phalanx	0	0.00	0.0	0	0.00	0.0
Third phalanx	0	0.00	0.0	0	0.00	0.0

Key: See Table 5.1a.
Note: See Table 5.1a.

the most frequent skeletal part of sheep & goat remains from Haimenkou, but as no VD data are available for the mandible of sheep, Figure 5.6 is problematic and should be interpreted with caution.

5.2.2.4 Large Cervids

As mentioned in chapter four, the large cervid category actually consists largely of the sambar (*Rusa unicolor*). Given its similar body size to the gaur, the skeletal part representation of the sambar is important for the purpose of comparison with the gaur.

Tables 5.4a–c and Figures 5.7a–b, which were derived from the tables, show the frequencies of skeletal part of large cervids (no figure was presented for the Late Phase as no data were available). Frequency of the skeletal parts of large cervids in Phase I was insufficient for analysis due to the very

Table 5.3b Skeletal Part Representation of Sheep & Goat from Haimenkou (Phase III and the Late Phase)

Skeletal Part	Phase III			Late Phase		
	MNE	MAU	%MAU	MNE	MAU	%MAU
Cranium	1	0.50	16.7	No data available in the Late Phase		
Mandible	6	3.00	100.0			
Atlas	0	0.00	0.0			
Axis	0	0.00	0.0			
Scapula	2	1.00	33.3			
Humerus, proximal	1	0.50	16.7			
Humerus, distal	2	1.00	33.3			
Radius, proximal	3	1.50	50.0			
Radius, distal	0	0.00	0.0			
Ulna, proximal	0	0.00	0.0			
Pelvis	2	1.00	33.3			
Sacrum	1	1.00	33.3			
Femur, proximal	1	0.50	16.7			
Femur, distal	2	1.00	33.3			
Patella	0	0.00	0.0			
Tibia, proximal	1	0.50	16.7			
Tibia, distal	0	0.00	0.0			
Carpal	0	0.00	0.0			
Metacarpal, proximal	0	0.00	0.0			
Metacarpal, distal	0	0.00	0.0			
Calcaneus	1	0.50	16.7			
Astragalus	0	0.00	0.0			
Tarsal, C+4	0	0.00	0.0			
Metatarsal, proximal	0	0.00	0.0			
Metatarsal, distal	1	0.50	16.7			
Metapodial, distal	1	0.50	16.7			
First phalanx	0	0.00	0.0			
Second phalanx	0	0.00	0.0			
Third phalanx	0	0.00	0.0			

Key: See Table 5.1a.
Note: See Table 5.1a.

small sample size. The skeletal part frequencies displayed in Phases II and III share many features in common, including the absence of the cranium, mandible, pelvis, sacrum, and proximal femur. Because relatively abundant antlers were recovered at Haimenkou, along with some loose teeth identified as sambar teeth, it is evident that sambar heads were brought back to the Haimenkou site after hunting. Also, the absence of the pelvis, sacrum, and proximal femur may not be the result of having been dumped at a kill site distant from the residential village, given that other large and heavy bones, such as the humerus and tibia, are present in the faunal assemblage. Moreover, small elements, such as calcaneus, astragalus, and phalanges, also appear, indicating whole bodies were transported back to the residential

Table 5.3c Skeletal Part Representation of Sheep & Goat from Haimenkou (All Phases)

Skeletal Part	Total		
	MNE	MAU	%MAU
Cranium	2	1.00	20.0
Mandible	10	5.00	100.0
Atlas	1	1.00	20.0
Axis	1	1.00	20.0
Scapula	3	1.50	30.0
Humerus, proximal	3	1.50	30.0
Humerus, distal	4	2.00	40.0
Radius, proximal	5	2.50	50.0
Radius, distal	5	2.50	50.0
Ulna, proximal	2	1.00	20.0
Pelvis	3	1.50	30.0
Sacrum	3	3.00	60.0
Femur, proximal	3	1.50	30.0
Femur, distal	6	3.00	60.0
Patella	0	0.00	0.0
Tibia, proximal	3	1.50	30.0
Tibia, distal	2	1.00	20.0
Carpal	0	0.00	0.0
Metacarpal, proximal	0	0.00	0.0
Metacarpal, distal	0	0.00	0.0
Calcaneus	1	0.50	10.0
Astragalus	1	0.50	10.0
Tarsal, C+4	0	0.00	0.0
Metatarsal, proximal	0	0.00	0.0
Metatarsal, distal	1	0.50	10.0
Metapodial, distal	1	0.50	10.0
First phalanx	0	0.00	0.0
Second phalanx	0	0.00	0.0
Third phalanx	0	0.00	0.0

Key: See Table 5.1a.
Note: See Table 5.1a.

village rather than some skeletal parts being discarded outside. Hence, the absence of these skeletal parts, including the cranium, mandible, pelvis, sacrum, and proximal femur, may have resulted from destruction during food preparation and tool manufacturing using bones as raw materials.

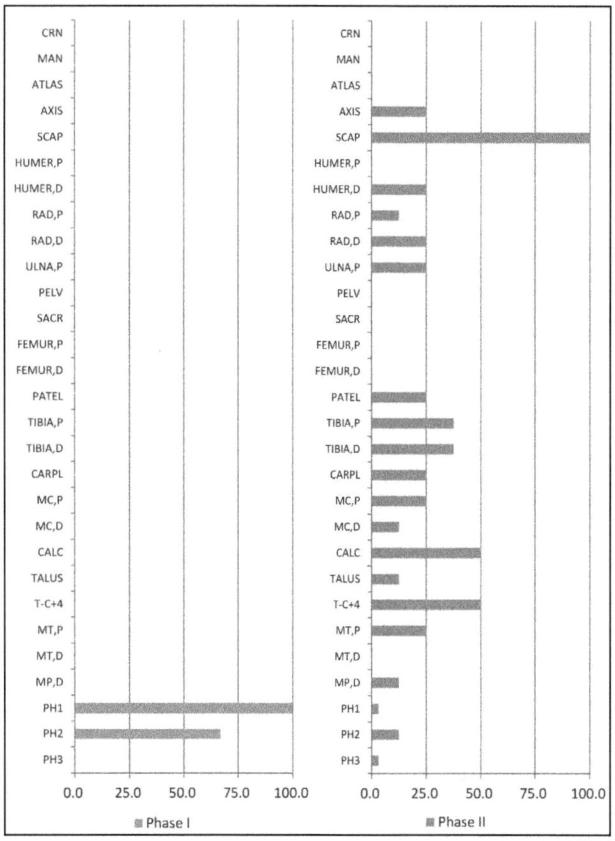

Figure 5.7a %MAU for Large Cervids from Haimenkou (Phases I and II)
Key: See Figure 5.1a. Note: See Figure 5.1a.

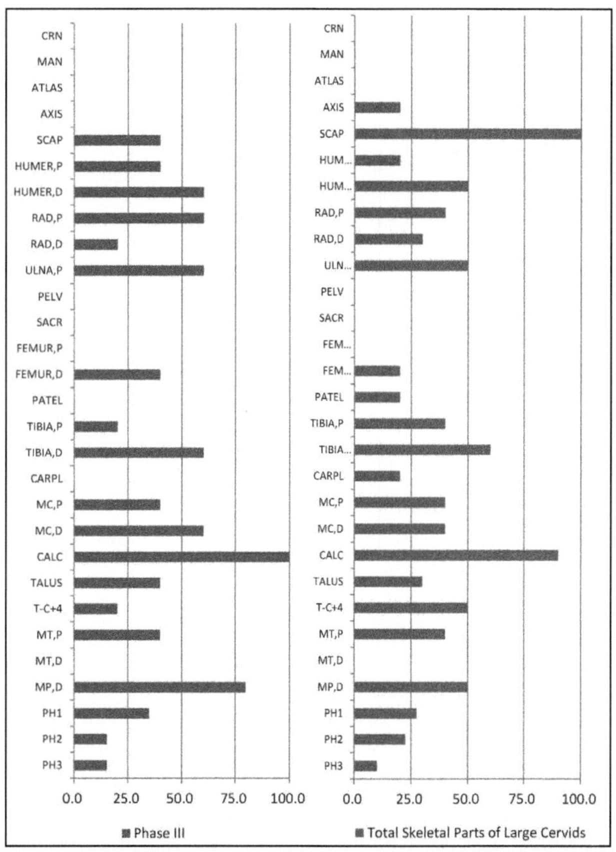

Figure 5.7b %MAU for Large Cervids from Haimenkou (Phase III and Total Skeletal Parts)
Key: See Figure 5.1a. Note: See Figure 5.1a.

Table 5.4a Skeletal Part Representation of Large Cervids from Haimenkou (Phases I and II)

Skeletal Part	Phase I			Phase II		
	MNE	MAU	%MAU	MNE	MAU	%MAU
Cranium	0	0.00	0.0	0	0.00	0.0
Mandible	0	0.00	0.0	0	0.00	0.0
Atlas	0	0.00	0.0	0	0.00	0.0
Axis	0	0.00	0.0	1	1.00	25.0
Scapula	0	0.00	0.0	8	4.00	100.0
Humerus, proximal	0	0.00	0.0	0	0.00	0.0
Humerus, distal	0	0.00	0.0	2	1.00	25.0
Radius, proximal	0	0.00	0.0	1	0.50	12.5
Radius, distal	0	0.00	0.0	2	1.00	25.0
Ulna, proximal	0	0.00	0.0	2	1.00	25.0
Pelvis	0	0.00	0.0	0	0.00	0.0
Sacrum	0	0.00	0.0	0	0.00	0.0
Femur, proximal	0	0.00	0.0	0	0.00	0.0
Femur, distal	0	0.00	0.0	0	0.00	0.0
Patella	0	0.00	0.0	2	1.00	25.0
Tibia, proximal	0	0.00	0.0	3	1.50	37.5
Tibia, distal	0	0.00	0.0	3	1.50	37.5
Carpal	0	0.00	0.0	2	1.00	25.0
Metacarpal, proximal	0	0.00	0.0	2	1.00	25.0
Metacarpal, distal	0	0.00	0.0	1	0.50	12.5
Calcaneus	0	0.00	0.0	4	2.00	50.0
Astragalus	0	0.00	0.0	1	0.50	12.5
Tarsal, C+4	0	0.00	0.0	4	2.00	50.0
Metatarsal, proximal	0	0.00	0.0	2	1.00	25.0
Metatarsal, distal	0	0.00	0.0	0	0.00	0.0
Metapodial, distal	0	0.00	0.0	1	0.50	12.5
First phalanx	3	0.38	100.0	1	0.13	3.1
Second phalanx	2	0.25	66.7	4	0.50	12.5
Third phalanx	0	0.00	0.0	1	0.13	3.1

Key: See Table 5.1a.
Note: See Table 5.1a.

The %MAU of large cervids is plotted against FUI and VD (volume density) values for caribou respectively in Figures 5.8 and 5.9, in which FUI values are taken from Metcalfe and Jones (1988: table 2). The VD values are from Lyman (1984: tables 6 and 10; 1994c: table 7.6, column deer).

The FUI values are taken from caribou rather than sambar. Since the caribou is also a kind of large cervid, the FUI values for caribou are appropriate when used in Figure 5.8. According to Figure 5.8, a not significant (R^2=0.0002) and very slightly positive relationship between FUI values and %MAU is evident. This indicates a rare relationship between food utility and the representation of skeletal parts of large cervids. This figure indicates that whole bodies of large cervids were carried back to the residential village after hunting.

Table 5.4b Skeletal Part Representation of Large Cervids from Haimenkou (Phase III and the Late Phase)

Skeletal Part	Phase III			Late Phase		
	MNE	MAU	%MAU	MNE	MAU	%MAU
Cranium	0	0.00	0.0	No data available in the Late Phase		
Mandible	0	0.00	0.0			
Atlas	0	0.00	0.0			
Axis	0	0.00	0.0			
Scapula	2	1.00	40.0			
Humerus, proximal	2	1.00	40.0			
Humerus, distal	3	1.50	60.0			
Radius, proximal	3	1.50	60.0			
Radius, distal	1	0.50	20.0			
Ulna, proximal	3	1.50	60.0			
Pelvis	0	0.00	0.0			
Sacrum	0	0.00	0.0			
Femur, proximal	0	0.00	0.0			
Femur, distal	2	1.00	40.0			
Patella	0	0.00	0.0			
Tibia, proximal	1	0.50	20.0			
Tibia, distal	3	1.50	60.0			
Carpal	0	0.00	0.0			
Metacarpal, proximal	2	1.00	40.0			
Metacarpal, distal	3	1.50	60.0			
Calcaneus	5	2.50	100.0			
Astragalus	2	1.00	40.0			
Tarsal, C+4	1	0.50	20.0			
Metatarsal, proximal	2	1.00	40.0			
Metatarsal, distal	0	0.00	0.0			
Metapodial, distal	4	2.00	80.0			
First phalanx	7	0.88	35.0			
Second phalanx	3	0.38	15.0			
Third phalanx	3	0.38	15.0			

Key: See Table 5.1a.
Note: See Table 5.1a.

Similarly to FUI values, the VD values for Figure 5.9 are also taken from modern caribou. There is still no significant relationship appearing between VD values and %MAU of the skeletal parts of large cervids (R^2 = 0.1675), but the trend is apparently positive. Like the gaur, the large cervid has a massive body and larger elements. This could decrease the loss of specimens due to inadequate sample collection and enhance the possibility of the appearance of small elements. This explains why Figure 5.9 shows the harder skeletal parts being present in higher frequencies.

In addition, large cervid bones at Haimenkou show high fragmentation. No complete long bones were revealed at Haimenkou, which may indicate that bodies of large cervids as well as gaur were dismembered into small chunks.

Table 5.4c Skeletal Part Representation of Large Cervids from Haimenkou (All Phases)

Skeletal Part	Total		
	MNE	MAU	%MAU
Cranium	0	0.00	0.0
Mandible	0	0.00	0.0
Atlas	0	0.00	0.0
Axis	1	1.00	20.0
Scapula	10	5.00	100.0
Humerus, proximal	2	1.00	20.0
Humerus, distal	5	2.50	50.0
Radius, proximal	4	2.00	40.0
Radius, distal	3	1.50	30.0
Ulna, proximal	5	2.50	50.0
Pelvis	0	0.00	0.0
Sacrum	0	0.00	0.0
Femur, proximal	0	0.00	0.0
Femur, distal	2	1.00	20.0
Patella	2	1.00	20.0
Tibia, proximal	4	2.00	40.0
Tibia, distal	6	3.00	60.0
Carpal	2	1.00	20.0
Metacarpal, proximal	4	2.00	40.0
Metacarpal, distal	4	2.00	40.0
Calcaneus	9	4.50	90.0
Astragalus	3	1.50	30.0
Tarsal, C+4	5	2.50	50.0
Metatarsal, proximal	4	2.00	40.0
Metatarsal, distal	0	0.00	0.0
Metapodial, distal	5	2.50	50.0
First phalanx	11	1.38	27.5
Second phalanx	9	1.13	22.5
Third phalanx	4	0.50	10.0

Key: See Table 5.1a.
Note: See Table 5.1a.

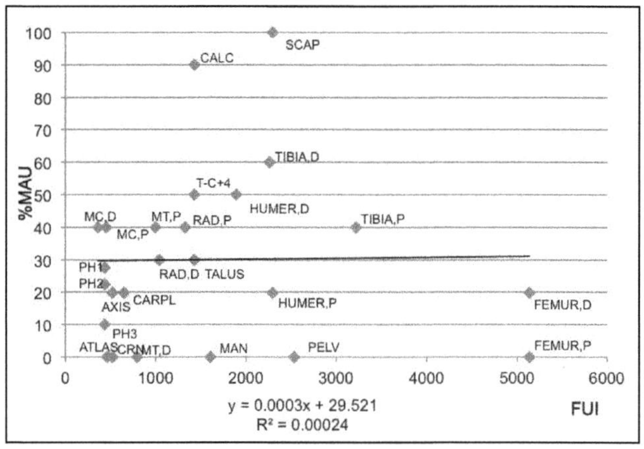

Figure 5.8 Scatterplot of %MAU of the Large Cervid Skeletal Parts from Haimenkou against FUI Values for Caribou

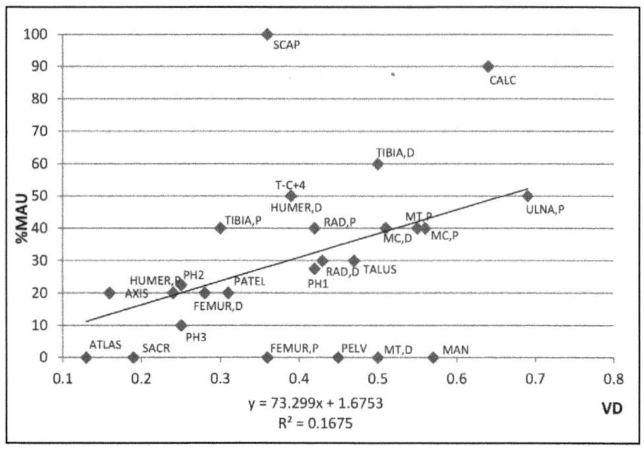

Figure 5.9 Scatterplot of %MAU of the Large Cervid Skeletal Parts from Haimenkou against VD Values for Caribou

5.2.2.5 Medium Cervids

Like large cervids, the medium cervid category may consist largely of one species, the red muntjac (*Muntiacus muntjak*). The deer belonging to this category have similar body sizes to sheep & goats, thus the comparison between them is useful.

The skeletal part representation of medium cervids can be observed in Tables 5.5a–c, from which Figures 5.10a–c are derived. Phase I cannot be analysed as only a few specimens of medium cervids were discovered in this phase. Figures 5.10a and 5.10b demonstrate that although there are a few disparities showing between Phases II and III, the main distribution of skeletal parts is the same. For example, no cranium or mandible parts were present in Phase III but they occurred in Phase II. However, more antlers were found in Phase III than in Phase II, when measuring based on NISP, MNI, and weight. This suggests that the human behaviour of bringing heads of medium cervids back to the residential settlement did not change from Phases II to III. Therefore, the absence of the cranium and mandible could result from destruction due to human activities and natural taphonomic processes. The relatively frequent presence of the calcaneus, along with the appearance of the astragalus, metapodial, and phalanges, indicates whole bodies were transported back to the site. The rare finds of small elements, such as phalanges, are possibly caused by inadequate sample collection, since they have a relatively high bone density and thus should have more chance of being preserved.

Table 5.5a Skeletal Part Representation of the Medium Cervids from Haimenkou (Phases I and II)

Skeletal Part	Phase I			Phase II		
	MNE	MAU	%MAU	MNE	MAU	%MAU
Cranium	0	0.00	0.0	1	0.50	33.3
Mandible	0	0.00	0.0	1	0.50	33.3
Atlas	0	0.00	0.0	0	0.00	0.0
Axis	0	0.00	0.0	0	0.00	0.0
Scapula	0	0.00	0.0	1	0.50	33.3
Humerus, proximal	0	0.00	0.0	0	0.00	0.0
Humerus, distal	2	1.00	100.0	0	0.00	0.0
Radius, proximal	0	0.00	0.0	2	1.00	66.7
Radius, distal	0	0.00	0.0	1	0.50	33.3
Ulna, proximal	0	0.00	0.0	2	1.00	66.7
Pelvis	0	0.00	0.0	1	0.50	33.3
Sacrum	0	0.00	0.0	0	0.00	0.0
Femur, proximal	0	0.00	0.0	2	1.00	66.7
Femur, distal	2	1.00	100.0	1	0.50	33.3
Patella	0	0.00	0.0	0	0.00	0.0
Tibia, proximal	0	0.00	0.0	2	1.00	66.7
Tibia, distal	0	0.00	0.0	3	1.50	100.0
Carpal	0	0.00	0.0	0	0.00	0.0
Metacarpal, proximal	0	0.00	0.0	3	1.50	100.0
Metacarpal, distal	0	0.00	0.0	2	1.00	66.7
Calcaneus	0	0.00	0.0	2	1.00	66.7
Astragalus	0	0.00	0.0	1	0.50	33.3
Tarsal, C+4	0	0.00	0.0	0	0.00	0.0
Metatarsal, proximal	0	0.00	0.0	1	0.50	33.3
Metatarsal, distal	0	0.00	0.0	1	0.50	33.3
Metapodial, distal	0	0.00	0.0	0	0.00	0.0
First phalanx	0	0.00	0.0	0	0.00	0.0
Second phalanx	0	0.00	0.0	1	0.13	8.3
Third phalanx	0	0.00	0.0	0	0.00	0.0

Key: See Table 5.1a.
Note: See Table 5.1a.

5.2.2.6 Small Cervids

A few specimens of small cervids were recovered at Haimenkou, and only Phases II and III have yielded their remains. Thus, this animal category is not appropriate for the analysis of skeletal part representation in this study. They are included here for comparison with other animal categories presented in the next sections.

Tables 5.6a–c show the frequency of recovered body parts of small cervids, and Figures 5.11a–b give a pictorial representation. Although only a small number of specimens are available, it is still clear that basic body parts of small cervids are present in the faunal assemblage from Haimenkou, as the head and the fore and hind upper limbs have been represented. Only the skeletal parts representing the lower limbs are missing, including the carpals,

Table 5.5b Skeletal Part Representation of Medium Cervids from Haimenkou (Phase III and the Late Phase)

Skeletal Part	Phase III			Late Phase		
	MNE	MAU	%MAU	MNE	MAU	%MAU
Cranium	0	0.00	0.0	0	0.00	0.0
Mandible	0	0.00	0.0	0	0.00	0.0
Atlas	0	0.00	0.0	0	0.00	0.0
Axis	2	2.00	66.7	0	0.00	0.0
Scapula	2	1.00	33.3	0	0.00	0.0
Humerus, proximal	2	1.00	33.3	0	0.00	0.0
Humerus, distal	5	2.50	83.3	0	0.00	0.0
Radius, proximal	2	1.00	33.3	0	0.00	0.0
Radius, distal	1	0.50	16.7	1	0.50	100.0
Ulna, proximal	1	0.50	16.7	0	0.00	0.0
Pelvis	2	1.00	33.3	0	0.00	0.0
Sacrum	1	1.00	33.3	0	0.00	0.0
Femur, proximal	5	2.50	83.3	0	0.00	0.0
Femur, distal	5	2.50	83.3	0	0.00	0.0
Patella	0	0.00	0.0	0	0.00	0.0
Tibia, proximal	2	1.00	33.3	0	0.00	0.0
Tibia, distal	4	2.00	66.7	0	0.00	0.0
Carpal	0	0.00	0.0	0	0.00	0.0
Metacarpal, proximal	3	1.50	50.0	0	0.00	0.0
Metacarpal, distal	1	0.50	16.7	0	0.00	0.0
Calcaneus	6	3.00	100.0	0	0.00	0.0
Astragalus	1	0.50	16.7	0	0.00	0.0
Tarsal, C+4	0	0.00	0.0	0	0.00	0.0
Metatarsal, proximal	1	0.50	16.7	0	0.00	0.0
Metatarsal, distal	0	0.00	0.0	0	0.00	0.0
Metapodial, distal	0	0.00	0.0	0	0.00	0.0
First phalanx	0	0.00	0.0	0	0.00	0.0
Second phalanx	0	0.00	0.0	0	0.00	0.0
Third phalanx	0	0.00	0.0	0	0.00	0.0

Key: See Table 5.1a.
Note: See Table 5.1a.

tarsals, and phalanges. This absence may also result from the inadequate sample collection at the site.

5.2.2.7 Comparison between Gaur and Large Cervids

The skeletal part representations of the gaur and large cervids show a number of similarities that could be explained given their similar body sizes. They have similar frequencies of limb bone recovery, in both upper and lower limbs, more so in the distal humerus than the proximal humerus. They also both exhibit high frequencies of lower limb bones, such as the calcaneus and distal metapodial.

The most noticeable disparity occurring between these two animal categories is the absence of certain skeletal parts in large cervids, including the cranium, mandible, pelvis,

Table 5.5c Skeletal Part Representation of Medium Cervids from Haimenkou (All Phases)

Skeletal Part	Total		
	MNE	MAU	%MAU
Cranium	1	0.50	12.5
Mandible	1	0.50	12.5
Atlas	0	0.00	0.0
Axis	2	2.00	50.0
Scapula	3	1.50	37.5
Humerus, proximal	2	1.00	25.0
Humerus, distal	7	3.50	87.5
Radius, proximal	4	2.00	50.0
Radius, distal	3	1.50	37.5
Ulna, proximal	3	1.50	37.5
Pelvis	3	1.50	37.5
Sacrum	1	1.00	25.0
Femur, proximal	7	3.50	87.5
Femur, distal	8	4.00	100.0
Patella	0	0.00	0.0
Tibia, proximal	4	2.00	50.0
Tibia, distal	7	3.50	87.5
Carpal	0	0.00	0.0
Metacarpal, proximal	6	3.00	75.0
Metacarpal, distal	3	1.50	37.5
Calcaneus	8	4.00	100.0
Astragalus	2	1.00	25.0
Tarsal, C+4	0	0.00	0.0
Metatarsal, proximal	2	1.00	25.0
Metatarsal, distal	1	0.50	12.5
Metapodial, distal	0	0.00	0.0
First phalanx	0	0.00	0.0
Second phalanx	1	0.13	3.1
Third phalanx	0	0.00	0.0

Key: See Table 5.1a.
Note: See Table 5.1a.

and sacrum bones. However, relatively abundant antlers and loose teeth of sambars are present, suggesting that heads of large cervids may have been carried back to the residential settlement. The lack of head bones in the large cervid category may be caused by human processing, such

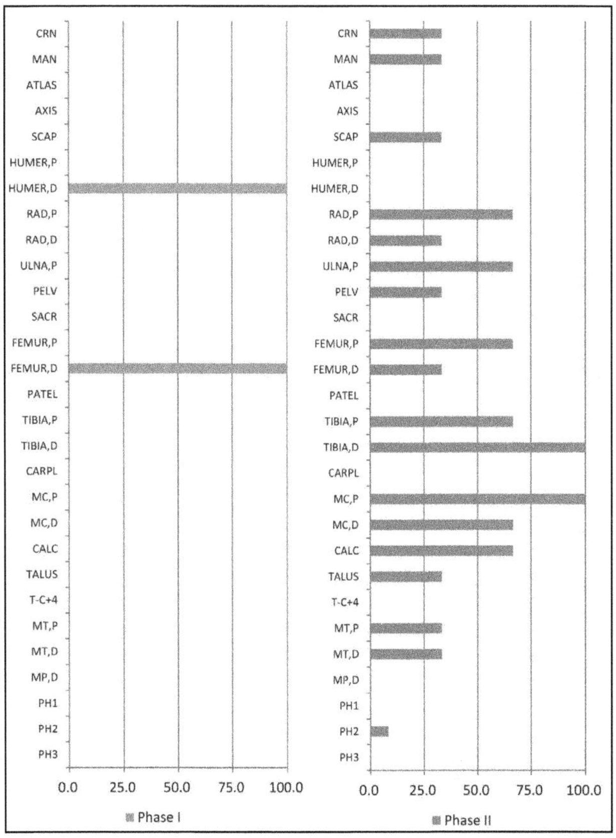

Figure 5.10a %MAU for Medium Cervids from Haimenkou (Phases I and II)
Key: See Figure 5.1a. Note: See Figure 5.1a.

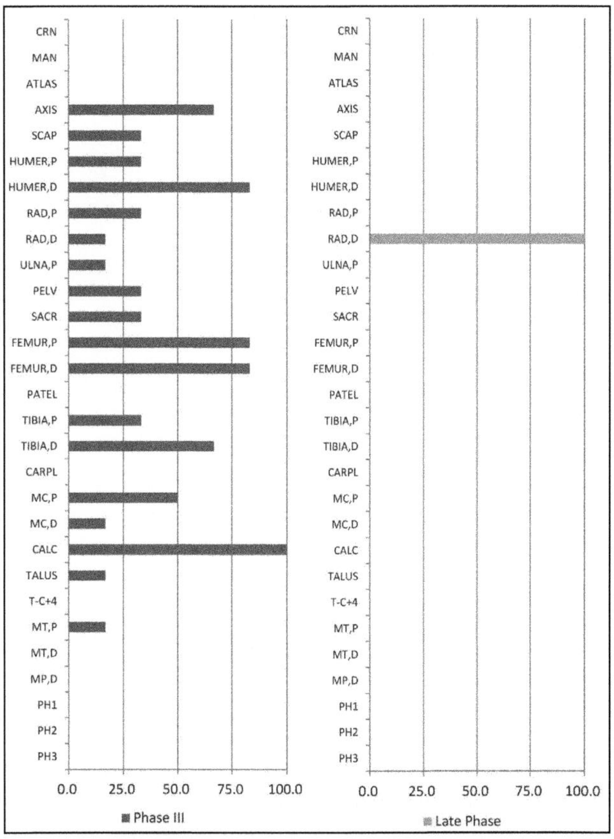

Figure 5.10b %MAU for Medium Cervids from Haimenkou (Phase III and the Late Phase)
Key: See Figure 5.1a. Note: See Figure 5.1a.

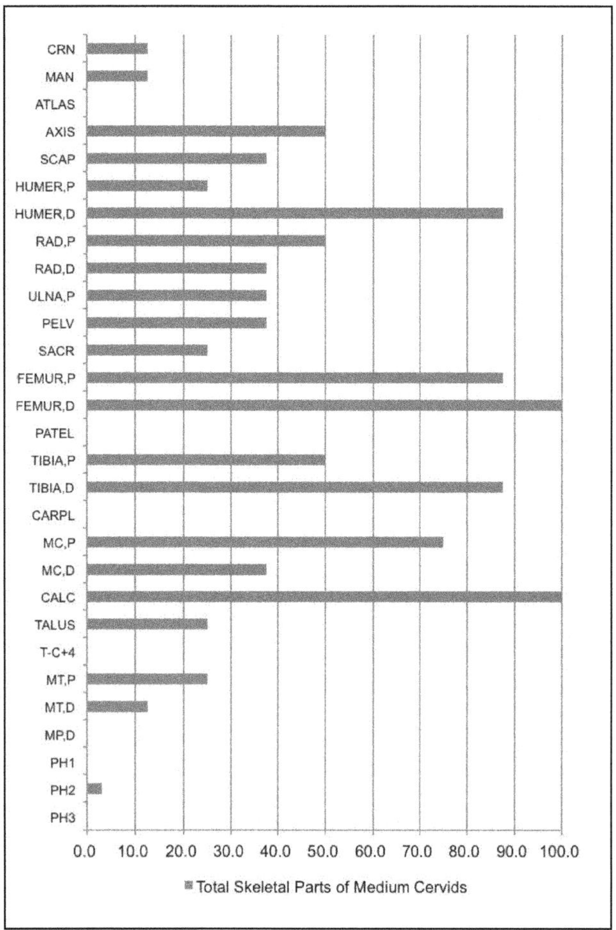

Figure 5.10c %MAU for Medium Cervids from Haimenkou (Total Skeletal Parts)
Key: See Figure 5.1a. Note: See Figure 5.1a.

as the crushing of skulls to get at the brains by Haimenkou inhabitants. This may have led to poor identifiability of the fragments of the cranium and mandible of large cervids. This does not explain the disappearance of the pelvis and sacrum of large cervids. However, since most of the massive upper limb bones of large cervids have been found at Haimenkou, it is unlikely that the pelvic bones were discarded at the kill site outside Haimenkou. Further research on this issue should be conducted with a larger sample size in future excavations.

5.2.2.8 Comparison between Large, Medium, and Small Cervids

The cranium and mandible bones from medium cervids were recovered from Haimenkou, whereas no cranium fragments from small cervids or large cervids were evident. No atlas was recovered from any of the three cervid categories, while axes of large and medium cervids were recovered. Therefore, it is probable that the heads of these cervids of different sizes were brought back to Haimenkou, the residential village, after hunting.

Another discrepancy between the large, medium, and small cervids is that a small number of complete long limb bones from medium and small cervids were recovered. This indicates that the long limb bones of large cervids may have been highly fragmented for food processing or marrow extraction. The pelvis and sacrum bones of medium cervids were present, which makes the representation of body parts in medium cervids more complete than in large cervids. Moreover, the small elements, such as the carpal and phalanges, were scant in medium cervid specimens and totally absent in small cervid specimens. This may be the result of inadequate sample collection.

5.2.2.9 Comparison between Pig, Sheep & Goat, and Medium Cervids

The pig and sheep & goat skeletal remains share some similarities, especially with regard to head bones. In both animal categories, the most frequent skeletal part occurring was the mandible. There were no carpals present in the skeletal part representation of either animal category. Among the other skeletal parts, some were better represented in pigs, while others were more frequent in sheep & goats. It is interesting that the representation of craniums among pigs was better than among sheep & goats. In practice, the cranial fragments of pigs were most frequent based on NISP, MNI, and bone weight among all the taxa found at Haimenkou. This phenomenon may result from the thickness of bones in pig craniums, which could lead to relatively high resistance to destruction, and thus the large number of identifiable specimens from pig craniums.

The most apparent discrepancy between these two domestic animals and medium cervids was in the frequency of mandibles, which was higher among pigs and sheep & goats, but infrequent among medium cervids. Another obvious dissimilarity was the representation of the calcaneus, which was highest in medium cervids but infrequent in pigs and sheep & goats.

Aside from the calcaneus and talus, other small elements were barely present in these three taxa. For example, phalanges rarely appear among pigs and medium cervids and were not recovered from sheep & goats. As mentioned before, this may be due to inadequate sample collection.

5.3 Bone Modification

The study of bone modification is considered a branch of taphonomy (Marshall, L. 1989:8), and there has been much written on the subject (e.g., Behrensmeyer 1978; Behrensmeyer, et al. 1986; Binford 1973; Bonnichsen and Sorg 1989; Brain 1981; Bunn and Kroll 1986; Fisher 1995; Gifford-Gonzalez 1989; Lyman 1994c; Marshall, L. 1989; Noe-Nygaard 1989; Olsen 1989; Russell 1995; Shipman 1981; Shipman, et al. 1984; Shipman and Rose 1983; White, Theodore 1952; 1953b; 1954; 1955; 1956). The study of bone modification is significantly useful when surveying diverse kinds of archaeological, palaeoanthropological, and taphonomic issues, such as subsistence adaptation, ritual behaviour, site formation process, and the evolution of economic and social pattern (Fisher 1995).

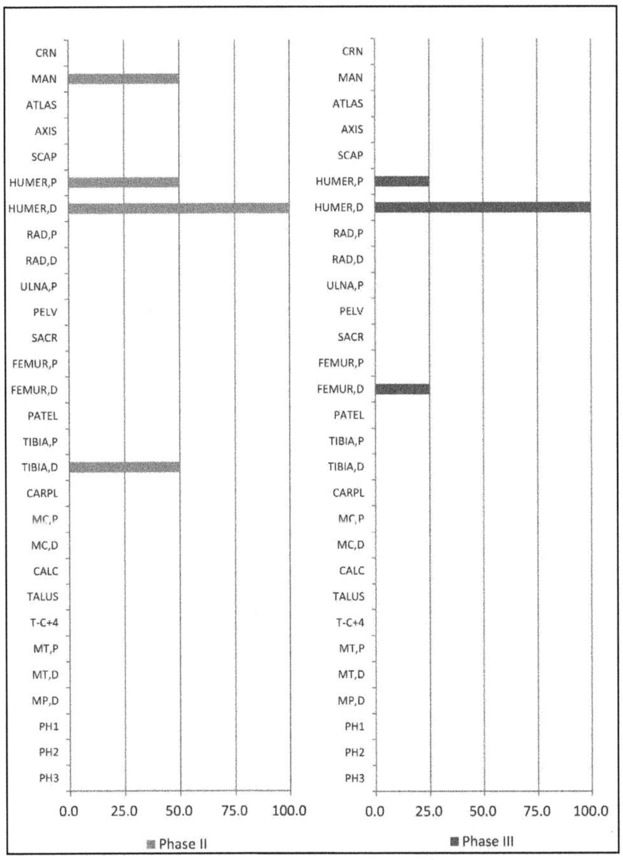

Figure 5.11a %MAU for Small Cervids from Haimenkou (Phases II and III)
Key: See Figure 5.1a. Note: See Figure 5.1a.

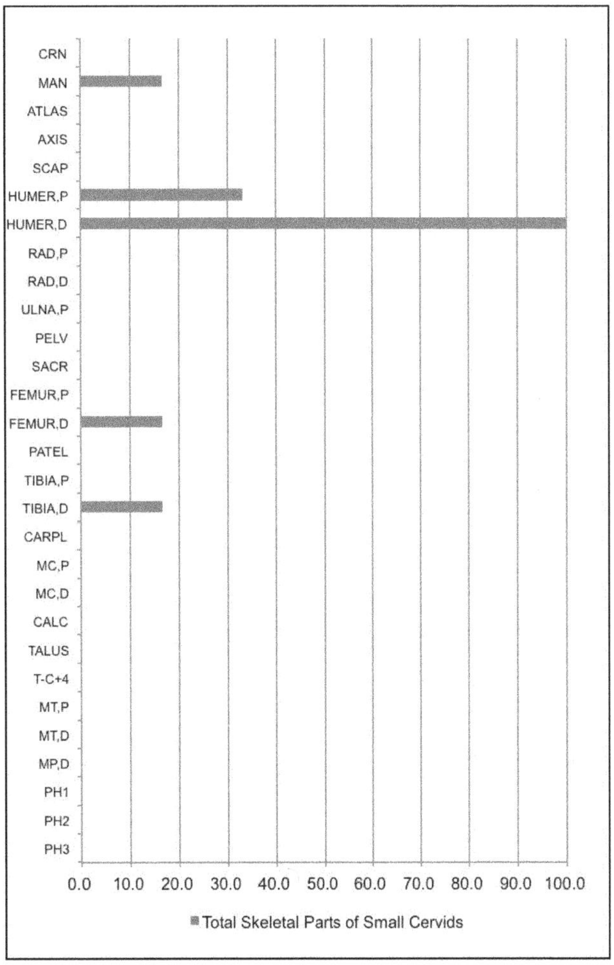

Figure 5.11b %MAU for Small Cervids from Haimenkou (Total Skeletal Parts)
Key: See Figure 5.1a. Note: See Figure 5.1a.

Bone modification can be defined as any alteration in size, structure, or texture of bone by an external agent (Marshall, L. 1989:8). This includes marks on bone fragments and breakage patterns (Reitz and Wing 2008:168). The term 'modification' here consists of two elements: the process and the pattern (Marshall, L. 1989:8). The process refers to the cause by which patterns come to be produced on a bone, also called the activity, agent, causal agent, causal factor, event, external agent, force, formation process, modifier, sequence of events, and strategy; the pattern, also called the effect, means the consequent alteration of the bone resulting from the process (Marshall, L. 1989:8).

The study of bone modification generally involves three operational steps (Marshall, L. 1989:11): a study of the patterns (e.g., Shipman 1981), an attempt to identify causal processes (e.g., Behrensmeyer, et al. 1986), and a comparison of observed patterns with experimental analogues for known processes (e.g., Gifford-Gonzalez 1989; Shipman, et al. 1984). To investigate the causal processes, two categories are recognised: processes conducted by humans and by non-humans. The human process refers to human behaviours that cause alterations of bone specimens. Analysis of human behaviours may be combined with the cultural aspect of zooarchaeology. The non-human taphonomic process in site formation concerns biological factors, namely the aspects of animals and nature that involve physical, chemical, geological, and mechanical processes (Marshall, L. 1989:11). To observe the resultant patterns, two categories are considered (Marshall, L. 1989:11). The basic types of modification patterns that reflect human activities are fractures and marks, characterised by cut, scrape, chop, hack, blow, saw, and burning marks, and so on (Marshall, L. 1989:11). These types of fractures and marks may be related to human behaviours including butchering, skinning, extracting soft tissue, cooking, and manufacturing tools (Shipman 1981; Shipman, et al. 1984; Shipman and Rose 1983). Patterns indicating animal behaviours can include gnaw marks, furrows, tooth grooves, and scrapes. Weathering, root etching, fluvial abrasion, and sedimentary abrasion are caused by natural forces (Marshall, L. 1989).

The bone modification methods discussed below includes cut and gnaw marks, burning, weathering, and pathology. The cut marks may reflect butchering and food preparation processes, while gnaw marks, burning, and weathering may provide lines of evidence for taphonomic processes after discard of the bones. Bones that seem to have been used as tools are excluded from this observation as these tools may reflect 'curate behaviour' (Binford 1973:242). According to Meadow (1980:68), such bones are 'rescued from immediate deposition and recycled through various cultural activities to be discarded at the end of their useful lives in a

Table 5.6a Skeletal Part Representation of Small Cervids from Haimenkou (Phases I and II)

Skeletal Part	Phase I			Phase II		
	MNE	MAU	%MAU	MNE	MAU	%MAU
Cranium	No data available in Phase I			0	0.00	0.0
Mandible				1	0.50	50.0
Atlas				0	0.00	0.0
Axis				0	0.00	0.0
Scapula				0	0.00	0.0
Humerus, proximal				1	0.50	50.0
Humerus, distal				2	1.00	100.0
Radius, proximal				0	0.00	0.0
Radius, distal				0	0.00	0.0
Ulna, proximal				0	0.00	0.0
Pelvis				0	0.00	0.0
Sacrum				0	0.00	0.0
Femur, proximal				0	0.00	0.0
Femur, distal				0	0.00	0.0
Patella				0	0.00	0.0
Tibia, proximal				0	0.00	0.0
Tibia, distal				1	0.50	50.0
Carpal				0	0.00	0.0
Metacarpal, proximal				0	0.00	0.0
Metacarpal, distal				0	0.00	0.0
Calcaneus				0	0.00	0.0
Astragalus				0	0.00	0.0
Tarsal, C+4				0	0.00	0.0
Metatarsal, proximal				0	0.00	0.0
Metatarsal, distal				0	0.00	0.0
Metapodial, distal				0	0.00	0.0
First phalanx				0	0.00	0.0
Second phalanx				0	0.00	0.0
Third phalanx				0	0.00	0.0

Key: See Table 5.1a.
Note: See Table 5.1a.

manner largely independent of the disposal of the unwanted by-products of meat preparation and consumption.'

5.3.1 Cut Marks

As mentioned above, the patterns related to human activities primarily consist of two types, fractures and marks. The fractures include longitudinal, stepped, saw tooth, depressed, V-shaped, flake, perpendicular, irregular, smooth, articular, and spiral fractures (Marshall, L. 1989:11-14). Types of marks that may be derived from human cultural activities include cuts, furrows, grooves, lines, pits, polish, scars, scrapes, scratches, and slices (Marshall, L. 1989:11-14). Fractures may be created

Table 5.6b Skeletal Part Representation of Small Cervids from Haimenkou (Phase III and the Late Phase)

Skeletal Part	Phase III			Late Phase		
	MNE	MAU	%MAU	MNE	MAU	%MAU
Cranium	0	0.00	0.0	No data available in the Late Phase		
Mandible	0	0.00	0.0			
Atlas	0	0.00	0.0			
Axis	0	0.00	0.0			
Scapula	0	0.00	0.0			
Humerus, proximal	1	0.50	25.0			
Humerus, distal	4	2.00	100.0			
Radius, proximal	0	0.00	0.0			
Radius, distal	0	0.00	0.0			
Ulna, proximal	0	0.00	0.0			
Pelvis	0	0.00	0.0			
Sacrum	0	0.00	0.0			
Femur, proximal	0	0.00	0.0			
Femur, distal	1	0.50	25.0			
Patella	0	0.00	0.0			
Tibia, proximal	0	0.00	0.0			
Tibia, distal	0	0.00	0.0			
Carpal	0	0.00	0.0			
Metacarpal, proximal	0	0.00	0.0			
Metacarpal, distal	0	0.00	0.0			
Calcaneus	0	0.00	0.0			
Astragalus	0	0.00	0.0			
Tarsal, C+4	0	0.00	0.0			
Metatarsal, proximal	0	0.00	0.0			
Metatarsal, distal	0	0.00	0.0			
Metapodial, distal	0	0.00	0.0			
First phalanx	0	0.00	0.0			
Second phalanx	0	0.00	0.0			
Third phalanx	0	0.00	0.0			

Key: See Table 5.1a.
Note: See Table 5.1a.

during butchering, dismemberment, extracting marrow and grease, cooking, and tool manufacture and use (Marshall, L. 1989:16). Cut and chop marks (furrows, grooves, lines, scrapes, scratches, and striations) are produced by humans with tools made of stone, bone, metal, and so on. A cut mark refers to an elongate, relatively narrow, linear striation, which is often V-shaped in the cross section and has flat sides. The cross-section shape and width of cut marks may vary (Fisher 1995:12). A chop mark refers to a broad, relatively short, and linear depression that generally has a V-shaped cross section (Fisher 1995:19; Noe-Nygaard 1989:473). In this section, cut and chop marks are the focus of discussion. They are attributed to the category 'cut marks' for the sake of simplification.

Table 5.6c Skeletal Part Representation of Small Cervids from Haimenkou (All Phases)

Skeletal Part	Total		
	MNE	MAU	%MAU
Cranium	0	0.00	0.0
Mandible	1	0.50	16.7
Atlas	0	0.00	0.0
Axis	0	0.00	0.0
Scapula	0	0.00	0.0
Humerus, proximal	2	1.00	33.3
Humerus, distal	6	3.00	100.0
Radius, proximal	0	0.00	0.0
Radius, distal	0	0.00	0.0
Ulna, proximal	0	0.00	0.0
Pelvis	0	0.00	0.0
Sacrum	0	0.00	0.0
Femur, proximal	0	0.00	0.0
Femur, distal	1	0.50	16.7
Patella	0	0.00	0.0
Tibia, proximal	0	0.00	0.0
Tibia, distal	1	0.50	16.7
Carpal	0	0.00	0.0
Metacarpal, proximal	0	0.00	0.0
Metacarpal, distal	0	0.00	0.0
Calcaneus	0	0.00	0.0
Astragalus	0	0.00	0.0
Tarsal, C+4	0	0.00	0.0
Metatarsal, proximal	0	0.00	0.0
Metatarsal, distal	0	0.00	0.0
Metapodial, distal	0	0.00	0.0
First phalanx	0	0.00	0.0
Second phalanx	0	0.00	0.0
Third phalanx	0	0.00	0.0

Key: See Table 5.1a.
Note: See Table 5.1a

Cut marks on different parts of animal skeletons may indicate activities such as slaughtering, scalping, skinning, dismembering, de-fleshing, filleting, and chopping carcasses into manageable lumps suitable for sharing, transport, and cooking (Fisher 1995; Marshall, L. 1989; Noe-Nygaard 1989). For the purpose of reconstructing the patterns of butchering practice and food preparation techniques, the location and frequency of cut marks on various skeletal parts were recorded and analysed in combination with knowledge of animal anatomy.

As only a few cut marks were observed on bones in the faunal assemblage from Haimenkou, the analysis of differences among phases is inappropriate. The number and distribution of cut marks found on specimens are summarised in Tables 5.7a and 5.7b. The taxa include gaur, pig, sheep & goat, dog, large cervids (including sambar and hog deer), and medium cervids (including red muntjac). Based on the tables, the description and analysis are conducted as follows from taxon to taxon.

5.3.1.1 Gaur

As Table 5.7a shows, cut marks on gaur bones are the most numerous type of human modification in the assemblage, although they are not so high in frequency. A total of eight gaur specimens were found with cut marks. These were located on the scapula (blade), humerus (distal shaft), radius (distal shaft), ilium (shaft), calcaneus (proximal end), and astragalus (see Figure 5.12). Most of these cut marks are chop marks, with broad, relatively short, linear depressions, and V-shaped cross sections. These marks may indicate human behaviours such as dismembering or chopping into smaller lumps rather than butchering or de-fleshing.

The high fragmentation of gaur specimens may lead to the lower relative frequency of cut marks in the total NISP of gaurs (2.4%). Also, highly fragmentary specimens are sometimes themselves the result of cutting or chopping. Moreover, highly fragmentary specimens are more vulnerable and may have lost marks during the long term taphonomic processes of deposition, burial, and recovery. Due to the low frequency of cut marks, the reconstruction of butchering is difficult. However, in combination with examination of high fragmentation, human activities such as dismembering, chopping into small lumps, and extraction of fat or grease, may be revealed. As mentioned in Section 5.2.2.1 under the discussion of skeletal part representation, the people of the Dulong minority in northwestern Yunnan Province retain an age-old custom of sacrificing semi-domestic gayals, the possible descendants of gaurs, to celebrate the New Year. During the celebration, the gayal's body is chopped into small lumps and shared among the whole village (Wang, W. and Long 2009:296-297). A similar practice may be reflected in the zooarchaeological record at Haimenkou. It may be assumed that after the killing or butchering, the gaur body would be divided into small lumps by chopping or some other way in order to share, transport, or cook.

5.3.1.2 Pig

Only one pig specimen with cut marks was recovered (Figure 5.13). These marks are multiple and relatively light cut marks on the distal shaft of a tibia. These cut marks may indicate

Table 5.7a Location and Frequency of Cut Marks on Bones of Gaur, Pig, and Sheep & Goat

Skeletal Part	Gaur			Pig			Sheep & Goat		
	N	%N	% NISP	N	%N	% NISP	N	%N	% NISP
Horn	0	0.0	0.0	—	—	—	0	0.0	—
Cranium	0	0.0	0.0	0	0.0	0.0	0	0.0	0.0
Mandible	0	0.0	0.0	0	0.0	0.0	0	0.0	0.0
Atlas	0	0.0	0.0	0	0.0	0.0	0	0.0	0.0
Axis	0	0.0	0.0	0	0.0	0.0	0	0.0	0.0
Vertebrae	0	0.0	0.0	0	0.0	0.0	0	0.0	—
Scapula	1	12.5	6.3	0	0.0	0.0	0	0.0	0.0
Humerus	1	12.5	11.1	0	0.0	0.0	1	33.3	14.3
Radius	2	25.0	15.4	0	0.0	0.0	0	0.0	0.0
Ulna	0	0.0	0.0	0	0.0	0.0	0	0.0	0.0
Pelvis	1	12.5	3.8	0	0.0	0.0	0	0.0	0.0
Sacrum	0	0.0	0.0	0	0.0	0.0	0	0.0	0.0
Femur	1	12.5	5.9	0	0.0	0.0	0	0.0	0.0
Patella	0	0.0	0.0	0	0.0	0.0	0	0.0	—
Tibia	0	0.0	0.0	1	100.0	1.6	2	66.7	40.0
Fibula	—	—	—	0	0.0	0.0	—	—	—
Carpal	0	0.0	0.0	0	0.0	—	0	0.0	—
Metacarpal	0	0.0	0.0	0	0.0	0.0	0	0.0	—
Calcaneus	1	12.5	7.1	0	0.0	0.0	0	0.0	0.0
Astragalus	1	12.5	10.0	0	0.0	0.0	0	0.0	0.0
Tarsal	0	0.0	0.0	0	0.0	—	0	0.0	—
Metatarsal	0	0.0	0.0	0	0.0	0.0	0	0.0	0.0
Metapodial	0	0.0	0.0	0	0.0	0.0	0	0.0	—
First phalanx	0	0.0	0.0	0	0.0	0.0	0	0.0	—
Second phalanx	0	0.0	0.0	0	0.0	0.0	0	0.0	—
Third phalanx	0	0.0	0.0	0	0.0	—	0	0.0	—
Total	8	100.0	2.4	1	100.0	0.1	3	100.0	4.9

Key: N = number of specimens bearing cut marks; %N = relative frequency of specimens bearing cut marks for each skeletal part to the total number of cut-mark-bearing specimens per animal taxon; %NISP = relative frequency of specimens bearing cut marks for each skeletal part to the total number of specimens identified for each skeletal part per animal taxon; — = no specimens available.

human activities like de-fleshing or filleting. The relative frequency of cut marks among pigs is the lowest among all the taxa bearing cut marks, which may be due to the few cut marks on pig bones as well as the great abundance of pig specimens revealed at Haimenkou. Compared with large-bodied gaurs, the bones of pigs are less fragmentary, implying a low frequency of activities such as hitting or fracturing.

5.3.1.3 Sheep & Goat

Three specimens were recovered with cut marks among sheep & goats, resulting in the highest relative frequency of specimens bearing cut marks to the total NISP of sheep & goat (4.9%). The three specimens bearing cut marks were all identified with certainty as sheep. These cut marks

Table 5.7b Location and Frequency of Cut Marks on Bones of Dog, Large Cervids, and Medium Cervids

Skeletal Part	Dog			Large Cervid			Medium Cervid		
	N	%N	% NISP	N	%N	% NISP	N	%N	% NISP
Antler	—	—	—	2	66.7	8.0	0	0.0	0.0
Cranium	0	0.0	—	0	0.0	—	0	0.0	0.0
Mandible	0	0.0	0.0	0	0.0	—	0	0.0	0.0
Atlas	0	0.0	0.0	0	0.0	—	0	0.0	—
Axis	0	0.0	—	0	0.0	0.0	0	0.0	0.0
Vertebrae	0	0.0	0.0	0	0.0	0.0	0	0.0	0.0
Scapula	0	0.0	—	0	0.0	0.0	0	0.0	0.0
Humerus	0	0.0	0.0	0	0.0	0.0	0	0.0	0.0
Radius	1	100.0	50.0	0	0.0	0.0	0	0.0	0.0
Ulna	0	0.0	0.0	0	0.0	0.0	0	0.0	0.0
Pelvis	0	0.0	—	0	0.0	—	0	0.0	0.0
Sacrum	0	0.0	—	0	0.0	—	0	0.0	0.0
Femur	0	0.0	0.0	0	0.0	0.0	1	50.0	6.3
Patella	0	0.0	—	0	0.0	0.0	0	0.0	—
Tibia	0	0.0	0.0	0	0.0	0.0	1	50.0	8.3
Fibula	0	0.0	—	—	—	—	—	—	—
Carpal	0	0.0	—	0	0.0	0.0	0	0.0	0.0
Metacarpal	0	0.0	0.0	1	33.3	11.1	0	0.0	0.0
Calcaneus	0	0.0	—	0	0.0	0.0	0	0.0	0.0
Astragalus	0	0.0	—	0	0.0	0.0	0	0.0	0.0
Tarsal	0	0.0	—	0	0.0	0.0	0	0.0	—
Metatarsal	0	0.0	—	0	0.0	0.0	0	0.0	0.0
Metapodial	0	0.0	—	0	0.0	0.0	0	0.0	—
First phalanx	0	0.0	—	0	0.0	0.0	0	0.0	—
Second phalanx	0	0.0	—	0	0.0	0.0	0	0.0	0.0
Third phalanx	0	0.0	—	0	0.0	0.0	0	0.0	—
Total	1	100.0	4.0	3	100.0	1.9	2	100.0	1.6

Key: See Table 5.7a.

were distributed on the humerus (distal end) and tibiae (distal shafts) (see Figure 5.14). The cut marks on the distal humerus were relatively light, which may suggest they relate to traces of de-fleshing or sinew removal. In contrast, the multiple cut marks on distal tibia shafts were short and relatively deep, with a V-shaped cross section. This type of butchery marks may indicate human behaviour such as chopping and dismembering.

5.3.1.4 Dog

As with pigs, there is only one specimen with cut marks present among dog remains. The multiple cut marks on a distal shaft seem short and relatively deep, which may indicate de-fleshing or tool manufacture (Figure 5.15).

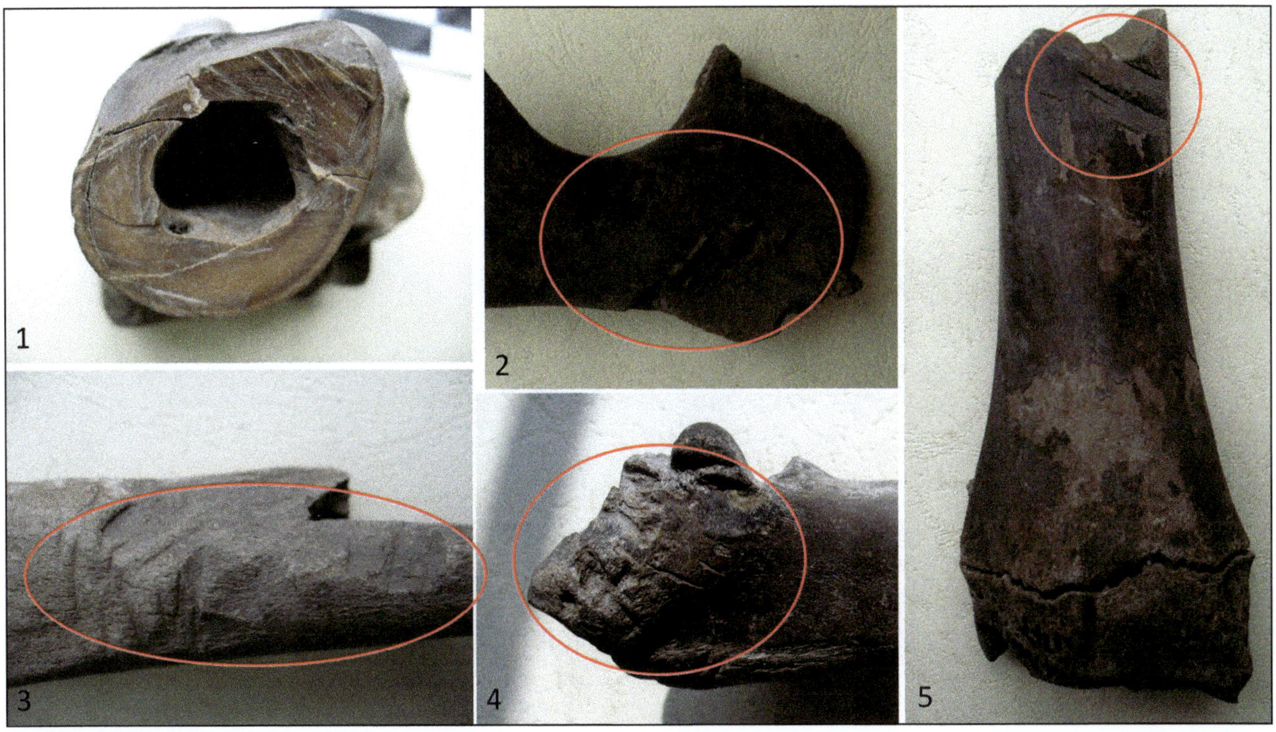

Figure 5.12 Cut Marks on Gaur Bones from Haimenkou (Photos by Juan Wang)
1: Distal Humerus (BON212, Level (6), Phase II), cut marks on breakage surface of shaft; 2: Ilium (BON255, Level (6), Phase II), cut marks on ilium wing; 3: Distal Femur (BON199, Level (6), Phase II), cut marks on middle, caudal shaft; 4: Calcaneus (BON373, Level (4), Phase III); 5: Distal Radius (BON147, Level (6), Phase II), cut marks on dorsal shaft.

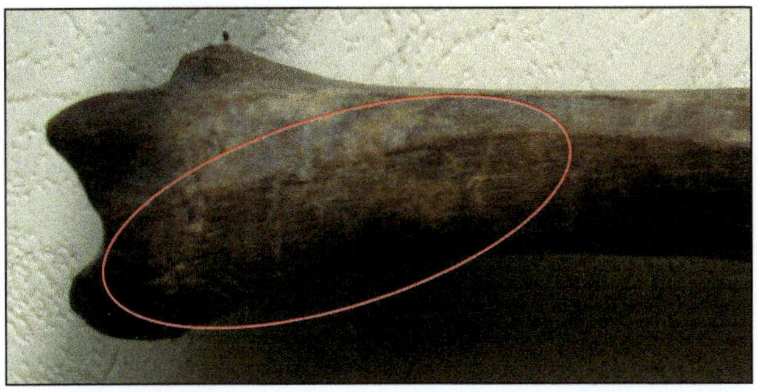

Figure 5.13 Cut Marks on Pig Bone from Haimenkou (Photo by Juan Wang)
Note: Distal Tibia (BON1148, Level (6), Phase II), cut marks on medial, distal shaft.

Dogs were still used as hunting partners in ethnic minorities groups in Yunnan several decades ago, and local customs forbade eating dogs (Li, G. and Lu 1987:172). Dogs also appeared to play the role of a hunting partner, guard, and shepherd dog in artefacts from the Shizhaishan site (*ca.* 2500–2100 B.P.) (Jiang, Z. 1999; Zhang, Zengqi 1998:157) in eastern Yunnan (see Figure 2.28). The evidence is scarce but it seems that dogs may have been eaten occasionally. It is difficult to tell whether they were also raised as food animals at Haimenkou.

5.3.1.5 Large Cervids

Three specimens bearing cut marks were recovered among large cervids from Haimenkou (Figure 5.16). They exhibited a low relative frequency of specimens bearing cut marks to the total number of identified specimens of large cervids (1.9%). Of the three specimens, two were from antlers and only one belonged to the post-cranial skeleton, a metacarpal. On the antlers, the cut marks were obviously a result of chopping, which suggests the antler may have been removed from the cranium for tool manufacture.

Figure 5.14 Cut Marks on Sheep Bones from Haimenkou (Photos by Juan Wang)
1: Distal Humerus (BON2052, Level (5), Phase III), cut marks on distal end; 2: Distal Tibia (BON365, Level (7), Phase II), cut marks on lateral, distal shaft; 3: Distal Tibia (BON586, Level (7), Phase II), cut marks on lateral, distal end.

Figure 5.15 Cut Marks on Dog Bone from Haimenkou (Photo by Juan Wang)
Note: Proximal Radius (BON 685, Level (8), Phase I), cut marks on medial, distal shaft.

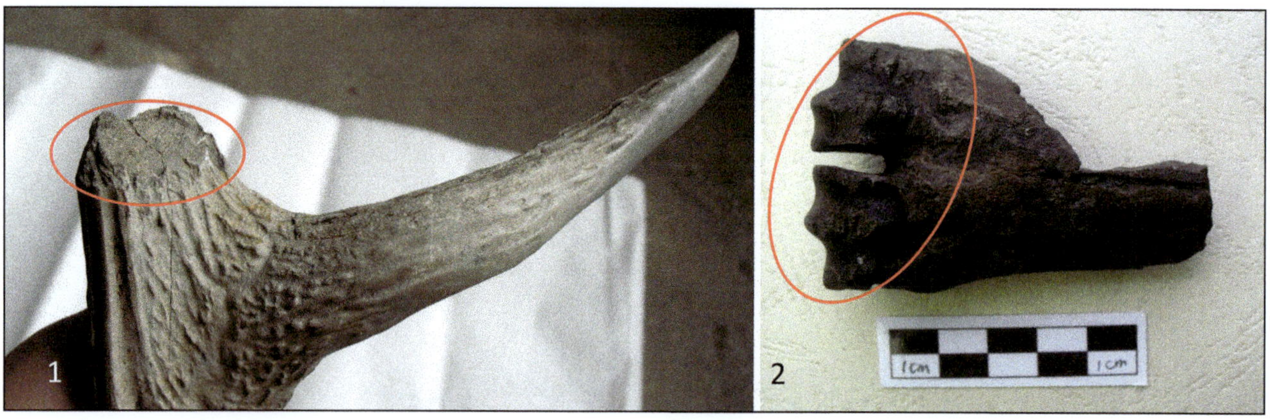

Figure 5.16 Cut Marks on Large Cervid Bones from Haimenkou (Photos by Juan Wang)
1: Antler (BON2054, Level (6), Phase II), cut marks on surface of tine breakage; 2: Distal Metacarpal (BON1333, Level (4), Phase III), cut marks on dorsal, distal end

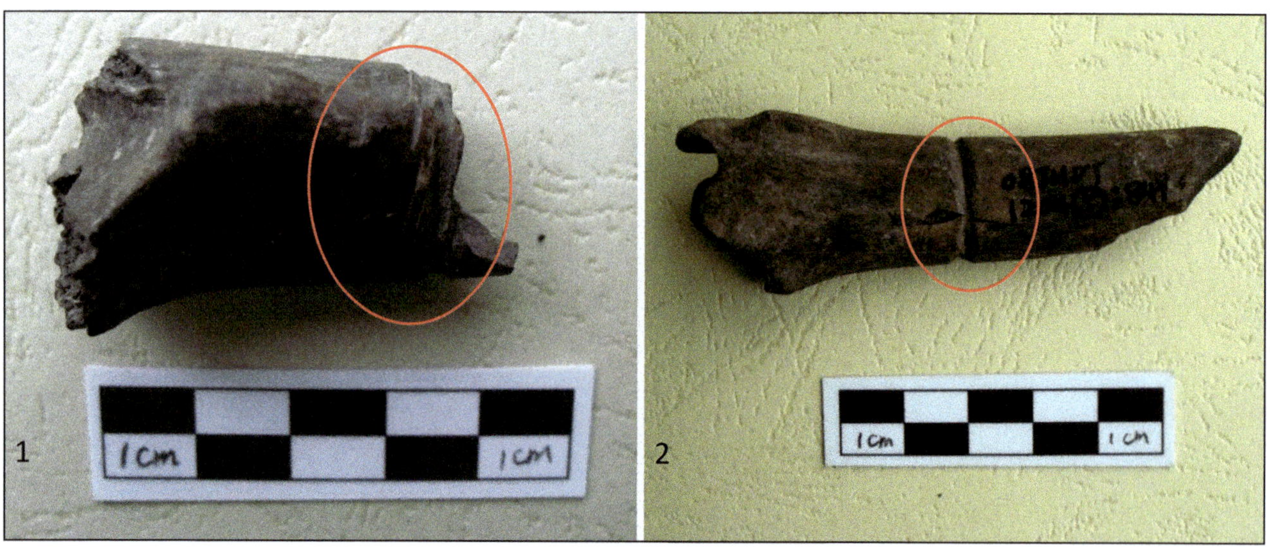

Figure 5.17 Cut Marks on Medium Cervid Bones from Haimenkou (Photos by Juan Wang)
1: Proximal Femur (BON924, Level (4), Phase III), cut marks on proximal shaft; 2: Distal Tibia (BON849, Level (4), Phase III), cut marks on distal shaft.

In contrast to the antlers, multiple relatively shallow cut marks were observed on the distal metacarpal, including the distal shaft and the lateral and medial condyles. These cut marks do not look like the result of chopping or dismembering, but are more similar to butchery practices including de-fleshing or sinew removal with a relatively sharp knife. Furthermore, given the location of the cut marks, skinning is also a possibility.

5.3.1.6 Medium Cervids

Two specimens of medium cervids were found with cut marks. One of these was located on the proximal shaft of a femur, while the other was evident on the distal shaft of a tibia (Figure 5.17). The cut marks on the proximal femur shaft were multiple and relatively deep. In association with the bone fracture condition, it may be assumed that the repeated cuttings were for the purpose of breaking the femur shaft. This activity may be conducted for marrow extraction or tool manufacture. The cut mark on the distal tibia shaft is a relatively deep groove that encircles the shaft. This cut mark was most probably caused by a sharp edged implement such as a knife. Furthermore, in association with the level and phase (Level (4), Phase III) of this distal tibia, it is possible that the knife which caused this cut mark was made of bronze. The purpose of this cut mark is unclear, but it may have a similar purpose to those on the proximal femur shaft, for extracting marrow or making a tool.

5.3.1.7 Summary

Cut marks are relatively rare on the specimens in the faunal assemblage from Haimenkou. However, this does not necessarily suggest the absence of the use of sharp edged implements during butchering and the sequential steps of food preparation process. As Shipmen and Rose (1983:86) proposed, it is possible to butcher an entire animal without leaving any cut marks on the bones. The reconstruction of the complete butchering process is difficult in this study due to the relative rarity of cut marks. Nevertheless, the human activities which may have been included in the butchering process were evident in the traces of cutting and chopping, such as de-fleshing, skinning, and dismembering.

5.3.2 Gnaw Marks

A variety of animals, especially carnivores and rodents, may leave tooth marks on bones when gnawing, chewing, or masticating. Carnivore gnawing could create marks such as striations, furrows, pits, punctures, ragged edges, chipped edges, sinuous edges, polish, conchoidal flake scars, and bone flakes (Fisher 1995:36). Rodent gnaw marks are characterised by parallel, multiple, broad, shallow, and flat-bottomed grooves produced by their chisel-like incisors (Fisher 1995:40; Marshall, L. 1989:19). In this section the gnaw marks created by carnivores are summarised and presented in Tables 5.8a–b with location and frequency, while the ones caused by rodents are summarised in Table 5.9.

As shown by Figures 5.8a–b, the gnaw marks produced by carnivores are concentrated on specimens of pigs, with the highest number of specimens bearing carnivore gnaw marks (26 specimens) and the highest proportion of specimens bearing carnivore gnaw marks in the total number of identified specimens per taxon (3.1%). Other taxa, including gaur, sheep & goat, large cervids, and medium cervids, also show carnivore gnaw marks in different numbers and relative frequencies. These carnivore gnaw marks may have been mostly created by dogs due to their diagnostic morphological characteristics (see Figure 5.18). These gnaw marks are widely distributed on several skeletal parts, including the axis, scapula, humerus, ulna, pelvis, femur, tibia, tarsals (especially the calcaneus), and phalanx. This may reflect a typical consumption of bones by dogs after human discard of the bones. The low frequency of these marks could not simply be explained by a low number of dogs living at Haimenkou during the long time span of human occupation. It is very likely the fragmentation of most specimens may have destroyed the evidence of gnaw marks. Furthermore, the bones that have undergone the process of gnawing and digestion by dogs would have a lesser chance to survive taphonomic processes. Given these considerations, the low frequency of dog gnaw marks is understandable.

Gnaw marks from rodents are less frequently observed than carnivore gnaw marks on the surface of specimens from Haimenkou (see Table 5.9). Only two animal categories exhibited rodent gnaw marks: pigs and medium cervids (Figure 5.19). The rodent gnaw marks are most frequently observed on the bone fragments of pigs (10 specimens). However, due to the high NISP of pigs, the relative frequency in this taxon (1.2%) is lower than among medium cervids (3.1%), which have four specimens with rodent gnaw marks. The rodent gnaw marks are scattered on some skeletal parts including the humerus, ulna, pelvis, femur, patella, astragalus, and phalanx. The tiny traces left by rodents seem to be erased more easily by external forces, such as sedimentary abrasion, than the relatively obvious marks caused by carnivores. This discrepancy in durability may have led to decreased evidence for rodent gnawing. Prior to recognition of rodent gnaw marks, it was unclear whether the rodent remains found at Haimenkou were contemporary with the Haimenkou people or were from disturbance to the layers during later periods. After recognition of rodent gnaw marks it was evident that rodents were present in or around the residential village at Haimenkou about 3,000 years ago. Therefore, the presence of both humans and rodents at the same time in Haimenkou may be inferred.

5.3.3 Burnt Bones

The burning degree (completely/partly calcined, completely/partly carbonised, burnt, etc.) of specimens was observed and recorded following the criteria of BONECODE (Meadow 1978; see also Appendix 1). All specimens that exhibit traces of burning were summarised and presented in Tables 5.10a–c.

Table 5.8a Location and Frequency of Carnivore Gnaw Marks on Bones of Gaur, Pig, and Sheep & Goat

Skeletal Part	Gaur			Pig			Sheep & Goat		
	N	%N	% NISP	N	%N	% NISP	N	%N	% NISP
Horn	0	0.0	0.0	—	—	—	0	0.0	—
Cranium	0	0.0	0.0	0	0.0	0.0	0	0.0	0.0
Mandible	0	0.0	0.0	0	0.0	0.0	0	0.0	0.0
Atlas	0	0.0	0.0	0	0.0	0.0	0	0.0	0.0
Axis	1	14.3	50.0	0	0.0	0.0	0	0.0	0.0
Vertebrae	0	0.0	0.0	0	0.0	0.0	0	0.0	—
Scapula	0	0.0	0.0	3	11.5	4.1	0	0.0	0.0
Humerus	0	0.0	0.0	5	19.2	6.9	0	0.0	0.0
Radius	0	0.0	0.0	0	0.0	0.0	0	0.0	0.0
Ulna	1	14.3	33.3	6	23.1	13.0	1	100.0	50.0
Pelvis	1	14.3	3.8	1	3.8	1.6	0	0.0	0.0
Sacrum	0	0.0	0.0	0	0.0	0.0	0	0.0	0.0
Femur	0	0.0	0.0	4	15.4	6.1	0	0.0	0.0
Patella	0	0.0	0.0	0	0.0	0.0	0	0.0	—
Tibia	1	14.3	8.3	7	26.9	11.5	0	0.0	0.0
Fibula	—	—	—	0	0.0	—	—	—	—
Carpal	0	0.0	0.0	0	0.0	—	0	0.0	—
Metacarpal	0	0.0	0.0	0	0.0	0.0	0	0.0	—
Calcaneus	2	28.6	14.3	0	0.0	0.0	0	0.0	0.0
Astragalus	0	0.0	0.0	0	0.0	0.0	0	0.0	0.0
Tarsal	1	14.3	20.0	0	0.0	—	0	0.0	—
Metatarsal	0	0.0	0.0	0	0.0	0.0	0	0.0	0.0
Metapodial	0	0.0	0.0	0	0.0	0.0	0	0.0	—
First phalanx	0	0.0	0.0	0	0.0	0.0	0	0.0	—
Second phalanx	0	0.0	0.0	0	0.0	0.0	0	0.0	—
Third phalanx	0	0.0	0.0	0	0.0	—	0	0.0	—
Total	7	100.0	2.1	26	100.0	3.1	1	100.0	1.6

Key: N = number of specimens bearing gnaw marks; %N = relative frequency of specimens bearing gnaw marks for each skeletal part to the total number of gnawing-mark-bearing specimens per animal taxon; %NISP = relative frequency of specimens bearing gnaw marks for each skeletal part to the total number of specimens identified for each skeletal part per animal taxon; — = no specimens available.

As Tables 5.10a–c show, burnt bones are commonly found among various animal taxa in the faunal assemblage from Haimenkou. There were relatively abundant amounts, high relative frequencies, and a wide distribution of various skeletal parts across the faunal assemblage. Seven taxa exhibited traces of burning, including gaur, pig, sheep & goat, dog, and large, medium, and small cervids (see Figure 5.20). The number of specimens exhibiting traces of burning varies from 100 (pig) to three (small cervid) among different taxa. Four taxa exhibit over 30 specimens and three exhibit 10 or fewer burnt specimens. Based on the NISP of each taxon, the frequencies of all taxa are relatively high, ranging from 32.8 percent (medium cervids) to 9.2 percent (gaur). Furthermore, the traces of burning are definitely the most frequent modifications in all categories of modification found at Haimenkou. These were mostly

Table 5.8b Location and Frequency of Carnivore Gnaw Marks on Bones of Large and Medium Cervids

Skeletal Part	Large Cervid			Medium Cervid		
	N	%N	% NISP	N	%N	% NISP
Antler	0	0.0	0.0	0	0.0	0.0
Cranium	0	0.0	—	0	0.0	0.0
Mandible	0	0.0	—	0	0.0	0.0
Atlas	0	0.0	—	0	0.0	—
Axis	0	0.0	0.0	1	33.3	50.0
Vertebrae	0	0.0	0.0	0	0.0	0.0
Scapula	0	0.0	0.0	1	33.3	25.0
Humerus	0	0.0	0.0	0	0.0	0.0
Radius	0	0.0	0.0	0	0.0	0.0
Ulna	2	66.7	40.0	0	0.0	0.0
Pelvis	0	0.0	—	1	33.3	33.3
Sacrum	0	0.0	—	0	0.0	0.0
Femur	0	0.0	0.0	0	0.0	0.0
Patella	0	0.0	0.0	0	0.0	—
Tibia	0	0.0	0.0	0	0.0	0.0
Fibula	—	—	—	—	—	—
Carpal	0	0.0	0.0	0	0.0	0.0
Metacarpal	0	0.0	0.0	0	0.0	0.0
Calcaneus	0	0.0	0.0	0	0.0	0.0
Astragalus	0	0.0	0.0	0	0.0	0.0
Tarsal	0	0.0	0.0	0	0.0	—
Metatarsal	0	%N	0.0	0	0.0	0.0
Metapodial	0	0.0	0.0	0	0.0	—
First phalanx	0	0.0	0.0	0	0.0	—
Second phalanx	1	33.3	11.1	0	0.0	0.0
Third phalanx	0	0.0	0.0	0	0.0	—
Total	3	100.0	1.9	3	100.0	2.3

Key: See Table 5.8a.

present on all skeletal parts, although this varies among different taxa. The traces of burning found on the bones of dogs may suggest that dogs were eaten by the prehistoric Haimenkou people, although they may not have been raised expressly for the purpose of meat production.

Completely calcined or carbonised specimens are rare among the burnt bones found at Haimenkou. Most burnt specimens only show some traces indicating exposure to fire for a brief time, which may be the result of human activities such as roasting. From the author's own experience, the roasting of animal body parts may sometimes leave no obvious traces of burning, such as colour changes or morphological alteration. Although micro morphological structural changes may be evident if proper equipment is used to examine the faunal remains (e.g., Shipman, et al. 1984), due to the lack of this equipment in the field, only macro traces of burning were observed during this study. However, based on the relatively abundant traces of burning, it appears that human behaviours, such as roasting or throwing bones into hearths after consuming edible parts, and whether intentional or unconscious, frequently occurred at Haimenkou.

Table 5.9 Location and Frequency of Rodent Gnaw Marks on Bones of Pig and Medium Cervids

Skeletal Part	Pig			Medium Cervid		
	N	%N	%NISP	N	%N	%NISP
Antler	—	—	—	0	0.0	0.0
Cranium	0	0.0	0.0	0	0.0	0.0
Mandible	0	0.0	0.0	0	0.0	0.0
Atlas	0	0.0	0.0	0	0.0	—
Axis	0	0.0	0.0	0	0.0	0.0
Vertebrae	0	0.0	0.0	0	0.0	0.0
Scapula	0	0.0	0.0	0	0.0	0.0
Humerus	1	10.0	1.4	2	50.0	20.0
Radius	0	0.0	0.0	0	0.0	0.0
Ulna	1	10.0	2.2	0	0.0	0.0
Pelvis	2	20.0	3.2	0	0.0	0.0
Sacrum	0	0.0	0.0	0	0.0	0.0
Femur	2	20.0	3.0	0	0.0	0.0
Patella	0	0.0	0.0	0	0.0	—
Tibia	2	20.0	100.0	0	0.0	0.0
Fibula	0	0.0	0.0	—	—	—
Carpal	0	0.0	—	0	0.0	0.0
Metacarpal	0	0.0	0.0	1	25.0	14.3
Calcaneus	0	0.0	0.0	0	0.0	0.0
Astragalus	2	20.0	18.2	0	0.0	0.0
Tarsal	0	0.0	—	0	0.0	—
Metatarsal	0	0.0	0.0	0	0.0	0.0
Metapodial	0	0.0	0.0	0	0.0	—
First phalanx	0	0.0	0.0	0	0.0	—
Second phalanx	0	0.0	0.0	1	25.0	100.0
Third phalanx	0	0.0	—	0	0.0	—
Total	10	100.0	1.2	4	100.0	3.1

Key: See Table 5.8a.

5.3.4 Weathering

Both chemical and mechanical processes may lead to the weathering of bones, causing effects such as cracking, splitting, exfoliation, disintegration, and decomposition (Behrensmeyer 1978; Fisher 1995).

In this study a simple weathering categorisation is used for describing the weathering characteristics of specimens. Two stages are applied, namely moderate and heavy weathering stages. The moderate stage corroborates with stages 1 and 2 in Behrensmeyer's (1978) criteria of bone weathering, while the heavy stage consists of stages 3 and 4 in the same criterion set. All specimens exhibiting weathering in these two stages are summarised and presented in Tables 5.11a–b.

After traces of burning, traces of weathering are the second most frequent type of modification found on specimens in the faunal assemblage from Haimenkou. Weathering traces were widely distributed among animal taxa and their skeletal parts (Figure 5.21), and occurred in

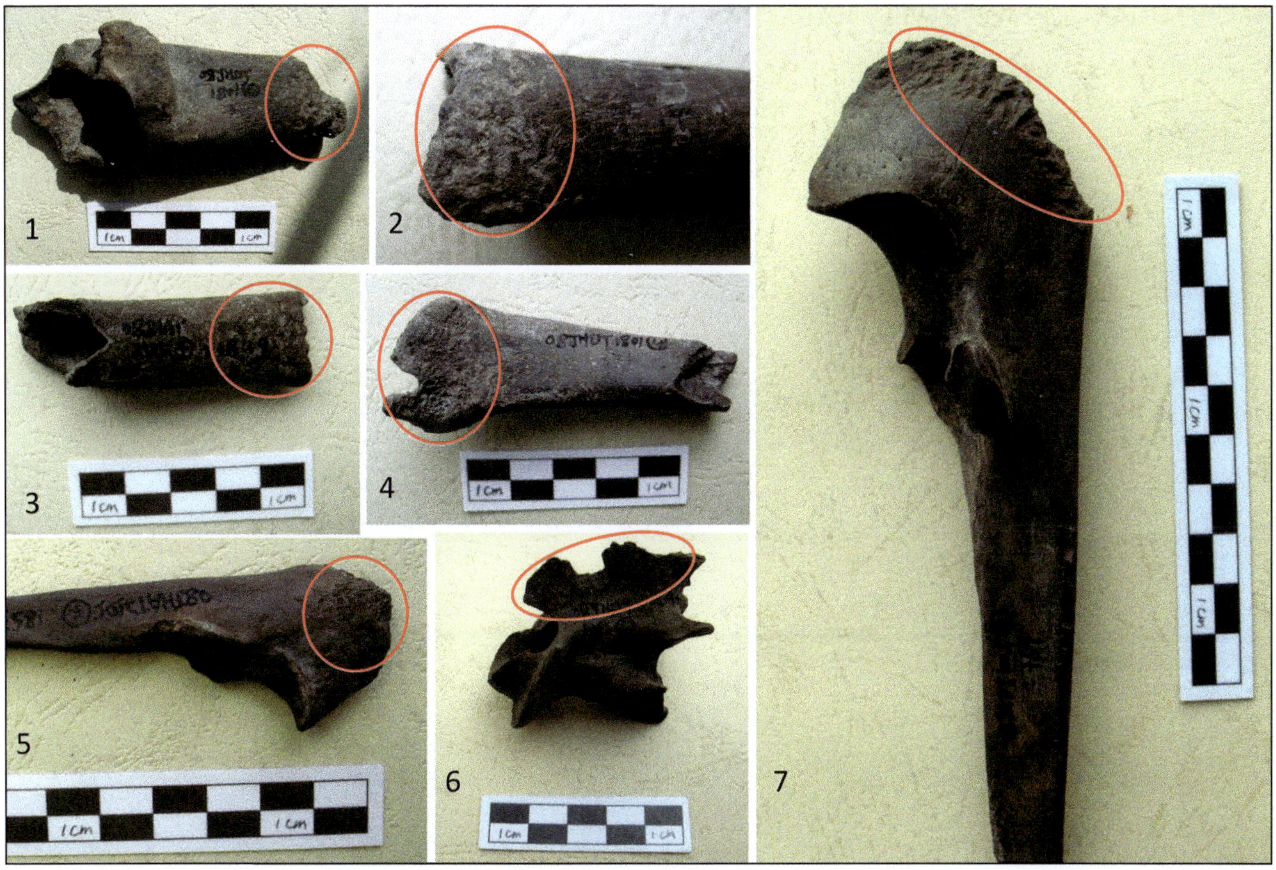

Figure 5.18 Carnivore Gnaw Marks on Animal Bones from Haimenkou (Photos by Juan Wang)
1: Calcaneus (Gaur, BON373, Level (4), Phase III), gnaw marks on tuber; 2: Distal Femur Shaft (Pig, BON923, Level (5), Phase III), gnaw marks on distal end of shaft; 3: Distal Humerus Shaft (Pig, BON1849, Level (7), Phase II), gnaw marks on distal end of shaft; 4: Distal Tibia (Pig, BON371, Level (4), Phase III), gnaw marks on distal end; 5: Ulna (Sheep, BON1853, Level (7), Phase II), gnaw marks on olecranon; 6: Axis (Medium Cervid, BON900, Level (5), Phase III), gnaw marks on spinous process; 7: Ulna (Large Cervid, BON769, Level (6), Phase II), gnaw marks on olecranon.

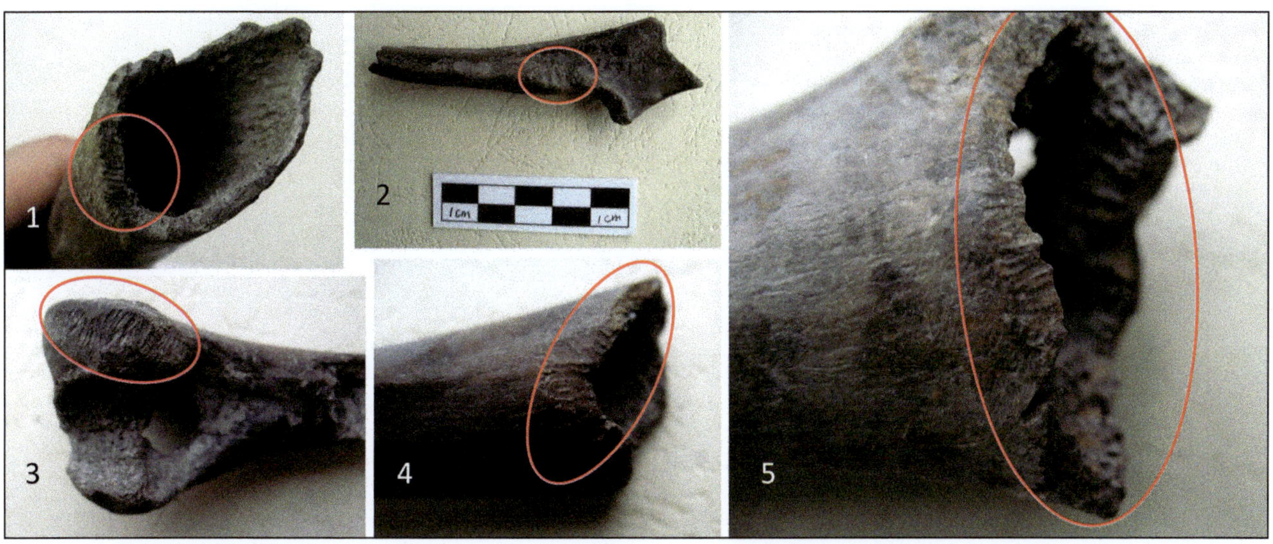

Figure 5.19 Rodent Gnaw Marks on Animal Bones from Haimenkou (Photos by Juan Wang)
1: Distal Humerus (Pig, BON216, Level (5), Phase III), gnaw marks on breakage surface of shaft; 2: Ulna (Pig, BON571, Level (4), Phase III), gnaw marks on coronoid process; 3: Distal Humerus (Medium Cervid, BON344, Level (4), Phase III), gnaw marks on distal end; 4: Tibia Shaft (Pig, BON583, Level (4), Phase III), gnaw marks on breakage surface of distal shaft; 5: Proximal Femur Shaft (Pig, BON844, Level (5), Phase III), gnaw marks on breakage surface of proximal shaft.

Table 5.10a Location and Frequency of Burning on Bones of Gaur and Pig

Skeletal Part	Gaur			Pig		
	N	%N	%NISP	N	%N	%NISP
Horn	2	6.5	15.4	—	—	—
Cranium	0	0.0	0.0	3	3.0	2.1
Mandible	0	0.0	0.0	5	5.0	4.5
Atlas	1	3.2	33.3	12	12.0	52.2
Axis	0	0.0	0.0	2	2.0	40.0
Vertebrae	7	22.6	9.6	2	2.0	2.3
Scapula	3	9.7	18.8	13	13.0	17.8
Humerus	0	0.0	0.0	4	4.0	5.6
Radius	0	0.0	0.0	7	7.0	21.9
Ulna	0	0.0	0.0	1	1.0	2.2
Pelvis	6	19.4	23.1	22	22.0	34.9
Sacrum	0	0.0	0.0	1	1.0	50.0
Femur	2	6.5	11.8	11	11.0	16.7
Patella	0	0.0	0.0	0	0.0	0.0
Tibia	1	3.2	8.3	12	12.0	19.7
Fibula	—	—	—	0	0.0	0.0
Carpal	1	3.2	7.7	0	0.0	—
Metacarpal	1	3.2	16.7	0	0.0	0.0
Calcaneus	1	3.2	7.1	1	1.0	12.5
Astragalus	0	0.0	0.0	2	2.0	18.2
Tarsal	1	3.2	20.0	0	0.0	—
Metatarsal	0	0.0	0.0	1	1.0	14.3
Metapodial	1	3.2	10.0	0	0.0	0.0
First phalanx	1	3.2	4.2	1	1.0	33.3
Second phalanx	1	3.2	5.0	0	0.0	0.0
Third phalanx	2	6.5	22.2	0	0.0	—
Total	31	100.0	9.2	100	100.0	11.9

Key: N = number of specimens bearing burning; %N = relative frequency of specimens bearing burning for each skeletal part to the total number of burning-bearing specimens per animal taxon; %NISP = relative frequency of specimens bearing burning for each skeletal part to the total number of specimens identified for each skeletal part per animal taxon; — = no specimens available.

relatively large numbers and high percentages. As displayed in Tables 5.11a–b, six taxa were found with traces of weathering, including gaur, pig, sheep & goat and, large, medium, and small cervids. Gaurs exhibited the highest relative frequency of weathering (10.7%) and the second largest number of specimens exhibiting weathering (n = 36), while pigs exhibited the largest number of specimens (n = 42) but show a lower relative frequency of weathering traces (5.0%). The number of weathered specimens in other taxa, including sheep & goat and large, medium and, small cervids, varies from seven to one, with a range of relative frequency from 8.3 percent to 4.4 percent. The traces of weathering were evident on every skeletal part of gaur, with a relatively narrower distribution among pigs. They were even less frequently observed in other taxa.

Table 5.10b Location and Frequency of Burning on Bones of Sheep & Goat and Dog

Skeletal Part	Sheep & Goat			Dog		
	N	%N	%NISP	N	%N	%NISP
Horn	0	0.0	—	—	—	—
Cranium	0	0.0	0.0	0	0.0	—
Mandible	0	0.0	0.0	1	25.0	25.0
Atlas	1	11.1	100.0	0	0.0	0.0
Axis	1	11.1	100.0	0	0.0	—
Vertebrae	0	0.0	—	0	0.0	0.0
Scapula	1	11.1	25.0	0	0.0	—
Humerus	0	0.0	0.0	0	0.0	0.0
Radius	0	0.0	0.0	1	25.0	50.0
Ulna	0	0.0	0.0	2	50.0	33.3
Pelvis	1	11.1	25.0	0	0.0	—
Sacrum	2	22.2	66.7	0	0.0	—
Femur	1	11.1	11.1	0	0.0	0.0
Patella	0	0.0	—	0	0.0	—
Tibia	1	11.1	20.0	0	0.0	0.0
Fibula	—	—	—	0	0.0	—
Carpal	0	0.0	—	0	0.0	—
Metacarpal	0	0.0	—	0	0.0	0.0
Calcaneus	1	11.1	100.0	0	0.0	—
Astragalus	0	0.0	0.0	0	0.0	—
Tarsal	0	0.0	—	0	0.0	—
Metatarsal	0	0.0	0.0	0	0.0	—
Metapodial	0	0.0	—	0	0.0	—
First phalanx	0	0.0	—	0	0.0	—
Second phalanx	0	0.0	—	0	0.0	—
Third phalanx	0	0.0	—	0	0.0	—
Total	9	100.0	14.8	4	100.0	16.0

Key: See Table 5.10a.

The general weathering phenomenon of gaur remains may suggest that this taxon is vulnerable to taphonomic process. This may be for a result of the high degree of fragmentation of gaur specimens. Also, as burning and boiling may facilitate weathering processes or result in effects that mimic these processes (Fisher 1995:32; White, Tim 1992:159), a high degree of gaur weathering may be indicative of cooking processes.

5.3.5 Pathology

The location and frequency of pathological abnormalities in the Haimenkou assemblage were recorded and summarised in Table 5.12. The nature and characteristics of bones exhibiting pathology were recorded following BONECODE (Meadow 1978; see also Appendix 1).

Table 5.10c Location and Frequency of Burning on Bones of Large, Medium, and Small Cervids

Skeletal Part	Large Cervid			Medium Cervid			Small Cervid		
	N	%N	% NISP	N	%N	% NISP	N	%N	% NISP
Antler	3	8.6	12.0	7	16.7	20.6	0	0.0	0.0
Cranium	0	0.0	—	0	0.0	0.0	0	0.0	—
Mandible	0	0.0	—	0	0.0	0.0	0	0.0	0.0
Atlas	0	0.0	—	0	0.0	—	0	0.0	—
Axis	0	0.0	0.0	2	4.8	100.0	0	0.0	—
Vertebrae	12	34.3	41.4	3	7.1	21.4	0	0.0	—
Scapula	3	8.6	30.0	1	2.4	25.0	0	0.0	—
Humerus	1	2.9	14.3	7	16.7	70.0	2	66.7	25.0
Radius	1	2.9	14.3	3	7.1	60.0	0	0.0	—
Ulna	0	0.0	0.0	0	0.0	0.0	0	0.0	—
Pelvis	0	0.0	—	2	4.8	66.7	0	0.0	—
Sacrum	0	0.0	—	1	2.4	100.0	0	0.0	—
Femur	0	0.0	0.0	4	9.5	25.0	0	0.0	0.0
Patella	2	5.7	100.0	0	0.0	—	0	0.0	—
Tibia	6	17.1	60.0	5	11.9	41.7	1	33.3	100.0
Fibula	—	—	—	—	—	—	—	—	—
Carpal	0	0.0	0.0	0	0.0	0.0	0	0.0	—
Metacarpal	3	8.6	33.3	1	2.4	14.3	0	0.0	—
Calcaneus	0	0.0	0.0	3	7.1	150.0	0	0.0	—
Astragalus	0	0.0	0.0	0	0.0	0.0	0	0.0	—
Tarsal	1	2.9	20.0	0	0.0	—	0	0.0	—
Metatarsal	0	0.0	0.0	2	4.8	50.0	0	0.0	—
Metapodial	0	0.0	0.0	0	0.0	—	0	0.0	—
First phalanx	1	2.9	9.1	0	0.0	—	0	0.0	—
Second phalanx	2	5.7	22.2	1	2.4	100.0	0	0.0	—
Third phalanx	0	0.0	0.0	0	0.0	—	0	0.0	—
Total	35	100.0	21.9	42	100.0	32.8	3	100.0	25.0

Key: See Table 5.10a.

A few specimens exhibited pathological abnormalities, evident as exostoses on bone shafts (the femur and metacarpal) or the dorsal patella, in sheep & goats and large cervids (Figure 5.22). These pathological abnormalities may indicate the relatively old age of animals.

5.4 Summary

Based on the skeletal part representation analysis of taxa in the faunal assemblage at Haimenkou, several deductions may be made. It appears that the representation of the skeletal part of each taxon is more closely related to the size of elements than to bone density. In other words, small elements of small sized animals are less frequent or absent from the archaeological record. Skeletal parts with a high bone density do not always exhibit high frequencies of this bone type in the zooarchaeological assemblage. The lower frequencies of the smaller bone elements may indicate inadequate sample collection during excavation and the need for an application of sieving. Furthermore,

Figure 5.20 Traces of Burning on Animal Bones from Haimenkou (Photos by Juan Wang)

1: Horn Core Fragment (Gaur, BON803, Level (5), Phase III), partly calcined, partly carbonised; 2: Distal Humerus (Pig, BON1625, Level (7), Phase II), partly calcined, partly carbonised; 3: Second Phalanx (Gaur, BON1876, Level (6), Phase II), carbonised; 4: Antler (Sambar, BON430, Level (6), Phase II), partly carbonised; 5: Mandible (Dog, BON1661, Level (4), Phase III), burnt; 6: Scapula (Medium Cervid, BON1295, Level (6), Phase II), burnt; 7: Distal Tibia (Small Cervid, BON1543, Level (6), Phase II), burnt.

Figure 5.21 Weathering on Animal Bones from Haimenkou (Photos by Juan Wang)

1: Mandible (Pig, BON1817, Level (6), Phase II), heavily weathered; 2: Ulna (Gaur, BON1913, Level (6), Phase II), heavily weathered; 3: Antler (Red Muntjac, BON2059, Level (6), Phase II), heavily weathered; 4: 2+3 Carpal (Gaur, BON115, Level (6), Phase II), moderately weathered; 5: Distal Femur (Small Cervid, BON1528, Level (4), Phase III), moderately weathered; 6: Proximal Femur (Sheep, BON1318, Level (6), Phase II), moderately weathered.

Table 5.11a Location and Frequency of Weathering on Bones of Gaur, Pig, and Sheep & Goat

Skeletal Part	Gaur			Pig			Sheep & Goat		
	N	%N	% NISP	N	%N	% NISP	N	%N	% NISP
Horn/Antler	0	0.0	0.0	—	—	—	0	0.0	—
Cranium	0	0.0	0.0	1	2.4	0.7	0	0.0	0.0
Mandible	1	2.8	14.3	1	2.4	0.9	0	0.0	0.0
Atlas	0	0.0	0.0	1	2.4	4.3	0	0.0	0.0
Axis	0	0.0	0.0	0	0.0	0.0	0	0.0	0.0
Vertebrae	2	5.6	2.7	0	0.0	0.0	0	0.0	—
Scapula	0	0.0	0.0	10	23.8	13.7	2	40.0	50.0
Humerus	2	5.6	22.2	4	9.5	5.6	1	20.0	14.3
Radius	2	5.6	15.4	0	0.0	0.0	0	0.0	0.0
Ulna	1	2.8	33.3	4	9.5	8.7	0	0.0	0.0
Pelvis	1	2.8	3.8	6	14.3	9.5	0	0.0	0.0
Sacrum	1	2.8	33.3	0	0.0	0.0	0	0.0	0.0
Femur	1	2.8	5.9	8	19.0	12.1	2	40.0	22.2
Patella	0	0.0	0.0	0	0.0	0.0	0	0.0	—
Tibia	2	5.6	16.7	5	11.9	8.2	0	0.0	0.0
Fibula	—	—	—	0	0.0	0.0	—	—	—
Carpal	7	19.4	53.8	0	0.0	—	0	0.0	—
Metacarpal	1	2.8	16.7	1	2.4	9.1	0	0.0	—
Calcaneus	2	5.6	14.3	0	0.0	0.0	0	0.0	0.0
Astragalus	1	2.8	10.0	1	2.4	9.1	0	0.0	0.0
Tarsal	1	2.8	20.0	0	0.0	—	0	0.0	—
Metatarsal	0	0.0	0.0	0	0.0	0.0	0	0.0	0.0
Metapodial	3	8.3	30.0	0	0.0	0.0	0	0.0	—
First phalanx	1	2.8	4.2	0	0.0	0.0	0	0.0	—
Second phalanx	7	19.4	35.0	0	0.0	0.0	0	0.0	—
Third phalanx	0	0.0	0.0	0	0.0	—	0	0.0	—
Total	36	100.0	10.7	42	100.0	5.0	5	100.0	8.2

Key: N = number of weathered specimens; %N = relative frequency of weathered specimens for each skeletal part to the total number of weathered specimens per animal taxon; %NISP = relative frequency of weathered specimens for each skeletal part to the total number of specimens identified for each skeletal part per animal taxon; — = no specimens available.

according to the general appearance of the skeletal parts of the main animal taxa—including gaur, pig, sheep & goat, and large, medium, and small cervids, whether domestic or wild—it is possible that all these were slaughtered at the Haimenkou site rather than at a killing site outside this residential village and then transported to Haimenkou, if such a site existed.

With respect to bone modification, traces of burning were most commonly encountered on specimens in the faunal assemblage from Haimenkou. This was followed by traces of weathering. Other types of modification, such as cut marks, and carnivore and rodent gnaw marks, were relatively rare in frequency. The frequent presence of burning may indicate common human activities, including food

Table 5.11b Location and Frequency of Weathering on Bones of Large, Medium, and Small Cervids

Skeletal Part	Large Cervid			Medium Cervid			Small Cervid		
	N	%N	% NISP	N	%N	% NISP	N	%N	% NISP
Antler	0	0.0	0.0	2	40.0	5.9	0	0.0	0.0
Cranium	0	0.0	—	0	0.0	0.0	0	0.0	—
Mandible	0	0.0	—	0	0.0	0.0	0	0.0	0.0
Atlas	0	0.0	—	0	0.0	—	0	0.0	—
Axis	0	0.0	0.0	0	0.0	0.0	0	0.0	—
Vertebrae	1	14.3	3.4	0	0.0	0.0	0	0.0	—
Scapula	3	42.9	30.0	1	20.0	25.0	0	0.0	—
Humerus	0	0.0	0.0	0	0.0	0.0	0	0.0	0.0
Radius	0	0.0	0.0	0	0.0	0.0	0	0.0	—
Ulna	0	0.0	0.0	0	0.0	0.0	0	0.0	—
Pelvis	0	0.0	—	0	0.0	0.0	0	0.0	—
Sacrum	0	0.0	—	0	0.0	0.0	0	0.0	—
Femur	1	14.3	50.0	1	20.0	6.3	1	100.0	100.0
Patella	0	0.0	0.0	0	0.0	—	0	0.0	—
Tibia	1	14.3	10.0	0	0.0	0.0	0	0.0	0.0
Fibula	—	—	—	—	—	—	—	—	—
Carpal	0	0.0	0.0	0	0.0	0.0	0	0.0	—
Metacarpal	0	0.0	0.0	0	0.0	0.0	0	0.0	—
Calcaneus	0	0.0	0.0	0	0.0	0.0	0	0.0	—
Astragalus	0	0.0	0.0	1	20.0	50.0	0	0.0	—
Tarsal	0	0.0	0.0	0	0.0	—	0	0.0	—
Metatarsal	1	14.3	25.0	0	0.0	0.0	0	0.0	—
Metapodial	0	0.0	0.0	0	0.0	—	0	0.0	—
First phalanx	0	0.0	0.0	0	0.0	—	0	0.0	—
Second phalanx	0	0.0	0.0	0	0.0	0.0	0	0.0	—
Third phalanx	0	0.0	0.0	0	0.0	—	0	0.0	—
Total	7	100.0	4.4	5	100.0	3.9	1	100.0	8.3

Key: See Table 5.11a.

preparations at the Haimenkou site. Food preparation may have included roasting and/or discarding of bones in the hearth after consumption. In some cases, the apparent weathering of the bones may also have resulted from human behaviours such as the food preparation and consumption mentioned above. Cut marks, although small in number, also reflect human activities related to food processing at Haimenkou. Gnaw marks made by carnivores (mainly dogs) and rodents indicate the presence of dogs and rodents at this time. It can also elucidate on the relationship between humans and animals at Haimenkou, including the keeping of domestic dogs.

In chapter six, the kill-off patterns for domestic animals will be analysed. The analysis will be based on tooth eruption and wear and epiphyseal fusion data.

Table 5.12 Location and Frequency of Pathological Abnormalities on Bones of Sheep & Goat and Large Cervids

Skeletal Part	Sheep & Goat			Large Cervid		
	N	%N	% NISP	N	%N	% NISP
Horn/Antler	0	0	—	0	0.0	0.0
Cranium	0	0.0	0.0	0	0.0	—
Mandible	0	0.0	0.0	0	0.0	—
Atlas	0	0.0	0.0	0	0.0	—
Axis	0	0.0	0.0	0	0.0	0.0
Vertebrae	0	0.0	—	0	0.0	0.0
Scapula	0	0.0	0.0	0	0.0	0.0
Humerus	0	0.0	0.0	0	0.0	0.0
Radius	0	0.0	0.0	0	0.0	0.0
Ulna	0	0.0	0.0	0	0.0	0.0
Pelvis	0	0.0	0.0	0	0.0	—
Sacrum	0	0.0	0.0	0	0.0	—
Femur	1	100.0	11.1	0	0.0	0.0
Patella	0	0.0	—	2	66.7	100.0
Tibia	0	0.0	0.0	0	0.0	0.0
Fibula	—	—	—	—	—	—
Carpal	0	0.0	—	0	0.0	0.0
Metacarpal	0	0.0	—	1	33.3	11.1
Calcaneus	0	0.0	0.0	0	0.0	0.0
Astragalus	0	0.0	0.0	0	0.0	0.0
Tarsal	0	0.0	—	0	0.0	0.0
Metatarsal	0	0.0	0.0	0	0.0	0.0
Metapodial	0	0.0	—	0	0.0	0.0
First phalanx	0	0.0	—	0	0.0	0.0
Second phalanx	0	0.0	—	0	0.0	0.0
Third phalanx	0	0.0	—	0	0.0	0.0
Total	1	100.0	1.6	3	100.0	1.9

Key: N = number of specimens bearing pathological abnormalities; %N = relative frequency of specimens bearing pathological abnormalities for each skeletal part to the total number of pathological-abnormalities- bearing specimens per animal taxon; %NISP = relative frequency of specimens bearing bearing for each skeletal part to the total number of specimens identified for each skeletal part per animal taxon; — = no specimens available.

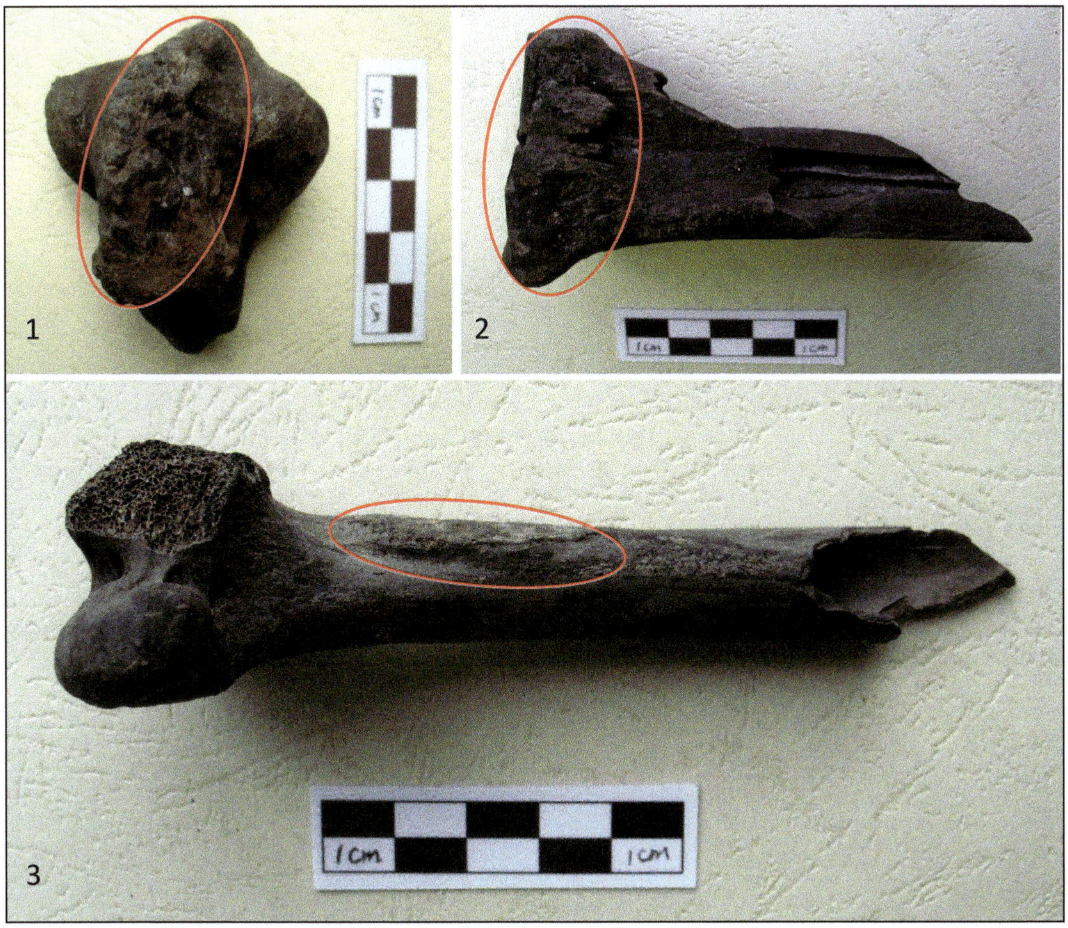

Figure 5.22 Pathological Abnormalities on Animal Bones from Haimenkou (Photos by Juan Wang)
1: Patella (Large Cervid, BON1532, Level (6), Phase II), exostosis on cranial surface; 2: Proximal Metacarpal (Large Cervid, BON192, Level (6), Phase II), exostosis on volar, proximal end; 3: Distal Femur (Sheep/Goat, BON1856, Level (6), Phase II), exostosis on lateral shaft.

6

Kill-off Patterns for Domestic Animals

6.1 Introduction

To investigate hunting practices, scavenging behaviours, and husbandry patterns of ancient humans, a number of zooarchaeologically identifiable bone attributes are useful. These include faunal identification data, relative abundance of taxa, skeletal part representation of taxa, and animal age profile. The age profile of a certain animal group may show a pattern different to the 'natural' attrition of a population of that species, namely death through old age, disease, and misadventure (O'Connor 2000:80). Based on the animal group age profile, a pattern dissimilar to the result caused by nature, such as a kill-off pattern, may represent selections made by humans. Selections made by hunters and farmers may further reflect the process of human decision-making in antiquity (O'Connor 2000:80). To investigate the kill-off patterns at a site, the construction of age profiles of the animal taxa is fundamental. It is also necessary to estimate the ages at which the animals died or were killed. The study on age at death has attracted researchers from a variety of fields, including veterinarians, zoologists, archaeologists, and zooarchaeologists. A number of publications have dealt with this issue from the early 19th century onwards (e.g., Amorosi 1989; Brown 1985; Bull and Payne 1982; Davis 1987; Deniz and Payne 1982; Grant 1975; 1982; Greenfield and Arnold 2008; Hillson 2005; Huidekoper 1891; Klein and Cruz-Uribe 1983; Klein, et al. 1981; Lauwerier 1983; Lewall and Cowan 1963; Matschke 1967; Mayhew 1849; Moran and O'Connor 1994; Morris 1972; Munson 2000; Noddle 1974; O'Connor 2000; Payne 1973; 1987; Prummel 1987a; b; 1988; 1989; Purdue 1983; Reitz and Wing 2008; Rolett and Chiu 1994; Schmid 1972; Silver 1970; Smith, R. 1956; Wilson, et al. 1982).

For domestic animals, the criteria used in selecting animals for slaughter may be indicated by the age estimation profiles. Thus, the background to the economic strategy may be revealed. The age at which animals are slaughtered depends on numerous factors, such as the relative value placed on the different products, the characteristics of the stock, a variety of environmental elements, and so forth (Payne 1973:281). Kill-off patterns are based on age profiles of certain animal groups, as particular animal species or breeds may reflect the range of factors mentioned above, especially their use in people's economic life. It is vital to estimate their age of slaughter, as different forms of animal consumption are characterised by different optimal slaughter ages (Uerpmann 1973:315). For instance, if meat production is the aim, most of the young males in a group are killed when they reach the optimum point in weight-gain, and only a few are kept for breeding. In Central and Western Europe, the optimal slaughter ages for animals raised for meat production were relatively constant (e.g., cattle 2.5–3.5 years, small ruminants 1–2 years, pigs 1.5 years). This is the point at which the 'rapid growth has ceased while the meat output no longer increased relative to the food input' (Uerpmann 1973:315-316). Therefore, if a sample is dominated by these age profiles, the animals were probably reared for meat (Uerpmann 1973:316).

6.2 Methods

Methods of age determination will be applied when constructing the age profiles for domestic animal taxa recovered from Haimenkou, and for presenting and comparing their kill-off patterns and survivorships. Conservatively speaking, the methods of age determination typically used may be considered estimates only, because they are not able to provide very precise timing but only a range or general time span. In other words, most of these methods provide relative results rather than absolute ones. In this situation, age profiles may be constructed on the basis of a faunal assemblage from an individual site, however comparisons among various faunal assemblages from different sites require more attention.

To date, there have been a number of methods, along with a variety of available data, to determine the age at death of an animal. These include changes in form and porosity of the bone (e.g., Prummel 1987a; b; 1988; 1989), tooth eruption and replacement sequences (e.g., Bull and Payne 1982; Deniz and Payne 1982; Greenfield and Arnold 2008; Hillson 2005), tooth wear (e.g., Klein, et al. 1981; Moran and O'Connor 1994; Payne 1973; 1987), incremental structures (e.g., Burke and Castanet 1995; Gordon 1991; Lowe 1967), epiphyseal fusion (e.g., Amorosi 1989; Bull and Payne 1982; Lewall and Cowan 1963; Noddle 1974; Smith, R. 1956), closure of cranial sutures (e.g., Chaplin 1971:81; Schmid 1972:73-74), and antler and horn development (e.g., Armitage and Clutton-Brock 1976; Chaplin 1971:89; Goss 1983; Schmid 1972:88-90). Although numerous data and discussions on various methods have been published, the different methods are not always appropriate for or applicable to a certain faunal assemblage. In this study, the methods employed include tooth eruption, tooth wear, and epiphyseal fusion. These will be introduced in the following text and described in detail in individual sections focusing on certain animal taxa.

6.2.1 Tooth Eruption and Wear

Sliver (1970:283) stated, 'the science of ageing an animal by the appearance of its teeth in life (or death) is probably as old as animal husbandry'. For the present zooarchaeological study, research on teeth eruption and wear may construct the most important foundation of the investigation for the age of animals at death. This is because: (1) teeth are the most durable and hard tissues in an animal body, with a usually relatively abundant amount; (2) changes in dentition may reflect advancing age closely and continuously in many ways, making teeth the most informative structures of a body (Klein, et al. 1981:2; Silver 1970:289-290). Information such as species, age, and approximate body size of an animal may be inferred from just a single tooth. However, more information may be extracted from complete or nearly complete tooth rows that are still embedded in maxillae or mandibles when excavated.

There are three general methods for estimating the ages of animals based on their teeth (Klein, et al. 1981:2). One method is to match tooth eruption and tooth wear stages in a dentition of unknown age with its closest counterpart in a series of dentitions of known age (e.g., Bull and Payne 1982; Deniz and Payne 1982; Grant 1975; 1982; Payne 1987). From the aspect of tooth wear stage, this method is mainly based on the observation of changes in the surface morphology related to the exposure of dentine when the enamel is worn away (O'Connor 2000:87). Another way is to count the increments (annuli) in dentine and enamel of teeth (e.g., Hillson 2005:159-168, 245-253; O'Connor 2000:80-83), for which a microscope is necessary. The third method is to measure a dental dimension, especially the crown height (e.g., Klein and Cruz-Uribe 1983; 1984-213; Klein, et al. 1981). This method is based on the assumption that there is a systematic relationship between increasing age and dimensional change (Klein, et al. 1981:2), and is particularly appropriate for herbivorous mammals with high-crowned teeth (O'Connor 2000:85). Since the decrease of tooth crown height is also caused by chewing attrition, the third method may be assigned to tooth wear. In this study, the main methods employed include tooth eruption and wear. Although the crown heights of teeth were measured in bovids and cervids as primary data, they will not be used here due to the small sample size.

Mammals usually have two sets of dentition, in which the deciduous teeth develop and erupt and then are replaced by permanent dentition (Hillson 2005; Reitz and Wing 2008:49). The sequences of teeth eruption and replacement are basically consistent for each animal taxon with advancing age, although the correlate age is variable and subject to many factors such as breed differences, diet composition, nutritional level, and environmental conditions (Chaplin 1971:78; Morris 1972:80). In this chapter, several protocols for recording tooth eruption and succession are applied comprehensively (Schmid 1972; Silver 1970), with details presented in relevant sections.

From the observation of animal tooth eruption and replacement, the age, especially the age of young animals, may be estimated. However, with only information on tooth eruption and succession the complete picture still may not be constructed, as teeth eruption only represent the life span of an animal after its full eruption of permanent teeth (Morris 1972:80). Thus this technique is usually employed in combination with another tooth method, tooth wear stages.

When teeth erupt through the alveolar bone of the maxilla or mandible and reach the occlusal plane of the adjacent teeth, teeth wear begins (O'Connor 2000:83-84). O'Connor (2000:86-87) has concluded that recording the tooth wear other than measuring tooth crown heights 'requires some form of classification of wear stages, and this is usually achieved by noting and utilising the sequence in which different cusps of each tooth come into wear, and by noting the typical sequence of exposure of dentine as the enamel covering is worn away'. In this study, a variety of systems for recording and coding tooth wear stage are used for different animal taxa, and detailed description will be presented in each animal taxon section respectively (e.g., Amorosi 1989; Bull and Payne 1982; Deniz and Payne 1982; Grant 1975; 1982; Greenfield and Arnold 2008; Moran and O'Connor 1994; Payne 1973; 1987; Rolett and Chiu 1994).

In ungulates, tooth eruption and succession may be employed as an accurate indicator for the age of younger individuals. In older individuals who already have full permanent dentition, tooth wear contributes more towards reflecting the age of older individuals (Morris 1972:81). A measure of crown height may be useful with regard to tooth wear, as this may provide a more accurate age estimate which cannot be obtained by merely observing occlusal surface wear pattern. For example, in sheep or goat, the tooth wear stage 'M1' (see Appendix 1) may remain for a relatively long time when applied to the lower M2 (second molar) (Payne 1973:289). Therefore, during this time span the wear pattern of the M2 will remain in stage 'M1' and remain relatively unchanged. However the attrition of the tooth crown will continue, leading to an apparent reduction in crown height which may be used to assess age more accurately.

As defined by Klein and Cruz-Uribe (1984:46), crown height is 'the minimum distance between the occlusal surface and the line separating the enamel of the crown from the dentine on the roots, measured on the buccal surface of mandible teeth and on the lingual surface of maxillary ones'. In this study, the crown height of teeth from bovids and cervids were measured, in the way that Klein and colleagues suggested (Klein, et al. 1981:6-7). However, due to the scarce finds of complete or nearly complete jaws with teeth embedded in bovids and cervids at Haimenkou, the crown height measurement was mostly taken on loose teeth (see Appendix 2). These were recovered in relatively small numbers and are thus hard to assign to individuals. Therefore, the measurement of crown height was unable to provide useful data for the present research. The data collected are presented in Appendix 2 and will not be analysed in this chapter.

The data for loose teeth should be recorded if possible, but it is better to extract information from a series of associated teeth which are still embedded in the maxilla or mandible rather than from loose and individual unassociated teeth (Reitz and Wing 2008:174). Following this suggestion, all of the teeth recovered, whether in alveoli or loose, were recorded in as much detail as possible. Nevertheless, the analysis is generally based on rows of teeth which are preserved as whole sets or at least left with some particular teeth that may reflect crucial information (e.g., a semi-mandible with only the third molar remaining).

Although tooth wear stage may be an effective indicator for the age of an animal after the completion of permanent teeth eruption, there is an obvious problem that may affect the results. That is, tooth wear will vary with the nature of diet (Morris 1972:79), especially the amount of grit consumed with food. General nutritional level and other factors may also affect tooth wear (Reitz and Wing 2008:174). Hence, unless there are lines of evidence from other sources (e.g., epiphyseal fusion), care should be taken when identifying tooth wear on zooarchaeological specimens (Silver 1970:290).

6.2.2 Epiphyseal Fusion

There are several centres of ossification in limb bones of mammals (O'Connor 2000:92; Silver 1970:283-284). In an immature animal, a typical limb bone consists of three parts: the diaphysis (shaft), and the proximal and distal epiphyses, which are attached to the diaphysis by the epiphyseal plate (an intervening layer of cartilage) (O'Connor 2000:92). A limb bone grows in length with new bone formation on either side of the epiphyseal plate (O'Connor 2000:92). When a mammal is growing to attain its adult size, the epiphyseal plates of different skeletal parts are replaced by real bone following a certain sequence, and finally most of the epiphyses fuse to the correlate diaphyses with a few minor exceptions (Reitz and Wing 2008:71; Silver 1970:284). Unlike animals with indeterminate growth (e.g., fish, shell, see Reitz and Wing 2008:73-79), mammals with determinate growth will stop growing in most body dimensions (height, length, and weight) when they reach the adult size, namely after most epiphyseal fusion is complete, although remoulding of bones will carry on for their whole life (Reitz and Wing 2008:70; Silver 1970:284-286).

The epiphyseal fusion in a animal skeleton occurs in stages over time, and follows a certain sequence in different animal taxa (Noddle 1974; Purdue 1983; Schmid 1972:75; Silver 1970:285-286; Wilson, et al. 1982). During this process, the epiphysis changes from the status of open, with the epiphysis separate from the diaphysis, to semi-fused, with the epiphysis and diaphysis partly united but the trabeculae interrupted. Finally the epiphysis fuses, and the epiphyseal line may still be visible or completely obscured in older individuals (Reitz and Wing 2008:173). If a young mammal dies, the epiphyseal plate generally decays more rapidly than the bone, meaning the epiphysis in its limb bones will separate from the associated diaphysis (O'Connor 2000:92). It is thus possible to observe an archaeological sample and record the frequency of specimens for which a particular epiphysis was fused to the diaphysis (without or with an epiphyseal line) or was unfused (O'Connor 2000:92).

As mentioned above, epiphyseal fusion occurs in a certain order. The age at which fusion takes place varies from element to element, species to species, and breed to breed, and is affected by environmental factors (Reitz and Wing 2008:72-73). For example, high planes of nutrition and sheltered conditions tend to accelerate epiphyseal fusion (Silver 1970:288), thus epiphyseal fusion may complete earlier in domestic animals than in wild or feral individuals (e.g., Noddle 1974). However, some elements consistently fuse early in the maturation process and others fuse very late. Although the chronological age at which epiphyseal fusion occurs varies, the order of fusion appears to be relatively constant among mammals, and may be employed to establish age classes. The age of a mammal may be inferred from these epiphyseal fusion based age classes (Noddle 1974; Purdue 1983; Reitz and Wing 2008:72; Schmid 1972:75; Silver 1970:285-286; Wilson, et al. 1982). For each element, it is possible to establish at least two age classes. Since many long bones have two epiphyses that fuse at different ages, one may actually infer three age classes from long bone epiphyseal fusion (Klein and Cruz-Uribe 1984:43). For example, in cattle the proximal epiphysis of the tibia fuses to the diaphysis later than the distal epiphysis, thus three age classes are formed: (1) neither epiphysis is fused, (2) distal is fused but proximal unfused, and (3) both epiphyses are fused. Because the distal epiphysis of the tibia tends to fuse at 24–30 months and the proximal epiphysis at 42–48 months, the age classes of the cattle tibia may be established as less than 24–30 months, 24–30 to 42–48 months, and more than 42–48 months (see Table 6.12). When the estimate of age from the tibia is used in conjunction with the age classes of other elements, the range of age assessed is more accurate. Therefore, for a complete skeleton or incomplete skeleton with crucial age-inferred elements, the age at death of a mammal may be estimated. By examining which epiphyses are fused and which are still unfused, the age can be bracketed: 'older than the earliest likely fusion timing for the last fused epiphysis, but younger than the latest likely fusion timing for the first unfused epiphysis' (O'Connor 2000:94).

Establishing the epiphyseal fusion sequence for mammalian taxa has consumed the time of many researchers (e.g., Lewall and Cowan 1963; Noddle 1974; Purdue 1983; Schmid 1972; Silver 1970; Smith, R. 1956). One way to arrive at such baseline data is to slaughter a known-aged animal then observe the fusion status in its bones, and another way is to X-ray the elements at regular intervals in an animal's life (O'Connor 2000:93). Needless to say this second method is less violent and destructive. It also has deficiencies, as different researchers have different criteria for reading radiographs when attempting to assess states of epiphyseal fusion (e.g., Lewall and Cowan 1963; Moran and O'Connor 1994; Smith, R. 1956). It goes without saying that determining age at death based on epiphyseal fusion is not an exact science.

Table 6.1 Age Stages for Pig and Sheep/Goat*

Animal	Infantile	Juvenile	Sub-adult	Adult
Pig	0–10 months	10–30 months	30–42 months	42+ months
Sheep/Goat	0–10 months	10–30 months	30–42 months	42+ months

Note: *Personal communication from Dr. Ajita K. Patel (May 5, 2010).

Table 6.2 Teeth Eruption Ages in Pig

Tooth	Eruption Timing		Age Class	Age Stage
	Silver(1970)	Schmid(1972)		
dI3	at Birth	at Birth	I	Infantile
dC	at Birth	at Birth		
dI1	4–14 days	at Birth		
dI2	6–12 weeks	12 weeks		
dP2	7–10 weeks	1 month		
dP3	1–5 weeks	at Birth		
dP4	2-7 weeks	at Birth		
P1	3.5–6.5 months	6 months	II	
M1	4–6 months	6 months		
M2	7–13 months	12 months	III	Juvenile
I3	8–12 months	9 months		
C	8–12 months	9 months		
P2	12–16 months	12 months	IV	
P3	12–16 months	12 months		
P4	12–16 months	15 months		
I1	12-17 months	12 months		
I2	17–20 months	15 months	V	
M3	17–22 months	15 months		

Key: d = deciduous; I = incisor; C = canine; P = premolar; M = molar.

There are several problems associated with determining age at death from epiphyseal fusion. Some of these problems may impact the present study in unknown ways. For example, ages of epiphyseal fusion among wild taxa, such as gaur, are not well understood (Greenfield and Arnold 2008:837; Klein and Cruz-Uribe 1984:43). And there are taphonomic consequences. Fused elements tend to be more durable than unfused elements, thus a bias against unfused elements, especially in the very young age classes, may occur (Greenfield and Arnold 2008:837). Consequently, epiphyseal fusion data are better be used in combination with other, potentially more precise methods, such as dental eruption and wear.

Most archaeological faunal assemblages contain disarticulated and fragmentary remains, while a complete or nearly complete skeleton is a pleasant but rarely occurring event (O'Connor 2000:94). In the absence of such optimal conditions, clarifying the age distribution in the remains belonging to a particular species involves listing the epiphyses recovered for each species, and tabulating the fused and unfused specimens in the sample (O'Connor 2000:94-95). The proportion of fused specimens for a particular epiphysis may then be taken to indicate 'the proportion of animals represented in the sample which survived beyond the typical age of fusion of that epiphysis' (O'Connor 2000:95), namely the survival rate.

6.3 Results and Analysis

In this chapter, two domestic animal taxa recovered from the faunal assemblage at Haimenkou were analysed,

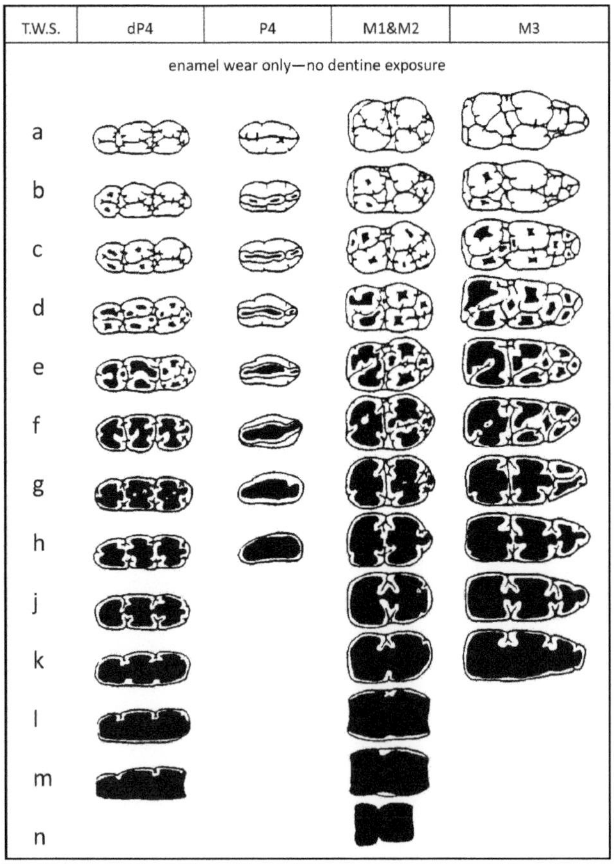

Figure 6.1 Mandibular Tooth Wear Stages for Pig
Key: T.W.S. = tooth wear stage; dP4 = deciduous fourth premolar; P4 = fourth premolar; M1 = first molar; M2 = second molar; M3 = third molar. Note: Modified from Grant (1982: figure 3); reproduced with permission of BAR Publishing, www.barpublishing.com

including pig and sheep & goat (all the caprine specimens are combined into one category in this chapter, as in chapter three, four, and five). Table 6.1 presents the age stages for these animal taxa examined in this study.

6.3.1 Pig

6.3.1.1 Method and Materials

Tooth Eruption and Wear

In the present study, dental attrition observations were based on Grant (1975; 1982) and Meadow (1978), while the teeth eruption timing was taken from Silver (1970: table G) and Schmid (1972: table X). Wear stages are shown in Table 6.2. All teeth, those in alveoli as well as isolated teeth, were observed and recorded. For this study, the permanent molars (upper and lower M1, M2, and M3) and the deciduous fourth premolar (upper and lower dP4) were recorded in Grant's style (see Figure 6.1), while other deciduous and permanent teeth were recorded following Meadow's methodology (1978). Breadth and length of all pig teeth were measured and recorded following von den Driesch (1976) (see Appendix 2).

Table 6.3 M.W.S. for Pig from Haimenkou

Age Class	(dP4)P4	M1	M2	M3
I	(1/2)	C		
I	(c/d)	V		
I	(d)	V		
I	(c)	V		
II	(c/d)	1/2		
II	(d)	U		
II	(j/k)	?	C	
II	(d/e)	U	C	
II	(d)	a	C	
II	(d/e)	b	C	
II	?	a	V	
II	?	b/c	V	
II	?	c	V	
III	?	d/e	1/2	
III	(k)	e	1/2	
III	3	c	U	
III	?	c/d	a	
III	?	e	a	
IV	(j)	c	a	C
IV	E	c	a	C
IV	E	c/d	b	C
IV	U	d	U	V
IV	a	b/c	a	V
IV	U	c/d	a	V
IV	U	d	a	V
IV	?	e	a	V
IV	?	?	b	V
IV	a	d	c	V
IV	a	e	c	V
IV	b	e	c	V
IV	b	e	c	V
IV	a	f	c	V
V	?	e/f	b	E
V	?	e	d	1/2
V	a/b	h	d/e	1/2
V	b	f	c/d	a
V	b	e	d/e	a
VI	d/e	h	e	c
VI	d	h	e	c
VI	?	h/j	f/g	c
VII	?	k/l	?	?
VII	?	?	?	e

Key: W.M.S. = mandible wear stage; ? = unclear.
Eruption stages (letters in uppercase and '1/2'): C = alveolar perforation in crypt visible; V = tooth visible in crypt but below alveolar rim; E = tooth protruding through bone; 1/2 = tooth half erupted; U = tooth almost at full height but unworn.
For other abbreviations, see Table 6.2.
Note: Tooth wear stages (letters in lowercase) are shown in Figure 6.1, following Grant (1982: figure 3)

Table 6.4 Criteria of Age Estimate for Pig Mandibles from Haimenkou Based on Tooth Eruption and Wear

Tooth Eruption	Age Class	Age Stage	Suggested Age	(dP4) P4	M1	M2	M3
	I	Infantile	0–4 months	(1/2, c, d)	C,V		
M1 erupting	II		4–7 months	(c, d, e, j, k)	1/2, U, a, b, c	C,V	
M2 erupting	III	Juvenile	7–12 months	(k)	c, d, e	1/2, U, a	
P4 erupting	IV		12–17 months	(j), E, U, a, b	c, d, e, f	U, a, b, c	C, V
M3 erupting	V		17–22 months	a, b	e, f, h	b, c, d, e	E, 1/2, a
	VI		22–30 months	d, e	h, j	e, f, g	c
	VII	Sub-adult	30+ months	—	k, l	—	e

Key: — = no data available.
For other abbreviations and symbols, see Table 6.3.

Table 6.2 gives the eruption timing of pig teeth that will be used for this research. Pig mandibles make up a relatively large proportion of the faunal assemblage from Haimenkou, hence the kill-off patterns for pigs were established on the basis of mandibles, although data were also recorded for maxillae containing teeth and loose teeth. Grant (1982:95) proposed the method known as 'mandible wear stage' (hereafter M.W.S.). M.W.S. of all pig mandibles containing at least one tooth were observed, recorded, and tabulated (see Table 6.3). These formed the age stage criteria data given in Table 6.4. Based on these criteria, the age at death of pigs was estimated (Table 6.5).

Based on the above data, the kill-off patterns for pigs were constructed for Phases I, II, and III. Complete (both sides preserved) and semi-mandibles were treated as single individuals.

Epiphyseal Fusion

According to Silver (1970: table A) and Schmid (1972: table IX), the criteria of age estimation for pig limb bones may be based on the ages at which epiphyses in different skeletal parts are fused (Table 6.6). Three age stages are presented: early, middle, and late, corresponding to Age Classes II, III, and IV. Results are presented in Table 6.7, showing the frequencies of fused and unfused skeletal parts of pigs from Haimenkou. This led to the construction of the survivorship curves for pigs from Phases I to III as shown in Figure 6.4. Again, there were no data for the Late Phase. In keeping with the natural conditions, the survival rate for new-borns was set at 100 percent.

6.3.1.2 Results and Analysis

Age Profiles Based on Tooth Eruption and Wear

Due to the small sample sizes from Phase I and the Late Phase (this phase is not included in the scope of this study due to its later date) at Haimenkou, only pig remains from Phases II and III were used to construct age profiles. However, Phase I and the Late Phase data were included in the overall age profile to illustrate the full range of age classes and stages for pigs at Haimenkou.

Figures 6.2a–c illustrate the distribution of age classes for pigs, based on dental eruption and wear. There are substantial peaks in Class IV in both Phases II and III, and in Phase II Class II is also well represented. The differences between these two distributions may indicate that between Phases II and III the timing for slaughter became more focused on Class IV (12–17 months) in Phase III. This may indicate a refinement of the husbandry practices in Phase III to optimize meat production. This is consistent with research by Uerpmann (1973:316) who states that slaughter between the ages of 12–17 months is optimal for pigs raised primarily for their meat.

Figures 6.3a–c give the age stage distribution based on dental eruption and wear using a reduced number of more general age categories such as infantile, juvenile, and sub-adult. The bias toward juvenile animals is clearer than in Figures 6.2a–c.

Grouping the data into stages (infantile, juvenile, sub-adult) rather than classes depicts the kill-off age at Haimenkou as relatively unambiguous. Most pigs were slaughtered during their juvenile stage, between 10 and 30 months. This is similar to pig meat production in the modern world, and the slaughter timing difference observed between Phases II and III may reflect a shift toward optimal pig husbandry techniques through time.

Survival Rate based on Epiphyseal Fusion

Despite the small sample of post-cranial pig remains from Phase I, it was possible to construct a survivorship curve for that period. However, it may be less statistically meaningful than for Phases II and III. The Late Phase yielded too few specimens to warrant a survivorship curve. As a

Table 6.5 Age Distribution of Pig Mandibles from Haimenkou Based on Tooth Eruption and Wear

Age Class/ Stage	Phase I		Phase II		Phase III		Total	
	N	%	N	%	N	%	N	%
I	0	0.0	2	7.1	3	8.8	5	7.9
II	1	100.0	7	25.0	6	17.6	14	22.2
Infantile	1	100.0	9	32.1	9	26.5	19	30.2
III	0	0.0	4	14.3	3	8.8	7	11.1
IV	0	0.0	7	25.0	12	35.3	19	30.2
V	0	0.0	6	21.4	5	14.7	11	17.5
VI	0	0.0	0	0.0	4	11.8	4	6.3
Juvenile	0	0.0	17	60.7	24	70.6	41	65.1
VII	0	0.0	2	7.1	1	2.9	3	4.8
Sub-adult	0	0.0	2	7.1	1	2.9	3	4.8

Key: N = number of pig mandibles.

Table 6.6 Criteria of Age Estimation for Pig Limb Bones from Haimenkou Based on Epiphyseal Fusion

Fusing Order	Epiphysis	Age of Fusion	Age Class	Age Stage
Early (12–18 months)	Scapula, d	12 months	II	Juvenile
	Humerus, d	12–18 months		
	Radius, p	12 months		
	Acetabulum	12 months		
	Phalanx 2, p	12 months		
Middle (24–30 months)	Tibia, d	24 months	III	Juvenile
	Fibula, d	30 months		
	Calcaneum	24–30 months		
	Metapodial, d	24–27 months		
	Phalanx 1, p	24 months		
Late (36–42 months)	Humerus, p	42 months	IV	Sub-adult
	Radius, d	42 months		
	Ulna, p	36–42 months		
	Ulna, d	36–42 months		
	Femur, p	42 months		
	Femur, d	42 months		
	Tibia, p	42 months		
	Fibula, p	42 months		

Key: p = proximal; d = distal.
Note: Based on data from Silver (1970: table A) and Schmid (1972: table IX).

Table 6.7a Frequencies of Fused and Unfused Skeletal Parts of Pig from Haimenkou (Phase I and II)

Age Class/ Stage	Epiphysis	Phase I			Phase II		
		N F	N UF	% F	N F	N UF	% F
II Juvenile (12–18 months)	Scapula, d	0	0	—	0	0	—
	Humerus, d	4	1	80.0	9	13	40.9
	Radius, p	0	0	—	9	4	69.2
	Acetabulum	1	0	100.0	13	2	86.7
	Phalanx 2, p	1	0	100.0	0	0	—
	Total	6	1	85.7	31	19	62.0
III Juvenile (24–30 months)	Tibia, d	1	2	33.3	4	9	30.8
	Fibula, d	0	0	—	0	1	0.0
	Calcaneum	0	0	—	0	2	0.0
	Metapodial, d	1	0	100.0	5	4	55.6
	Phalanx 1, p	0	0	—	0	1	0.0
	Total	2	2	50.0	9	17	34.6
IV Sub-adult (36–42 months)	Humerus p	0	1	0.0	1	10	9.1
	Radius, d	0	0	—	0	8	0.0
	Ulna, p	0	0	—	1	8	11.1
	Ulna, d	0	0	—	0	4	0.0
	Femur, p	0	2	0.0	2	10	16.7
	Femur, d	0	1	0.0	0	14	0.0
	Tibia, p	0	1	0.0	3	7	30.0
	Fibula, p	0	0	—	1	2	33.3
	Total	0	5	0.0	8	63	11.3

Key: N = number; F = fused epiphysis; UF = unfused epiphysis; % F = proportion of fused specimens; — = no data available. For other abbreviations, see Table 6.6.

result, this section focuses on Phases II and III, as was the case with the dental observations. As with the previous dental section, the total limb bone specimens were used to construct a complete picture of the survivorship curve for pigs from Haimenkou.

Figures 6.4a–d depicts the survival curves for pigs from Haimenkou based on epiphyseal fusion. According to Figures 6.4b and 6.4c, the survivorship curves show a similar trend. Both figures suggest that over 60 percent of the pigs survived to a juvenile stage, Class II, while many of them were slaughtered before the sub-adult stage. Only about 11 percent survived the beginning of the sub-adult Stage. Figures 6.4b–c accord well with the kill-off pattern seen in tooth eruption and tooth wear results. The same evidence appears in the evidence between Phases II and III as seen in the tooth eruption and wear results. The survival rate at Class III is higher in Phase III (52.0%) than in Phase II (34.6%). This may signal a refinement of husbandry techniques or a shift to a greater emphasis on optimizing meat production.

In summary, the survivorship curves for pigs based on epiphyseal fusion generally concur with the same husbandry strategies and kill-off patterns reflected by tooth eruption and wear stages. Pigs were mainly slaughtered during their juvenile period, and most were slaughtered intensively in a certain time span for optimal meat production and harvested.

6.3.2 Sheep & Goat

6.3.2.1 Method and Materials

Tooth Eruption and Wear

In the same manner as in previous chapters, the specimens found at Haimenkou which were identified as sheep or

Table 6.7b Frequencies of Fused and Unfused Skeletal Parts of Pig from Haimenkou (Phase III and the Late Phase)

Age Class/ Stage	Epiphysis	Phase III			Late Phase		
		N F	N UF	% F	N F	N UF	% F
II Juvenile (12–18 months)	Scapula, d	0	2	0.0	0	0	—
	Humerus, d	8	7	53.3	1	0	100.0
	Radius, p	10	3	76.9	2	0	100.0
	Acetabulum	13	6	68.4	1	1	50.0
	Phalanx 2, p	1	0	100.0	0	0	—
	Total	32	18	64.0	4	1	80.0
III Juvenile (24–30 months)	Tibia, d	8	4	66.7	0	1	0.0
	Fibula, d	0	0	—	0	0	—
	Calcaneum	1	4	20.0	0	0	—
	Metapodial, d	3	3	50.0	0	0	—
	Phalanx 1, p	1	1	50.0	0	0	—
	Total	13	12	52.0	0	1	0.0
IV Sub-adult (36–42 months)	Humerus p	0	2	0.0	0	0	—
	Radius, d	0	4	0.0	0	0	—
	Ulna, p	0	7	0.0	0	0	—
	Ulna, d	0	1	0.0	0	0	—
	Femur, p	2	9	18.2	0	0	—
	Femur, d	3	13	18.8	0	0	—
	Tibia, p	1	11	8.3	0	0	—
	Fibula, p	0	0	—	0	0	—
	Total	6	47	11.3	0	0	—

Key: See Table 6.7a.

sheep/goat are grouped in the category 'sheep & goat' for convenience. Since only sheep may be confidently identified in the faunal assemblage from Haimenkou, all data used here for ageing is specific to sheep rather than goats.

Methods for recording the tooth eruption and wear of sheep & goats were taken from Payne (1973) and Meadow (1978), while the eruption timing shown by Table 6.8 was after Silver (1970: table B) and Schmid (1972: table X). All the caprine teeth were observed and recorded, in the same way as they were for pigs and cattle in the previous sections, with the coding system proposed by Meadow (1978: figure 1) (see Figures 6.5a–d). Breadth and length of all the caprine teeth were measured and recorded, and the height of all isolated teeth was also measured and recorded (see Appendix 2) after von den Driesch (1976) and Klein and colleagues (1981: figure 1).

A relatively small sample size of sheep & goat mandibles from Haimenkou makes the M.W.S. method ineffective. Thus, the age stage criteria for sheep & goat mandibles (see Table 6.9) were taken from Payne (1973: table 1). Based on these criteria, the age distribution of sheep & goat mandibles are outlined in Table 6.10. Due to the absence of available materials in Phase I and the Late Phase, no table for these two phases were presented. Therefore, the kill-off patterns for sheep & goats were only established for Phases II and III. All sheep & goat mandibles found at Haimenkou were semi-mandibles (left or right), with no evident pairing present. Hence, they were all considered to be different individuals, with kill-off patterns built by age class and age stage respectively (Figures 6.6 and 6.7).

Epiphyseal Fusion

The method and process used to describe epiphyseal fusion were the same ones used for pigs, with Table 6.11 presenting the criteria of age estimation for sheep & goats based on a proposed epiphyseal fusion sequence. The frequencies of fused and unfused skeletal parts of sheep & goats

Table 6.7c Frequencies of Fused and Unfused Skeletal Parts of Pig from Haimenkou (Total)

Age Class/ Stage	Epiphysis	Total		
		N F	N UF	% F
II Juvenile (12–18 months)	Scapula, d	0	2	0.0
	Humerus, d	22	21	51.2
	Radius, p	21	7	75.0
	Acetabulum	28	9	75.7
	Phalanx 2, p	2	0	100.0
	Total	73	39	65.2
III Juvenile (24–30 months)	Tibia, d	13	16	44.8
	Fibula, d	0	1	0.0
	Calcaneum	1	6	14.3
	Metapodial, d	9	7	56.3
	Phalanx 1, p	1	2	33.3
	Total	24	32	42.9
IV Sub-adult (36–42 months)	Humerus p	1	13	7.1
	Radius, d	0	12	0.0
	Ulna, p	1	15	6.3
	Ulna, d	0	5	0.0
	Femur, p	4	21	16.0
	Femur, d	3	28	9.7
	Tibia, p	4	19	17.4
	Fibula, p	1	2	33.3
	Total	14	115	10.9

Key: See Table 6.7a.

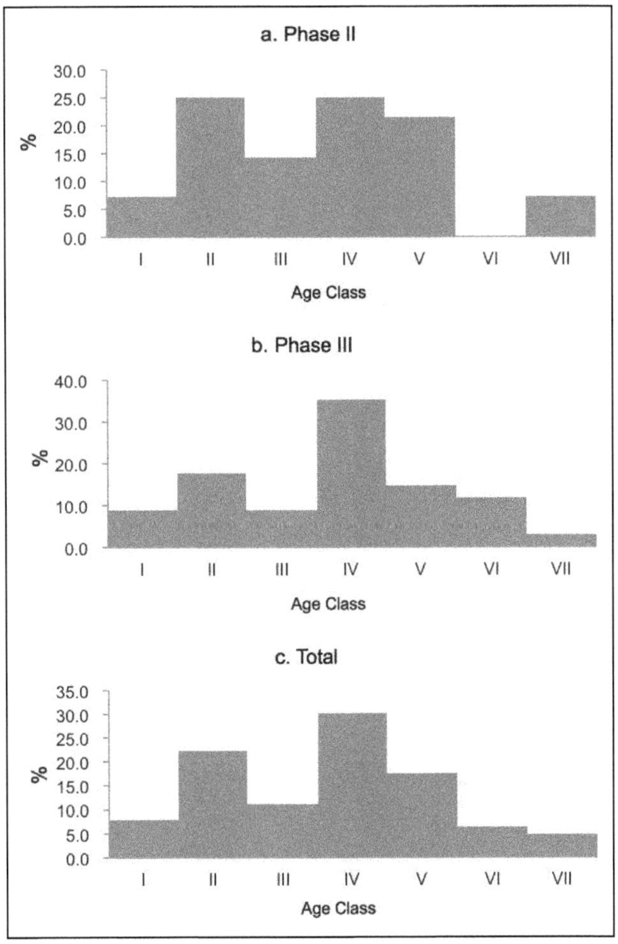

Figure 6.2a–c Age Profiles for Pig Based on Tooth Eruption and Wear (by Age Class)

were given in Tables 6.13a–b, reflecting the survival rates for sheep & goats. The Late Phase was absent from Table 6.12a–b, as there were no useful skeletal parts. Because of the small sample size of sheep & goats in each phase, the survivorship curve was illustrated on the basis of the total data from all phases, as shown by Figures 6.6a–c.

6.3.2.2 Results and Analysis

Age Profiles based on Tooth Eruption and Wear

The sample size of sheep & goat specimens from each phase is too small to establish a meaningful kill-off pattern. Hence, all useful specimens were grouped together to construct the total age profile. This reflected a relatively accurate age distribution for sheep & goats based on the available evidence.

Figure 6.7c describes a complete picture of the age distribution of sheep & goats from Haimenkou based on teeth eruption and teeth wear stages. In this figure, the ages of sheep & goats at their death were distributed across all life stages including Infantile, Juvenile, and Adult Stages, but mainly at the Adult Stage. A clearer pattern emerges in Figure 6.6c, where sheep & goats appear to have been mostly slaughtered at the Class VIII (42.9%) stage. This indicates slaughter at an age between 6 and 8 years old, which is quite old for a husbanded caprine. The secondary intensive slaughter timing is at Class IV (28.6%). This indicates slaughter at an age range between 1 to 2 years, which is the optimal slaughter age for small ruminants (Uerpmann 1973:316).

As discussed before, pigs from Haimenkou were probably reared for meat production. Figure 6.6c seems to suggest a kill-off pattern for sheep & goats that differs from those for pigs. The considerably higher proportion of adult animals (6–8 years old) among sheep & goats may indicate they were used for other purposes than meat, possibly for a secondary product such as wool or milk. Davis (1987:158) proposes that milk and wool production among sheep and goats does not decline considerably until they are several years old (4–7 years). Therefore, as suggested by Payne (1973:281), if these secondary products are the aims, a greater number of adult animals will be present in the age profiles. In addition, the relatively large proportion of

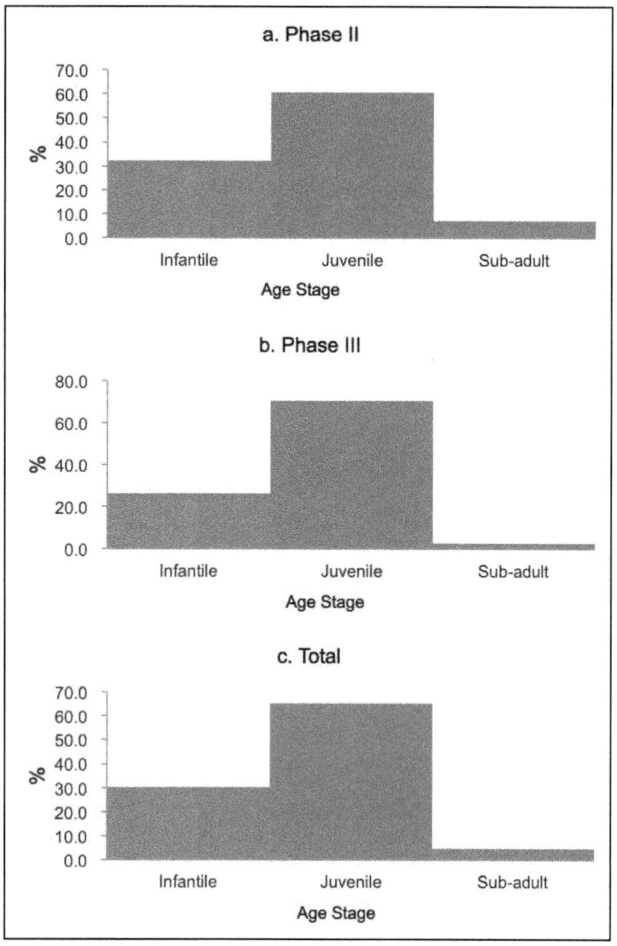

Figure 6.3a–c Age Profiles for Pig Based on Tooth Eruption and Wear (by Age Stage)

Table 6.8 Tooth Eruption Ages in Sheep

Tooth	Eruption Timing		Age Stage
	Silver (1970)	Schmid (1972)	
dI1	at Birth–1 week	at Birth	Infantile
dI2	at Birth–1 week	at Birth	
dI3	at Birth–2 weeks	at Birth	
dC	at Birth–3 weeks	at Birth	
dP2	at Birth–6 weeks	at Birth	
dP3	at Birth–6 weeks	at Birth	
dP4	at Birth–6 weeks	at Birth	
M1	3 months (lower)	6 months	
M2	9–12 months	24 months	Juvenile
I1	12–18 months	15 months	
I2	18–24 months	21 months	
P2	21–24 months	18 months	
P3	21–24 months	21 months	
P4	21–24 months	24 months	
M3	18–24 months	21 months	
I3	27–36 months	33 months	Sub-adult
C	33-48 months	45 months	Adult

Key: See Table 6.2

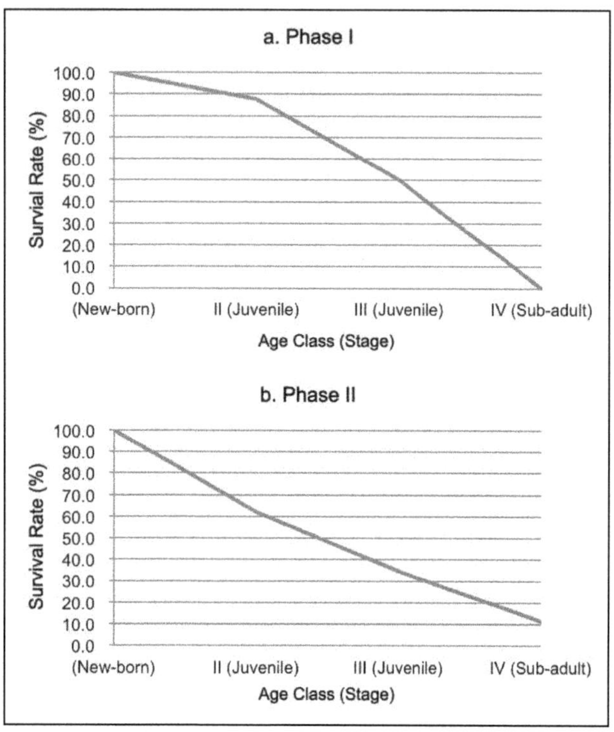

Figure 6.4a–b Survivorship Curves for Pigs Based on Epiphyseal Fusion (Phases I and II)

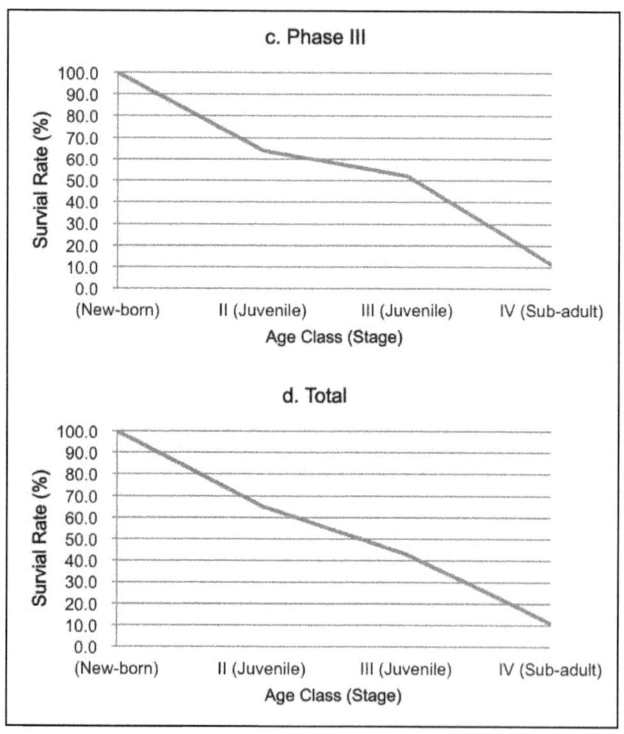

Figure 6.4c–d Survivorship Curves for Pigs Based on Epiphyseal Fusion (Phase III and Total)

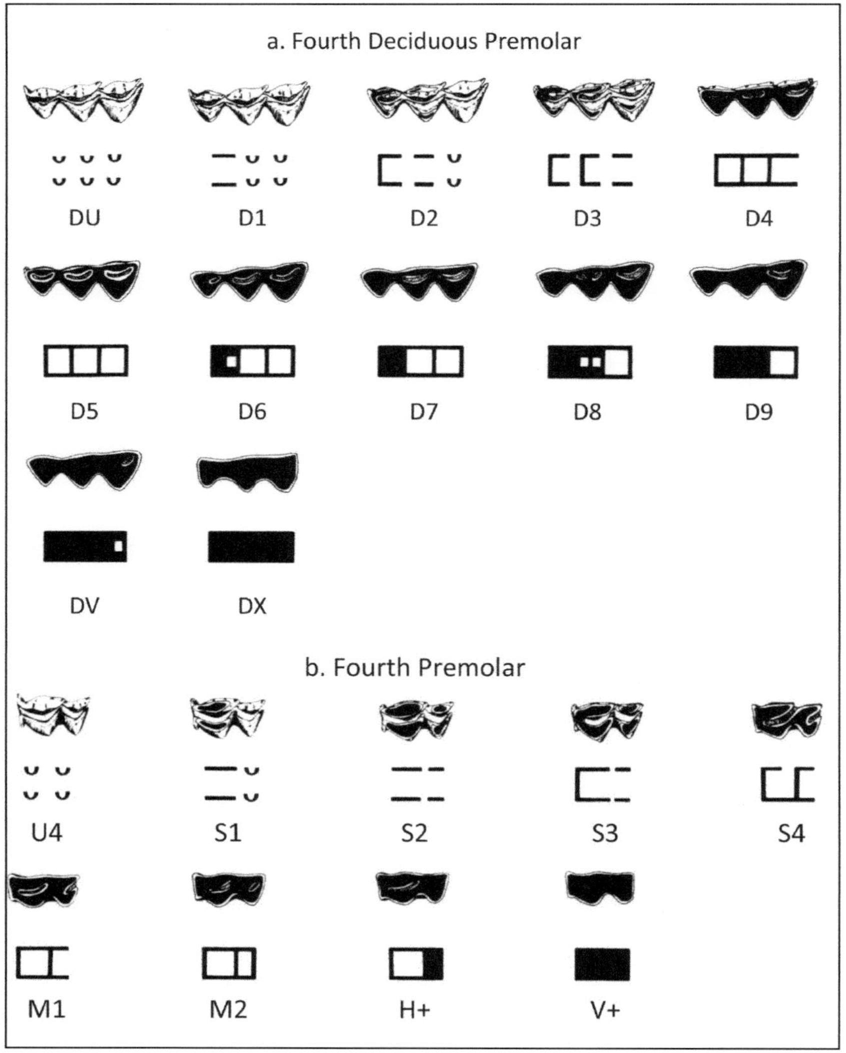

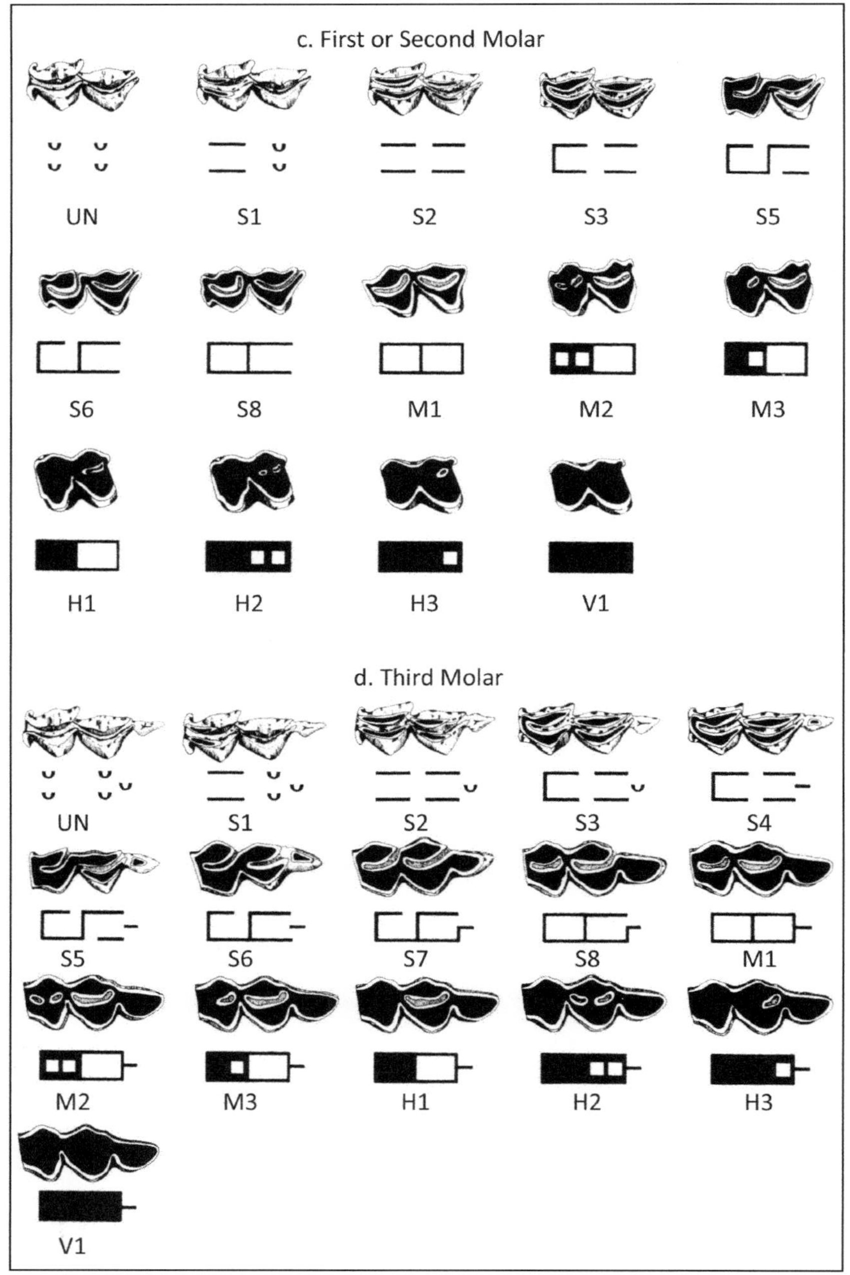

Figure 6.5a–d Mandibular Tooth Wear Stages for Sheep and Goat
Note: Modified from Payne (1973: figures 6, 8, 9, and 10), reproduced with permission of the British Institute at Ankara; for tooth wear codes, see BONECODE (Meadow 1978: figure 1), see also Appendix 1.

Table 6.9 Criteria of Age Estimate for Sheep & Goats at Haimenkou Based on Tooth Eruption and Wear

Tooth Eruption and Wear	Suggested Age	Age Class	Age Stage
dP4 still unworn	0–2 months	I	Infantile
dP4 in wear, M1 unworn	2–6 months	II	
M1 in wear, M2 unworn	6–12 months	III	Juvenile
M2 in wear, M3 unworn	1–2 years	IV	
M3 in wear, posterior cusp unworn	2–3 years	V	
Posterior cusp of M3 in wear, M3 pre-(M1)	3–4 years	VI	Sub-adult
M3 (M1), M2 (M1)	4–6 years	VII	Adult
M3 (M1), M2 post-(M2)	6–8 years	VIII	
M3 post-(M2)	8–10 years	IX	

Key: See Table 6.2; abbreviations in brackets refer to the codes proposed by Meadow (1978: table 1), see also in Figure 6.5a–d.
Note: Based on Payne (1973: table 1).

Table 6.10 Age Distribution of Sheep & Goat Mandibles from Haimenkou Based on Tooth Eruption and Wear

Age Class/ Stage	Phase II		Phase III		Total	
	N	%	N	%	N	%
I	0	0.0	1	33.3	1	14.3
II	1	25.0	0	0.0	1	14.3
Infantile	1	25.0	1	33.3	2	28.6
III	0	0.0	0	0.0	0	0.0
IV	2	50.0	0	0.0	2	28.6
V	0	0.0	0	0.0	0	0.0
Juvenile	2	50.0	0	0.0	2	28.6
VI	0	0.0	0	0.0	0	0.0
Sub-adult	0	0.0	0	0.0	0	0.0
VII	0	0.0	0	0.0	0	0.0
VIII	1	25.0	2	66.7	3	42.9
IX	0	0.0	0	0.0	0	0.0
Adult	1	25.0	2	66.7	3	42.9

Key: N = number of sheep & goat mandibles.

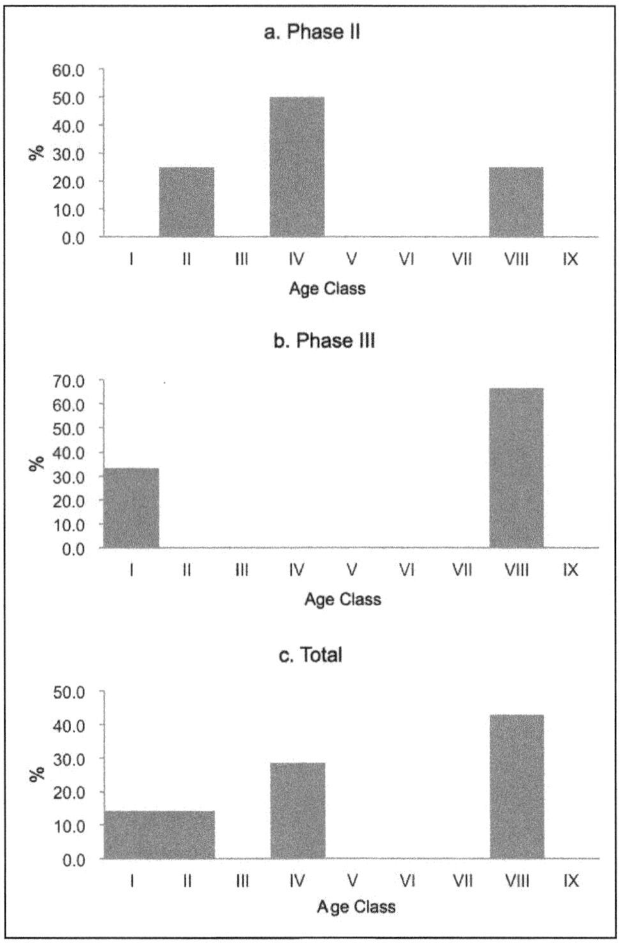

Figure 6.6a–c Age Profiles for Sheep & Goat Based on Tooth Eruption and Wear (by Age Class)

specimens at Class IV may imply the existence of some type of meat production. If meat is the primary objective, most young males are slaughtered when they reach the optimum point of weight-gain, and only a few are kept for breeding (Payne 1973:281). For meat production young females may or may not be killed depending on circumstances; however, mainly young females may be kept for maintaining the breeding stock, and for secondary products including milk and wool (Payne 1973:281). Consequently, from the age profiles based on tooth eruption and wear, it appears that there may have been multiple uses of sheep & goats at Haimenkou: meat, milk, and wool.

Table 6.11 Criteria of Age Estimation for Sheep & Goat from Haimenkou Based on Epiphyseal Fusion

Fusing Order	Epiphysis	Age of Fusion	Age Class	Age Stage
Early (3-16 months)	Scapula, d	6–8 months	I	Infantile
	Humerus ,d	3–10 months		
	Radius ,p	3–10 months		
	Acetabulum	6–10 months		
	Phalanx 1, p	6–16 months		
	Phalanx 2, p	6–16 months		
Middle (16-36 months)	Tibia, d	15–24 months	II	Juvenile
	Metapodial, d	18–28 months		
	Calcaneum	30–36 months	III	Sub-adult
Late (36-42 months)	Humerus, p	36–42 months	IV	
	Radius, d	36–42 months		
	Ulna, p	36–42 months		
	Ulna, d	42 months		
	Femur, p	30–42 months		
	Femur, d	36–42 months		
	Tibia, p	36–42 months		

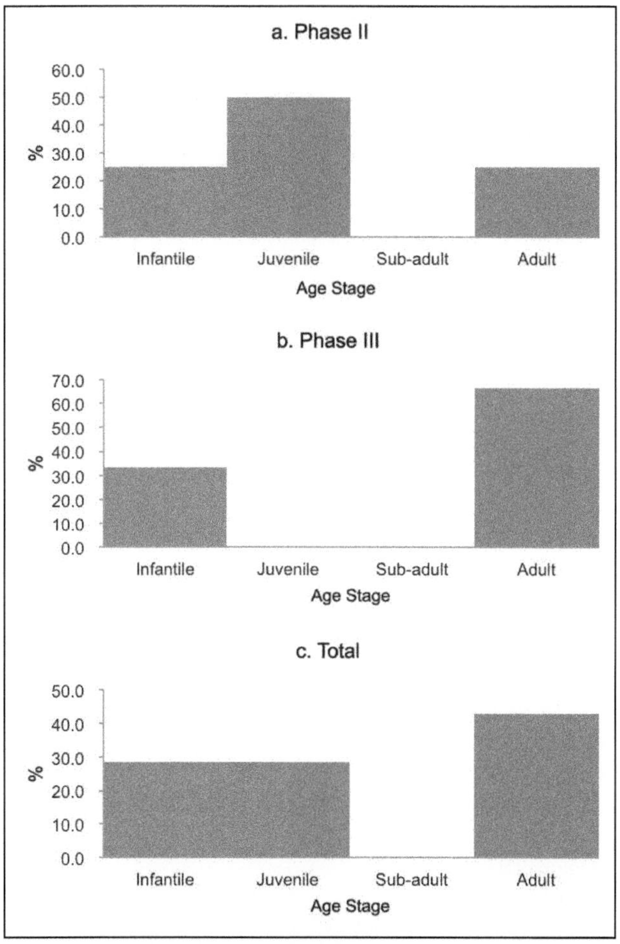

Figure 6.7a–c Age Profiles for Sheep & Goat Based on Tooth Eruption and Wear (by Age Stage)

Key: See Table 6.6.
Note: See Table 6.6.

Survival Rate based on Epiphyseal Fusion

Due to a lack of specimens available for establishing the survivorship curve for each phase, all the skeletal part specimens were put together to increase the sample size, and only the survivorship curve for the total number of specimens is shown by Figure 6.8.

The kill-off pattern for sheep and goats indicated in Figure 6.8 exhibits some similarities with Figure 6.7, which presents a high proportion of adult animals. According to Figure 6.8, more than half the sheep & goats survived the beginning of the Sub-adult Stage (i.e., Classes

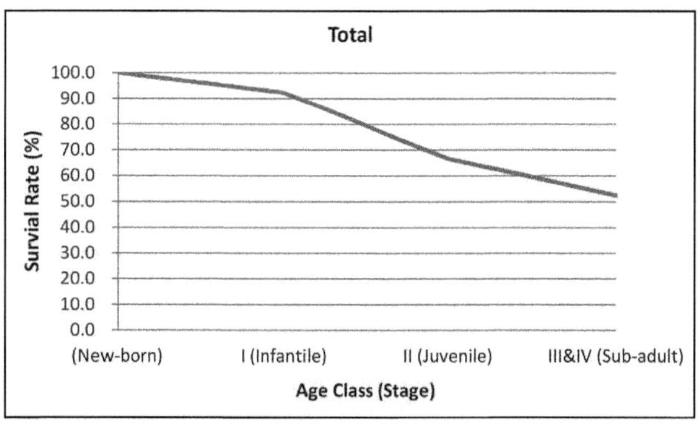

Figure 6.8 Survivorship Curve for Sheep & Goat Based on Epiphyseal Fusion (Total)

Table 6.12a Frequencies of Fused and Unfused Skeletal Parts of Sheep & Goat from Haimenkou (Phase I–II)

Age Class/Stage	Epiphysis	Phase I			Phase II		
		N F	N UF	% F	N F	N UF	% F
I Infantile (3–16 months)	Scapula, d	0	0	—	0	0	—
	Humerus, d	0	0	—	2	0	100.0
	Radius, p	0	0	—	2	0	100.0
	Acetabulum	0	0	—	1	0	100.0
	Phalanx 1, p	0	0	—	0	0	—
	Phalanx 2, p	0	0	—	0	0	—
	Total	0	0	—	5	0	100.0
II Juvenile (16–28 months)	Tibia, d	0	0	—	2	0	100.0
	Metapodial, d	0	0	—	0	0	—
	Total	0	0	—	2	0	100.0
III & IV Sub-adult (30–42 months)	Calcaneum	0	0	—	0	0	—
	Humerus, p	0	0	—	1	1	50.0
	Radius, d	1	0	100.0	1	3	25.0
	Ulna, p	0	0	—	1	0	100.0
	Ulna, d	0	0	—	0	0	—
	Femur, p	0	0	—	1	1	50.0
	Femur, d	0	0	—	2	2	50.0
	Tibia, p	0	0	—	1	1	50.0
	Total	1	0	100.0	7	8	46.7

Key: See Table 6.7a.

III and IV). This is a very high survival rate and is considerably different from that shown by pigs which were raised primarily for meat. However, based on the survival rates at the Infantile and Juvenile Stages, about 30 percent of sheep & goats were slaughtered between the Infantile and Juvenile Stages. Moreover, because of the wide range of early fusion timing for sheep (3–16 months, see Table 6.11), the animals that passed the beginning of the Infantile Stage may have reached the Juvenile Stage (10–30 months). Thus, it is possible that the animals slaughtered before reaching the Sub-adult Stage actually belonged to the Juvenile Stage, fitting the optimal slaughter age for small ruminants (i.e., 1–2 years, see Uerpmann 1973:316). Therefore, the kill-off pattern reflected by the survivorship curve based on epiphyseal fusion may match with the kill-off pattern indicated by the age profile based on tooth eruption and wear. Both of these indicated that the sheep & goats may have been reared for meat, milk, and wool production.

6.4 Summary

The results and analysis of the kill-off patterns for domestic animals recovered at the Haimenkou site offers new interpretations of animal husbandry and exploitation carried out by the Haimenkou people around 3,000 years ago. A variety of husbandry strategies were indicated among domestic animals, including pigs and sheep & goats.

Pigs were unquestionably the most important domestic animal in the daily life of the Haimenkou people, as the analyses and discussion in the previous chapters suggest. As described in this chapter, most individuals were slaughtered at a juvenile age which indicated they were kept for meat. The pig husbandry techniques appear to have substantially improved from Phases II to III, as reflected by more intensive slaughter ages for optimal meat products. Hence, it is probable that pigs were raised for meat products and thus can be considered food animals. This

Table 6.12b Frequencies of Fused and Unfused Skeletal Parts of Sheep & Goat from Haimenkou (Phase III and Total)

Age Class/Stage	Epiphysis	Phase III			Total		
		N F	N UF	% F	N F	N UF	% F
I Infantile (3–16 months)	Scapula, d	0	1	0.0	0	1	0.0
	Humerus, d	2	0	100.0	4	0	100.0
	Radius, p	3	0	100.0	5	0	100.0
	Acetabulum	2	0	100.0	3	0	100.0
	Phalanx 1, p	0	0	—	0	0	—
	Phalanx 2, p	0	0	—	0	0	—
	Total	7	1	87.5	12	1	92.3
II Juvenile (16–28 months)	Tibia, d	0	0	—	2	0	100.0
	Metapodial, d	0	1	0.0	0	1	0.0
	Total	0	1	0.0	2	1	66.7
III & IV Sub-adult (30–42 months)	Calcaneum	1	0	100.0	1	0	100.0
	Humerus, p	0	1	0.0	1	2	33.3
	Radius, d	0	0	—	2	3	40.0
	Ulna, p	0	0	—	1	0	100.0
	Ulna, d	0	0	—	0	0	—
	Femur, p	1	0	100.0	2	1	66.7
	Femur, d	1	1	50.0	3	3	50.0
	Tibia, p	0	0	—	1	1	50.0
	Total	3	2	60.0	11	10	52.4

Key: See Table 6.7a.

contrasts with the use of sheep & goats which were probably raised for meat and secondary products such as milk and wool. This conclusion is based on the distinctive kill-off pattern for sheep and goats at Haimenkou, which indicates that a larger proportion of caprine animals were not killed until relative old age.

The next chapter is the final chapter which will provide a general discussion and outline of conclusions drawn from the zooarchaeological study of the Haimenkou site. Directions for future research will also be outlined.

7

Discussion and Conclusion

7.1 Introduction

The main aims of the Haimenkou faunal assemblage study were to investigate human-animal relationships, the subsistence economy of the Haimenkou people, and cultural connections linking Haimenkou to other regions in prehistoric China.

Analyses were conducted from several aspects: the environmental setting and archaeological background of the Haimenkou site (chapter two); the outline of the procedures for faunal analysis (chapter three); the range and relative importance of identified taxa at the site (chapter four); skeletal part representation and bone modification (chapter five); and kill-off patterns for domestic animals (chapter six). In this final chapter, the results obtained from these previous chapters will be reviewed and summarised. Tentative conclusions will be given, and issues and directions for future research will be discussed.

7.2 Human-Animal Relationships at Haimenkou

In addition to a few specimens of birds, fish, and molluscs, mammal bones dominated the entire faunal assemblage from Haimenkou. A total of 19 mammal taxa were identified in this study (the less diagnostic categories such as 'large cervid' and 'small carnivore' were not included in this count). Among these animals, three mammal taxa were identified as domesticates, including pig, sheep & goat, and dog. The possibility of gaur domestication cannot be absolutely excluded.

7.2.1 Domestic Animals

7.2.1.1 Pig

Pigs were considered domesticated due to their small body size and the obvious human involvement in their kill-off patterns, as suggested by age profiles and survivorship curves.

As mentioned in chapter four, many Neolithic sites across mainland China yielded pig remains, and there is still a wide distribution of wild boars in Yunnan today (Smith, A. and Xie 2008: 452). Therefore, although several specimens belonging to wild boars were present in the faunal assemblage, the domestic pigs recovered may not have been domesticated locally. The origin of these pigs is not yet clear. An analysis of measurement of pig teeth indicates that pig remains recovered from Haimenkou were domesticated prior to their arrival at Haimenkou with their keepers. The reduction in size of pig teeth also indicates that the domestication process continued after their arrival at Haimenkou.

7.2.1.2 Sheep & Goat

The remains of some caprine animals were diagnostically identifiable as sheep. Less diagnostic remains were identified as sheep/goat. The absence of confirmed goat bones from Haimenkou and the presence of sheep in other sites belonging to later time periods in Yunnan make it more likely that the remains identified as sheep/goat from Haimenkou were actually sheep. Therefore, in this study the remains from sheep and sheep/goat were analysed in a combined sheep & goat category.

Based on the similar bone size and features to modern sheep, as well as the absence of reports on their wild ancestors in this region or vicinity, the sheep & goat remains recovered were identified as domestic animals.

Sheep and goats may have been introduced to Haimenkou in Phase II (*ca.* 3700–3200 B.P.) from northwestern China (Qinghai and Gansu Provinces), along with cultural artefacts such as the *Qiang* style double-handled pottery jars and botanical remains such as wheat (see Section 7.4 below).

7.2.1.3 Dog

The dog remains recovered at Haimenkou were morphologically similar to the skeleton of a modern Chinese rural dog from the comparative reference collection. The dog remains at Haimenkou were characterised by their small size, similar to that of a red fox. They were smaller than the wolf, dhole, Chinese rural dog, and dogs recovered from the Neolithic sites in China other than in Yunnan. Other sites in Yunnan dating from the Neolithic Period to the Bronze Age also yielded remains of this small-sized dog. The dog recovered at Haimenkou, therefore, may have belonged to a unique small-sized breed that was confined to Yunnan.

These small-sized dogs were present throughout the whole period of human occupation at Haimenkou (Phases I–III, *ca.* 5000–2400 B.P.). It is possible that these dogs were brought to Haimenkou by their owners from the beginning of settlement. This small-sized dog may also have been kept by inhabitants of the middle Lancang River Valley, in southwestern Yunnan, from the Shang dynasty onwards (*ca.* 3600–3100 B.P.). The similarity of dog remains in these two regions indicates the possibility of cultural connections or communication between these regions, as will be discussed in Section 7.4.

7.2.2 Wild Mammals, Birds, and Others

The wild animals revealed at Haimenkou included a high proportion of mammals, some birds, fish, and molluscs.

7.2.2.1 Wild Mammals

In the faunal assemblage from Haimenkou a total of 16 taxa of wild mammals were identified. Among these taxa, the gaur showed the highest relative importance. This was followed by cervids, including the sambar, hog deer, red muntjac, Reeve's muntjac, and forest musk deer. Other wild mammal taxa included wild boar, macaque, bear, red fox, leopard cat, masked palm civet, hare, bamboo rat, porcupine, and rat. Wild mammal taxa made up a smaller part of the whole faunal assemblage based on NISP (41.7%), but a relatively larger proportion based on MNI (52.6%) and weight (69.2%).

The diversity of wild mammal taxa suggested that, along with domestic animals, these taxa contributed to the range of food resources used by the Haimenkou people. Aside from this relationship between hunters and game, a symbiotic relationship between humans and wild animals was also indicated by rat remains recovered at the site, as well as by the rat gnaw marks widely observed on animal bones in the faunal assemblage.

7.2.2.2 Birds and Others

A very small percentage of bird specimens were recovered from Haimenkou, along with an even smaller proportion of fish and molluscs. Based on the high fragmentation of bird specimens and their small number and the lack of a comparative collection, most of these specimens were only identifiable to genus or family. The fish bones were not able to be identified even to family taxonomic level.

As a typical waterside site, it is abnormal that such a small sample of birds, fish, and freshwater molluscs were recovered from Haimenkou. This phenomenon is probably due to the recovery bias discussed in chapter two.

7.3 Subsistence Economy of the Haimenkou People

As stated above, there was a mixed spectrum of animal taxa present at prehistoric Haimenkou. The diversity of animal taxa indicated that these faunal remains represent a wide range of food sources as well as secondary products in the subsistence economy of Haimenkou. Although the results of specific analyses of cultural artefacts and botanic evidence from Haimenkou have not been presented, based on the present data the subsistence strategy of the Haimenkou people appears to have been a mixture of multi-cropping, gathering, animal husbandry, hunting, and fishing.

7.3.1 Domestic Animals

It is apparent that domestic animals made up a significant proportion in daily diet of the ancient people throughout the occupation at Haimenkou in prehistory (Phases I–III, *ca.* 5000–2400 B.P.), with a slight and gradual increase in relative importance through time.

7.3.1.1 Pig

Pigs were the most common and probably most important domestic animal in the subsistence economy of the Haimenkou people, as suggested by all quantification measures. The NISP, MNI, and weight of pig remains all show similar positive trends from Phases I to III. It is clear that the abundance of pigs was concentrated in Phases II and III, which represent the period ranging from the early to middle Bronze Ages in Yunnan Province. The considerable increase in numerical values by NISP, MNI, and weight may suggest a marked development in the husbandry techniques of pigs corresponding with the transition from the late Neolithic Period to the early Bronze Age. The slower growth of these relative proportions may indicate increased exploitation of other animals at Haimenkou in Phases II and III.

Both age profiles and survivorship curves regarding the kill-off pattern for pigs suggested that most individuals were slaughtered before reaching their sub-adult age stage, with the intensive slaughter age between 1 and 2 years. This kill-off pattern fits the optimal slaughter ages for pigs raised for meat production today (Uerpmann 1973:316). Furthermore, based on age profiles, the slaughter ages for pigs tended to be more concentrated in Phase III than in Phase II. This is at the chronological age of 12–17 months which fits the optimal slaughter age—1.5 years—for pigs reared for meat production (Uerpmann 1973:316). This tendency recurs in survivorship curves. Consequently, this phenomenon may suggest an improvement in pig exploitation strategies through time with more advanced exploitation patterns applied in later phases.

7.3.1.2 Sheep & Goat

Sheep & goat remains contributed a relatively small proportion to the faunal assemblage at Haimenkou. This is similar to the proportion of caprine bones recovered from other Neolithic sites in Yunnan (e.g., Dadunzi). The small proportion indicates the low status of sheep & goats in the Haimenkou people's subsistence economy, and may be associated with their origin as introduced and then small-group animals. As with pig remains, the greatest abundance of sheep & goat remains was concentrated in Phases II and III. It is possible that the sole sheep & goat specimen unearthed in Phase I (Level (8)) actually belonged to Phase II. On the other hand and in contrast to pig remains, trends for sheep & goat from Phase II to III were negative by both amount and relative proportion, in all analysed measures. Inadequate sample collection may have led to this phenomenon. If there was no recovery bias, an inappropriate husbandry strategy of sheep & goats may have reduced the herd size and been responsible for the decline of caprine animals.

As there were small sample sizes for each phase, the kill-off patterns for sheep & goats were based on the total specimens from all phases. As indicated by age profiles and the survivorship curve, in sheep & goats more individuals survived to old age. This is different from the kill-off pattern for pigs. This kill-off pattern may indicate that secondary products, such as milk and wool, were part of the objective of husbandry. At the same time, a certain proportion of animals were slaughtered at young ages, fitting the optimal slaughter ages for small ruminants (Uerpmann 1973:316). It seems that sheep & goats may have been reared for multiple aims: meat, milk, and wool.

7.3.1.3 Dog

Dog remains make up the minimum proportion of total specimens of domestic animals at Haimenkou based on NISP, MNI, and weight. This pattern is consistent throughout the three phases, although the small sample sizes may be caused partly by a recovery bias. It seems that these animals were not regularly consumed at Haimenkou, as only one specimen exhibited cut marks that may have come from tool manufacture. In association with the cultural evidence of bronze figures from later Bronze Age sites in Yunnan, the unique small-sized dog breed revealed at Haimenkou may have been raised as a sheepdog or hunting partner rather than for meat production. It is probable that dogs did not contribute much to the daily diet of the Haimenkou people, but with their help people may have found it easier to access other food resources.

7.3.2 Wild Animals

Wild animals recovered from Haimenkou also occupied a very important place in the subsistence economy of the Haimenkou people. Compared with the limited domestic animal taxa revealed at Haimenkou, the taxa and amount of wild animals recovered signal the complexity of food resources and the hunting ability of the Haimenkou people in ancient times.

The gaur was the second most abundant mammal species after pigs at Haimenkou. This was based on NISP during Phases I to III, although it was the most abundant species based on weight in all periods. Among the wild animals, gaur specimens outweighed the total specimens from other wild animals based on both NISP and weight. This suggests that it contributed a very large proportion to the daily diet of the Haimenkou people, and indicates that gaurs were the most important wild animals at prehistoric Haimenkou.

As with pig values, the numerical values of the gaur increase from Phases I to III by all methods of analysis. The marked growth in numbers based on NISP, MNI, and weight, especially between Phases I and II, suggests a prominent increase in the number of gaurs acquired through hunting. This may indicate enhanced hunting activities of the Haimenkou people. On the other hand, the relative proportions show a slight fall in Phase II based on NISP, MNI, and weight. This slight decrease may correspond with the growing importance of domestic pigs, in accordance with the shift from the late Neolithic Period to the early Bronze Age.

Other wild animals also display similar trends to gaurs based on NISP, MNI, and weight numerical values. As with domestic animals and gaurs, the amount of other wild animals dramatically increased from Phases I to III, and a slight growth continued between Phases II and III. This phenomenon indicates a marked development of human hunting activities occurring at the transition from the Neolithic Period to the Bronze Age.

7.4 Communication Networks Linking Haimenkou and Other Regions in Prehistoric China

As mentioned in chapter two, the Haimenkou site is located at the bottom of the 'San Jiang Bing Liu [Three Parallel Rivers of Yunnan]' region. This included the Nu (Salween), Lancang (Mekong), and Jinsha (Yangtze) Rivers which flow from north to south. Given this pivotal location, Haimenkou probably played an important role in cultural communication between this and other regions in southwestern China, such as Sichuan Province and southwestern and eastern Yunnan. In addition, the site probably had some type of communication or contact with distant regions such as Qinghai and Gansu Provinces in northwestern China.

7.4.1 Northwestern China

Chapter two described some of the artefacts and botanic remains recovered from Haimenkou which indicate influences from northwestern China. The *Qiang* style double-handled pottery jars, small-sized bronzes, and wheat all exhibited strong influence from northwestern China (Qinghai and Gansu Provinces) in Phases II and III (*ca.* 3700–2400 B.P.) This may indicate cultural communication and/or human migration occurring and continued between these two regions from Phase II through multiple routes, such as Tibet and Sichuan. This influence is suggested by the presence of sheep & goats in the faunal remains from Haimenkou, as sheep and goats are generally considered to be domestic animals first introduced to northwestern China from West Asia. As stated in Section 7.3.1.2, it is probable that sheep & goats were introduced to Haimenkou from the beginning of Phase II (*ca.* 3700 B.P.) along with the artefacts and botanic remains representing northwestern Chinese styles. The reduction in faunal group size of the sheep & goats in Phase III may be due to the unfamiliarity of the Haimenkou people with the husbandry of these exotic animals.

7.4.2 Southwestern and Eastern Yunnan

Connections between Haimenkou and the middle Lancang River Valley, which is in southwestern Yunnan, were discussed in chapter two. These are evident from the similarities in the shape of bronze *yue* axes recovered from Haimenkou to stone *yue* axes recovered from the Manghuai site. These connections recurred in the analysis of the animal remains from Haimenkou, based on the study of unique small-sized dog breed. According to ancient Chinese records and modern evidence, these small dogs may have been kept by people who inhabited the middle Lancang River Valley from the Shang dynasty (*ca.* 3600–3100 B.P.) and may still be found in the same region today. Evidence from artistic data indicates that this dog breed may have been raised as a sheepdog and used for hunting by the Shizhaishan people who lived in eastern Yunnan in the later Bronze Age (*ca.* 2500–2100 B.P.). It is difficult to demonstrate the routes and nature of the communication among these regions in prehistoric Yunnan, however, connections may be inferred based on the presence of this unique small-sized dog breed.

7.5 Directions for Future Research

The analysis of the faunal assemblage from Haimenkou revealed the mixed spectrum of animal resources in the subsistence economy of the Haimenkou people in prehistoric Yunnan (*ca.* 5000–2400 B.P.). It suggested a considerable improvement in animal husbandry and hunting occurred from Phase I, the late Neolithic Period, to Phases II, the early Bronze Age, and moved on through Phase III, the middle Bronze Age. The faunal analysis also suggested that there were multiple connections between Haimenkou and other regions in prehistoric China during the time periods under study.

The entire area of the Haimenkou site was estimated to be over 50 thousand square metres (5 ha.) with approximately 20 thousand square metres (2 ha.) making up the central area of the site. However, in the last excavation season in 2008 only 1,395 square metres were excavated. There is still a vast area yet to be excavated, which may provide more valuable information on a number of issues. There are problems evident in the understanding of the site, especially those caused by the faunal specimen recovery bias. Consequently, there are several points of interest that were unable to be investigated in this book and which should be investigated by future research.

Firstly, an advanced sample collection method is required for the next excavation. As a site adjacent to Jianhu Lake and yielding abundant stone and pottery implements for fishing and hunting, Haimenkou was expected to unearth a large quantity of remains from waterfowls, fish, and molluscs. However, this expectation was not met in the results of the zooarchaeological study of the faunal assemblage from the third excavation season. These problems are partly a result of the recovery bias noted, which led directly to the dearth of specimens available for a variety of analyses. The recovery bias at Haimenkou, typically the inadequate sample collection, not only generally obscured the existence of fish and molluscs, but also reduced the sample sizes of birds and small elements from mammals. During the next excavation season, more attention should be paid to adequate sample collection, and hand-collection should be assisted by water-sieving or other methods.

Secondly, the study of bone tools is a point for future research. In addition to the study of animal remains, other projects based on plants, stone implements, and tools made of various materials are under way. The results from these works may provide significant information and shed light on the zooarchaeological study. Of these the study of bone tools may be supplementary to the study of animal exploitation.

Thirdly, a number of scientific analyses may be conducted in the future, especially on wild animals. For example, an analysis on wild mammals' teeth and bone collagen may show up patterns of seasonal subsistence through isotope analysis. This could indicate the timing of wild animal hunting, in specific seasons or all year round. In addition, observations of micro morphological structural changes in burnt bones may further the study of burning degree and diet.

Finally, it is crucial to pay close attention to future excavations in Yunnan Province and other relevant regions in China or in other neighbouring countries, especially countries in Southeast Asia including Vietnam, Laos, and Myanmar. As only a few excavation reports of sites in Yunnan include zooarchaeological data, faunal comparison and analysis were obfuscated in this study. The similarities and discrepancies among archaeological sites located in China or in countries of Southeast Asia, in the same or neighbouring regions, may provide more information on a variety of issues, including site formation processes, development of animal hunting techniques, origins of domestic animals, local animal domestication processes, cultural contacts, and movement of humans and animals through Yunnan and neighbouring areas.

Bibliography

Amorosi, Thomas, 1989, A Postcranial Guide to Domestic Neo-Natal and Juvenile Mammals: The Indentification and Aging of Old World Species. Oxford, England: B.A.R.

An, Jiayuan, and Fengxiang Long, 1993, Dongwu Kaoguxue zai Meiguo [Zooarchaeology in the U.S.A]. Wenwu Tiandi [Cultural Relics World] (1):44-45.

Archaeological Team of Yunnan Provincial Museum, 1977, Yunnan Yun Xian Manghuai Xinshiqi Shidai Yizhi Diaocha [Survey at the Neolithic Manghuai Site in Yun County, Yunnan]. Kaogu [Archaeology] (3):176-177+214.

Archaeological Team of Yunnan Provincial Museum, 1983a, Yunnan Deqin Xian Nagu Shiguan Mu [Stone Graves at Nagu in Deqin County, Yunnan]. Kaogu [Archaeology] (3):220-225.

Archaeological Team of Yunnan Provincial Museum, 1983b, Yunnan Deqin Xian Shidi Gumu [Ancient Graves at Shidi in Deqin County, Yunnan]. Kaogu [Archaeology] (3):275-276.

Archaeological Team of Yunnan Provincial Museum, 1983c, Yunnan Ninglang Xian Daxingzhen Gumuqun [Ancient Graves at Daxingzhen in Ninglang County, Yunnan]. Kaogu [Archaeology] (3):226-232.

Armitage, P. L., and Juliet Clutton-Brock, 1976, A System for Classification and Description of the Horn Cores of Cattle from Archaeological Sites. Journal of Archaeological Science 3(4):329-348.

Bai, Yunxiang, 2002, Zhongguo Zaoqi Tongqi yu Qingtongqi de Qiyuan [A Study on Chinese Primal Bronze Wares and the Origin of Bronze Wares]. Dongnan Wenhua [Southeast Culture] (7):25-37.

Behrensmeyer, Anna K., 1978, Taphonomic and Ecologic Information from Bone Weathering. Paleobiology 4(2):150-162.

Behrensmeyer, Anna K., Kathleen D. Gordon, and Glenn T. Yanagi, 1986, Trampling as a Cause of Bone Surface Damage and Pseudo-Cutmarks. Nature 319(6056):768-771.

Binford, Lewis Roberts, 1973, Interassemblage Variability—The Mousterian and the 'Functional' Argument. *In* Explanation of Culture Change: Models in Prehistory. C. Renfrew, ed. Pp. 227-253. London: Duckworth.

Binford, Lewis Roberts, 1978, Nunamiut Ethnoarchaeology. New York: Academic Press.

Binford, Lewis Roberts, 1981, Bones: Ancient Men and Modern Myths. New York: Academic Press.

Binford, Lewis Roberts, 1984, Faunal Remains from Klasies River Mouth. New York: Academic Press.

Boessneck, Joachim, 1970, Osteological Differences between Sheep (*Ovis aries* Linne) and Goat (*Capra hirus* Linne). *In* Science in Archaeology: A Survey of Progress and Research. D. Brothwell and E. Higgs, eds. Pp. 331-358. New York: Praeger.

Boessneck, Joachim, and Angela von den Driesch, 1973, Die jungpleistozänen Tierknochenfunde aus der Brillenhöhle. *In* Das Paläolithikum der Brillenhöhle bei Blaubeuren (Schwäbische Alb). G. Riek, ed. Forschungen und Berichte zur Vor-und Frühgeschichte in Baden-Württemberg. Stuttgart: Landesdenkmalamt Baden-Württemberg, Verlag Müller and Gräff.

Bökönyi, Sandor, 1970, A New Method for the Determination of the Number of Individuals in Animal Bone Material. American Journal of Archaeology 74(3):291-292.

Bökönyi, Sandor, 1974, History of Domestic Mammals in Central and Eastern Europe. Budapest: Akademiai Kiado.

Bökönyi, Sandor, 1989, Definitions of Animal Domestication. *In* The Walking Larder: Patterns of Domestication, Pastoralism, and Predation. J. Clutton-Brock, ed. Pp. 22-27. London: Unwin Hyman.

Bonnichsen, Robson, and Marcella H. Sorg, eds., 1989, Bone Modification. Orono, Me.: Centre for the Study of the First Americans, Institute for Quaternary Studies, University of Maine.

Brain, C. K., 1981, The Hunters or the Hunted? An Introduction to African Cave Taphonomy. Chicago: University of Chicago Press.

Brown, George T., 1985, Dentition as Indicative of the Age of the Animals of the Farm. London: J. Murray.

Bull, Gail, and Sebastian Payne, 1982, Tooth Eruption and Epiphyseal Fusion in Pigs and Wild Boar. *In* Ageing and Sexing Animal Bones from Archaeological Sites. B. Wilson, C. Grigson, and S. Payne, eds. Pp. 55-71. Oxford, England: B.A.R.

Bunn, Henry T., and Ellen M. Kroll, 1986, Systematic Butchery by Plio/Pleistocene Hominids at Olduvai Gorge, Tanzania. Current Anthropology 27(5):431-442.

Burke, A., and J. Castanet, 1995, Histological Observations of Cementum Growth in Horse Teeth and Their Application to Archaeology. Journal of Archaeological Science 22(4):479-493.

Cai, Dawei, Lu Han, Xiaolei Zhang, Hui Zhou, and Hong Zhu, 2007, DNA Analysis of Archaeological Sheep Remains from China. Journal of Archaeological Science 34(9):1347-1355.

Cai, Dawei, Zhuowei Tang, Quanjia Chen, Lu Han, and Hui Zhou, 2010, Zhongguo Mianyang Qiyuan de Fenzi Kaoguxue Yanjiu [Molecular Archaeological Study on the Origin of Chinese Domestic Sheep]. *In* Bianjiang Kaogu Yanjiu [Research of China's Frontier Archaeology]. Research Centre for Chinese Frontier Archaeology of Jilin University, ed. Pp. 291-300. Beijing: Science Press.

Cai, Dawei, Zhuowei Tang, Lu Han, Camilla F. Speller, Dongya Y. Yang, Xiaolin Ma, Jian'en Cao, Hong Zhu, and Hui Zhou, 2009, Ancient DNA Provides New Insights into the Origin of the Chinese Domestic Horse. Journal of Archaeological Science 36(3):835-842.

Cannon, Michael D., 1999, A Mathematical Model of the Effects of Screen Size on Zooarchaeological Relative Abundance Measures. Journal of Archaeological Science 26(2):205-214.

Cannon, Michael D., 2001, Archaeofaunal Relative Abundance, Sample Size, and Statistical Methods. Journal of Archaeological Science 28(2):185-195.

Casteel, Richard W., 1972, Some Biases in the Recovery of Archaeological Fauna Remains. Proceedings of the Prehistoric Society 38:383-388.

Casteel, Richard W., 1976, Comparison of Column and Whole Unit Samples for Recovering Fish Remains. World Archaeology 8(2):192-196.

Casteel, Richard W., 1977, Characterization of Faunal Assemblages and the Minimum Number of Individuals Determined from Paired Elements: Continuing Problem in Archaeology. Journal of Archaeological Science 4(2):125-134.

Casteel, Richard W., 1978, Faunal Assemblages and the 'Wiegemethode' or Weight Method. Journal of Field Archaeology 5(1):71-77.

Chang, Kwang-chih, 1983, Zhongguo Qingtong Shidai [The Bronze Age of China]. Beijing: SDX Joint Publishing Company.

Chang, Kwang-chih, 1997, Ancient China. *In* Food in Chinese Culture: Anthropological and Historical Perspectives. K. C. Chang and E. N. Anderson, eds. Pp. 23-52. Taipei: SMC Pub. Inc.

Chaplin, Raymond Edwin, 1971, The Study of Animal Bones from Archaeological Sites. London: Seminar Press.

Chen, Quanjia, 1993, Nong'an Zuojiashan Yizhi Dongwu Guge Jianding ji Hengji Yanjiu [Identification and Mark Analysis of Animal Bones from the Zuojiashan Site in Nong'an County]. *In* Qingguo Ji [Collection of Green Fruits]. Department of Archaeology of Jilin University, ed. Pp. 57-71. Beijing: Knowledge Publishing House.

Chen, Wenhua, 2002, Nongye Kaogu [Agricultural Archaeology]. Beijing: Cultural Relics Press.

Chen, Yifeng, Xiangnian Shan, Xiaomei Cao, Chongyun Li, and Liuyong Yin, 1978, Yunnan Yeniu de Ranseti [Somatic Chromosome of *Bos gaurus readei*]. Yichuan Xuebao [Acta Genetica Sinica] 5(3):249-251.

Cheng, Qingtai, 1988, Fulu Er: Sanlihe Yizhi Chutu de Yugu, Yulin Jianding Baogao [Appendix II: Identification of the Fish Remains and Scales from the Sanlihe Site]. *In* Jiao Xian Sanlihe [Sanlihe Site in Jiao County]. The Institue of Archaeology, Chinese Academy of Social Science, ed. Beijing: Cultural Relics Press.

Chou, Shunwu, 1992, Yunnan Province. *In* China Provincial Geography. 1 edition. Pp. 384-401. Beijing: Foreign Languages Press.

Clark, J., and K. K. Kietzke, 1967, Palaeoecology of the Lower Nodular Zone, Brule Formation, in the Big Badlands of South Dakota. Fieldiana Memoirs 5:111-137.

Clason, A. T., and W. Prummel, 1977, Collecting, Sieving and Archaeozoological Research. Journal of Archaeological Science 4(2):171-175.

Clutton-Brock, Juliet, 1970, The Origins of the Dog. *In* Science in Archaeology: A Survey of Progress and Research. D. Brothwell and E. Higgs, eds. Pp. 303-309. New York: Praeger.

Clutton-Brock, Juliet, 1981, Domesticated Animals from Early Times. Austin: University of Texas Press.,

Clutton-Brock, Juliet, 1984, Dog. *In* Evolution of Domesticated Animals. I. L. Mason, ed. Pp. 198-211. London: Longman.

Clutton-Brock, Juliet, 1999, A Natural History of Domesticated Mammals. Cambridge, U.K. : Cambridge University Press.

Codification Committee of Annals of Jianchuan County, Yunnan Province, ed., 1999, Jianchuan Xian Zhi [Jianchuan County Annals]. Kunming: The Nationalities Publishing House of Yunnan.

Committee for the Preservation of Ancient Monuments of Jiangxi Province, 1963, Jiangxi Wannian Dayuan Xianrendong Dongxue Yizhi Shijue [Excavation at the Xianrendong Cave in Wannian County, Jiangxi Province]. Kaogu Xuebao [Acta Archaeologia Sinica] (1):1-16.

Committee for the Preservation of Ancient Monuments of Sichuan Province, and Culture Centre of Maowen County, 1983, Sichuan Sheng Maowen Qiangzu Zizhixian Shiguan Fajue Baogao [Excavation report of Stone Graves in Maowen *Qiang* Autonomous County, Sichuan Province]. *In* Wenwu Ziliao Conghan [Series of Cultural Relics Data]. ECCR, ed. Pp. 34-55, Vol. 7. Beijing: Cultural Relics Press.

Committee for the Preservation of Ancient Monuments of Tibet Autonomous Region, and Department of History, Sichuan University, eds., 1985, Changdu Karuo [Karuo: A Neolithic Site in Tibet]. Beijing: Cultural Relics Press.

Cruz-Uribe, Kathryn, and Richard G. Klein, 1986, Pascal Programs for Computing Taxonomic Abundance in Samples of Fossil Mammals. Journal of Archaeological Science 13(2):171-187.

Cucchi, T., A. Hulme-Beaman, J. Yuan, and K. Dobney, 2011, Early Neolithic Pig Domestication at Jiahu, Henan Province, China: Clues from Molar Shape Analyses Using Geometric Morphometric Approaches. Journal of Archaeological Science 38(1):11-22.

Cui, Jianfeng, and Xiaohong Wu, 2008, Qian Tongweisu Kaogu Yanjiu [Archaeological Study of Lead Isotope]. Beijing: Cultural Relics Press.

Cultural Relics Management, Aba Tibetan Autonomous Prefecture, 1987, Sichuan Li Xian Jiashan Shiguan Zang Fajue Qingli Baogao [Excavation Report of Stone Graves at Jiashan in Li County, Scihuan]. *In* Nanfang Minzu Kaogu [Southern Ethnology and Archaeology]. Sichuan University Museum and Chinese Association for the Study of Ancient Bronze Drums, eds. Pp. 211-236, Vol. 1. Chengdu: Sichuan University Press.

Cultural Relics Management, Aba Tibetan and Qiang Autonomous Prefecture, Sichuan Provincial Institute of Cultural Relics and Archaeology, Chengdu Municipal Institute of Cultural Relics and Archaeology, and Culture and Sport Bureau, Maerkang County, 2007, Sichuan Maerkang Xian Haxiu Yizhi Diaocha Jianbao [Investigation Report of the Haxiu Site in Maerkang County, Sichuan]. Sichuan Wenwu [Sichuan Cultural Relics] (4):8-15.

d'Alpoim Guedes, Jade, 2011, Millets, Rice, Social Complexity, and the Spread of Agriculture to the Chengdu Plain and Southwest China. Rice 4(3):104-113.

Davis, Simon J. M., 1981, The Effects of Temperature Change and Domestication on the Body Size of Late Pleistocene to Holocene Mammals of Israel. Paleobiology 7(1):101-114.

Davis, Simon J. M., 1987, The Archaeology of Animals. Conn: Yale University Press.

Deng, Hui, Jing Yuan, Guoding Song, Changsui Wang, and Masaki Eda, 2013, Zhongguo Gudai Jiaji de Zaitantao [The Re-examination of the Domestic Chicken in Ancient China]. Kaogu [Archaeology] (6):83-96.

Deniz, Esref, and Sebastian Payne, 1982, Eruption and Wear as in the Mandibular Dentition as a Guide to Ageing Turkish Angora Goats. *In* Ageing and Sexing Animal Bones from Archaeological Sites. B. Wilson, C. Grigson, and S. Payne, eds. Pp. 155-206. Oxford, England: B.A.R.

Ducos, Pierre, 1978, 'Domestication' Definition and Methodological Approaches to Its Recognition in Faunal Assemblages. *In* Approaches to Faunal Analysis in the Middle East. R. H. Meadow and M. A. Zeder, eds. Pp. 53-56. Cambridge, Mass: Peabody Museum of Archaeology and Ethnology, Harvard University.

Ducos, Pierre, 1989, Defining Domestication: A Clarification. *In* The Walking Larder: Patterns of Domestication, Pastoralism, and Predation. J. Clutton-Brock, ed. Pp. 28-30. London: Unwin Hyman.

Editorial Committee of Breeds Records for Livestock and Poultry in Yunnan Province, ed., 1987 Yunnansheng Jiachu Jiaqin Pinzhongzhi [Breeds Records for Livestock and Poultry in Yunnan Province]. Kunming: Yunnan Science and Technology Press.

Editorial Committee of Fauna Sinica, Academia Sinica, ed., 1963, Zhongguo Jingji Dongwu Zhi: Niaolei [Economic Fauna of China: Birds]. Beijing: Science Press.

Editorial Section of the 'Bovine Breeds in China', ed., 1986, Zhongguo Niu Pinzhongzhi [Bovine Breeds in China]. Shanghai: Shanghai Kexue Jishu Chubanshe [Shanghai Scientific and Technical Publishers].

Editorial Section of the 'Pig Breeds in China', ed., 1986, Zhongguo Zhu Pinzhongzhi [Pig Breeds in China]. Shanghai: Shanghai Kexue Jishu Chubanshe [Shanghai Scientific and Technical Publishers].

Epstein, H., 1984 Pig. *In* Evolution of Domesticated Animals. I. L. Mason, ed. Pp. 145-162. London: Longman.

Fang, Guoyu, 2001, Yunnan Junxian Zhidu Liangqian Nian [2,000 Years of the Prefecture System in Yunnan]. *In* Fang Guoyu Wenji [Collected Works of Guoyu Fang]. C. Lin, ed. Pp. 31-42. Kunming: Yunnan Education Publishing House.

Fieller, N. R. J., and A. Turner, 1982, Number Estimation in Vertebrate Samples. Journal of Archaeological Science 9(1):49-62.

Fisher, John W. Jr. , 1995, Bone Surface Modifications in Zooarchaeology. Journal of Archaeological Method and Theory 2(1):7-68.

Fitch, J. E., 1966, Additional Fish Remains, Mostly Otoliths, from a Pleistocene Deposit at Playa del Rey, California. Contributions in Science, Los Angeles County Museum 119:1-16.

Flad, Rowan K., Jing Yuan, and Shuicheng Li, 2009, Lun Zhongguo Ganqing Diqu Xinshqi Shidai Jiayang Dongwu de Laiyuan ji Tezheng [On the Source and Features of the Neolithic Domestic Animals in the Gansu and Qinghai Region, China]. Kaogu [Archaeology] (5):80-86.

Flannery, Kent V., 1961, Skeletal and Radiocarbon Evidence of the Origins of Pig Domestication. M. A. Thesis, University of Chicago.

Flannery, Kent V., 1969, The Animal Bones. *In* Prehistory and Human Ecology of the Deh Luran Plain: An Early Village Sequence from Khuzistan, Iran. F. Hole, K. V. Flannery, and J. A. Neely, eds. Pp. 262-330. Ann Arbor: University of Michigan Museum and Anthropology.

Flannery, Kent V., 1983, Early Pig Domestication in the Fertile Crescent: A Retrospective Look. *In* The Hilly Flanks and Beyond: Essays on the Prehistory of Southwestern Asia Presented to Robert J. Braidwood, November 15, 1982. J. T. C. Young, P. E. L. Smith, and P. Mortensen, eds. Studies in Ancient Oriental Civilization, Vol. 36. Chicago: The Oriental Institute of the University of Chicago.

Fujian Provincial Museum, 1976, Minhou Tanshishan Yizhi Diliuci Fajue Baogao [Sixth Season Excavations at Tanshishan in Minhou County, Fujian Province]. Kaogu Xuebao [Acta Archaeologica Sinica] (1):83-119.

Gansu Archaeological Team, Institute of Archaeology, Chinese Academy of Science, 1975, Gansu Yongjing Qinweijia Qijia Wenhua Mudi [Excavation of a Qijia Culture Cemetery at Qinweijia in Yongjing County, Gansu Province]. Kaogu Xuebao [Acta Archaeologia Sinica] (2):57-96.

Gansu Provincial Archaeological Team, Department of Archaeology, Peking University, and Culture Centre of Xihe County, 1987, Gansu Xihe Lanqiao Siwa Wenhua Muzang [Excavation of Siwa Culture Cemetery at Lanqiao in Xihe County, Gansu]. Kaogu [Archaeology] (8):678-691.

Gansu Provincial Institute of Cultural Relics and Archaeology, 1990, Yongchang Sanjiaocheng yu Hamadun Shajing Wenhua Yicun [Remains of the Shajing Culture at Sanjiaocheng and Hamadun in Yongchang County, Gansu]. Kaogu Xuebao [Acta Archaeologia Sinica] (2):205-237+271-276.

Gansu Provincial Institute of Cultural Relics and Archaeology, and Archaeology Laboratory of North China, Jilin University, eds., 1998, Minle Donghuishan Kaogu—Siba Wenhua Mudi de Jieshi yu Yanjiu [Archaeological Analysis of the Donghuishan Site in Minle County—Reveal and Research of the Siba Culture Cemetery]. Beijing: Science Press.

Gansu Squad, Yellow River Reservoir Archaeological Team, 1962, Gansu Linxia Jijiachuan Yizhi Fajue Jianbao [Brief Report of the Excavation at the Jijiachuan Site in Linxia City, Gansu]. Kaogu [Archaeology] (2):69-71.

Gao, Yaoting, Changkun Lu, Jie Zhang, and Song Wang, 1962, Yunnan Xishuangbanna Shoulei Diaocha Baogao [Mammals of the Hsi-Shuan-Pan-Na Area in Southern Yunnan]. Dongwu Xuebao [Acta Zoologica Sinica] 14(2):180-196.

Gifford-Gonzalez, D. P., 1989, Ethnographic Analogues for Interpreting Modified Bones: Some Cases from East Africa. *In* Bone Modification. R. Bonnichsen and M. H. Sorg, eds. Pp. 179-246. Orono, Me.: Centre for the Study of the First Americans, Institute for Quaternary Studies, University of Maine.

Giuffra, E., J. M. H. Kijas, V. Amarger, Ö. Carlborg, J.-T. Jeon, and L. Andersson, 2000, The Origin of the Domestic Pig: Independent Domestication and Subsequent Introgression. Genetics 154(4):1785-1791.

Gordon, Bryan C., 1991, Archaeological Seasonality Using Incremental Structures in Teeth. New York: Zooarchaeological Research News.

Goss, Richard J., 1983, Deer Antlers: Regeneration, Function, Evolution. New York: Academic Press.

Gould, Richard A., and Patty Jo Watson, 1982, A Dialogue on the Meaning and Use of Analogy in Ethnoarchaeological Reasoning. Journal of Anthropological Archaeology 1(4):355-381.

Grant, Annie, 1975, Appendix B: The Use of Tooth Wear as a Guide to the Age of Domestic Animals—A Brief Explanation. *In* Excavations at Portchester Castle, Vol. I, Roman. B. W. Cunliffe, ed. Pp. 437-450. London: Society of Antiquaries of London: Distributed by Thames and Hudson.

Grant, Annie, 1982, The Use of Tooth Wear as a Guide to the Age of Domestic Ungulates. *In* Ageing and Sexing Animal Bones from Archaeological Sites. B. Wilson, C. Grigson, and S. Payne, eds. Pp. 91-108. Oxford, England: B.A.R.

Gray, Annie P., 1972, Mammalian Hybrids: A Check-List with Bibliography. Slough: Commonwealth Agricultural Bureaux.

Grayson, Donald K., 1973, On the Methodology of Faunal Analysis. American Antiquity 38(4):432-439.

Grayson, Donald K., 1979, On the Quantification of Vertebrate Archaeofaunas. Advances in Archaeological Method and Theory 2:199-237.

Grayson, Donald K., 1984, Quantitative Zooarchaeology: Topics in the Analysis of Archaeological Faunas. Orlando: Academic Press.

Greenfield, Haskel J., and Elizabeth R. Arnold, 2008, Absolute Age and Tooth Eruption and Wear Sequences in Sheep and Goat: Determining Age-at-Death in Zooarchaeology Using a Modern Control Sample. Journal of Archaeological Science 35(4):836-849.

Grzimek, Bernhard, ed., 1990, Grzimek's Encyclopedia of Mammals. Volume 5. New York: McGraw-Hill.

He, Zhanxing, Kaixing Qu, Xiping Yuan, Yehua Sang, Wenzhang Ma, Jicai Zhang, and Ziliang Li, 2009, Fenghuangshan Qiandi Baohuzhong Daeniu de Waimao Tezheng ji Zhuyao Xixing [Appearance Characteristics and Major Behaviour of Gayal (*Bos frontalis*) on the Conservation *ex situ* in Phoenix Mountain]. Yunnan Nongye Daxue Xuebao [Journal of Yunnan Agricultural University] 24(2):225-230.

Henan Provincial Institute of Cultural Relics and Archaeology, ed., 1999, Wuyang Jiahua [Jiahu Site in Wuyang County]. Volume 1. Beijing: Science Press.

Higgs, E. S., and M. R. Jarman, 1972, The Origins of Animal and Plant Husbandry. *In* Papers in Economic Prehistory. E. S. Higgs, ed. Pp. 3-13. Cambridge: Cambridge University Press.

Higham, C. F. W., 1975, Non Nok Tha: The Faunal Remains from the 1966 and 1968 Excavations at Non Nok Tha, Northeastern Thailand. Dunedin, N.Z.: University of Otago, Dept. of Anthropology.

Hillson, Simon, 1992, Mammal Bones and Teeth: An Introductory Guide to Methods of Identification. London: Institute of Archaeology, University College London.

Hillson, Simon, 2005, Teeth. Cambridge: Cambridge University Press.

Horton, D. R., 1984, Minimum Numbers: A Consideration. Journal of Archaeological Science 11(3):255-271.

Howard, Hildegarde, 1930, A Census of the Pleistocene Birds of Rancho La Brea from the Collections of the Los Angeles Museum. Condor 32:81-88.

Huang, Huaixin, 2006, Yizhoushu Jiaobu Zhuyi [Collation and Annotation of the Superfluous (Chapters of the) Book of Zhou]. Xi'an: Sanqin Publishing House.

Huang, Wanbo, and Hongxiang Ji, 1963, Jiangxi Wannian Xianrendong Quanxinshi Dongxue Duiji [Note on Holocene Xianrendong Cave Deposits of Wannian, Jiangxi]. Gu Jizhui Dongwu yu Gu Renlei [Certebrata Palasiatica] 7(3):263-277.

Huang, Yunping, 1996, Neimenggu Zhukaigou Yizhi Shougu de Jianding yu Yanjiu [Identification and Study of the Animal Bones from the Zhukaigou Site, Inner Mongolia]. Kaogu Xuebao [Acta Archaeologia Sinica] (4):515-536.

Huangyuan County Museum, Qinghai Provincial Archaeological Team, and History Laboratory, Qinghai Provincial Academy of Social Science, 1985, Qinghai Huangyuan Xian Dahuazhongzhuang Kayue Wenhua Mudi Fajue Baogao [Excavation of the Kayue Culture Cemetery at Dahuazhongzhuang in Huangyuan County, Qinghai]. Kaogu yu Wenwu [Archaeology and Cultural Relics] (5):11-34.

Huidekoper, Rush Shippen, 1891, Age of the Domestic Animals Being a Complete Treatise on the Dentition of the Horse, Ox, Sheep, Hog, and Dog, and on the Various Other Means of Determining the Age of These Animals. Philadelphia: F.A. Davis.

Institute of Archaeology, Chinese Academy of Social Science, ed., 1991, Zhongguo Kaoguxue zhong Tanshisi Niandai Shujuji (1965–1991) [Radiocarbon Dates in Chinese Archaeology (1965–1991)]. Beijing: Cultural Relics Press.

Institute of Archaeology, Chinese Academy of Social Science, ed., 2003a, Zhongguo Kaoguxue—Liangzhou Juan [Chinese Archaeology—Western Zhou and Eastern Zhou]. Beijing: China Social Science Press.

Institute of Archaeology, Chinese Academy of Social Science, ed., 2003b, Zhongguo Kaoguxue—Xia Shang Juan [Chinese Archaeology—Xia and Shang]. Beijing: China Social Science Press.

James, Steven R., 1997, Methodological Issues Concerning Screen Size Recovery Rates and Their Effects on Archaeofaunal Interpretations. Journal of Archaeological Science 24(5):385-397.

Jarman, M. R., and P. F. Wilkinson, 1972, Criteria of Animal Domestication. *In* Papers in Economic Prehistory. E. S. Higgs, ed. Pp. 83-96. Cambridge: Cambridge University Press.

Jia, Lanpo, and Zhenbiao Zhang, 1989, Henan Xichuan Xiawanggang Yizhi zhong de Dongwuqun [Fauna from Xiawanggang in Xichuan County, Henan Province]. *In* Xichuan Xiawanggang [Xiawanggang Site in Xichuan]. Pp. 429-439. Beijing: Cultural Relics Press.

Jiang, Leping, and Li Liu, 2006, New Evidence for the Origins of Sedentism and Rice Domestication in the Lower Yangzi River, China. Antiquity 80(308):355-361.

Jiang, Xiaochun, 2010, Zhongguo Qingtong Shidai Qishi Shijian Kao [Study on the Beginning of Bronze Age in China]. Kaogu [Archaeology] (6):556-562.

Jiang, Zhilong, 1999, New Excavations at the Bronze Age Site of Shizhaishan, Yunnan, and Implications for the Archaeology of the Ancient Dian kingdom. Bulletin of the Indo-Pacific Prehistory Association 18:117-120.

Jin, Guiyun, 2007, Zhongguo Zaoqi Xiaomai de Kaogu Faxian yu Yanjiu [Archaeological Reveal and Studies of Wheat in Early China]. Nongye Kaogu [Agricultural Archaeology] (4):11-20.

Kan, Yong, 1998, Shilun Yunnan Xinshiqi Wenhua [Discussion on the Neolithic Cultures in Yunnan]. *In* Yunnan Kaogu Wenji [Collected Works on Yunnan Cultural Relics and Archaeology]. D. Wang, M. Xiao, and F. Yang, eds. Pp. 17-40. Kunming: The Nationalities Publishing House of Yunnan.

Klein, Richard G., 1989, Why Does Skeletal Part Representation Differ between Smaller and Larger Bovids at Klasies River Mouth and Other Archeological Sites? Journal of Archaeological Science 16(4):363-381.

Klein, Richard G., and Kathryn Cruz-Uribe, 1983, The Computation of Ungulate Age (Mortality) Profiles from Dental Crown Heights. Paleobiology 9(1):70-78.

Klein, Richard G., and Kathryn Cruz-Uribe, 1984, The Analysis of Animal Bones from Archeological Sites. Chicago: University of Chicago Press.

Klein, Richard G., Cornelia Wolf, Leslie G. Freeman, and Kathryn Allwarden, 1981, The Use of Dental Crown Heights for Constructing Age Profiles of Red Deer and Similar Species in Archaeological Samples. Journal of Archaeological Science 8(1):1-31.

Kunming Institution of Zoology, Chinese Academy of Science, ed., 1999, Wild Life in Yunnan. Beijing: China Forestry Publishing House.

Laboratory of Institute of Archaeology, Chinese Academy of Science, 1972, Fangshexing Tansu Niandai Ceding Baogao (Er) [Report of Carbon-14 Dates (2)]. Kaogu [Archaeology] (5):56-58.

Laboratory of Institute of Archaeology, Chinese Academy of Science, 1974, Fangshexing Tansu Niandai Ceding Baogao (San) [Report of Carbon-14 Dates (3)]. Kaogu [Archaeology] (5):333-338.

Laboratory of Institute of Archaeology, Chinese Academy of Science, 1977, Fangshexing Tansu Niandai Ceding Baogao (Si) [Report of Carbon-14 Dates (4)]. Kaogu [Archaeology] (3):200-204.

Laboratory of Institute of Archaeology, Chinese Academy of Social Science, 1978, Fangshexing Tansu Niandai Ceding Baogao (Wu) [Report of Carbon-14 Dates (5)]. Kaogu [Archaeology] (4):280-27+243.

Laboratory of Institute of Archaeology, Chinese Academy of Social Science, 1984, Fangshexing Tansu Niandai Ceding Baogao (Shiyi) [Report of Carbon-14 Dates (11)]. Kaogu [Archaeology] (7):649-653.

Laboratory of Institute of Archaeology, Chinese Academy of Social Science, 1990, Fangshexing Tansu Niandai Ceding Baogao (Shiqi) [Report of Carbon-14 Dates (17)]. Kaogu [Archaeology] (7):663-668.

Lan, Hong, Xikun Xiong, Shiying Lin, Aihua Liu, and Liming Shi, 1993, Yunnan Huangniu he Daeniu mtDNA Duotaixing Yanjiu [Mitochondrial DNA Polymorphism of Cattle (*Bos taurus*) and Mithun (*Bos frontalis*) in Yunnan Province]. Yichuan Xuebao [Acta Genetica Sinica] 20(5):419-425.

Larson, Greger, Keith Dobney, Umberto Albarella, Meiying Fang, Elizabeth Matisoo-Smith, Judith Robins, Stewart Lowden, Heather Finlayson, Tina Brand, Eske Willerslev, Peter Rowley-Conwy, Leif Andersson, and Alan Cooper, 2005, Worldwide Phylogeography of Wild Boar Reveals Multiple Centers of Pig Domestication. Science 307(5715):1618-1621.

Lauwerier, Roel C. G. M., 1983, Pigs, Piglets and Determining the Season of Slaughtering. Journal of Archaeological Science 10(5):483-488.

Lawrence, B., 1967, Early Domestic Dogs. Zeitschrift für Säugetierkunde 32(1):44-59.

Leonard, Jennifer A., Robert K. Wayne, Jane Wheeler, Raúl Valadez, Sonia Guillén, and Carles Vilà, 2002, Ancient DNA Evidence for Old World Origin of New World Dogs. Science 298(5598):1613-1616.

Lewall, E. F., and I. McT. Cowan, 1963, Age Determination in Black-Tail Deer by Degree of Ossification of the Epiphyseal Plate in the Long Bones. Canadian Journal of Zoology 41(4):629-636.

Li, Genpan, and Xun Lu, 1987, Zhongguo Nanfang Shaoshu Minzu Yuanshi Nongye Xingtai [Primitive Agriculture Patterns of Ethnic Minority Groups in South China]. Beijing: Agriculture Press.

Li, Jiarui, 1965, Yunnan Xiangyun Dabona Tongguan Mu Shidai de Shangxiaxian [Dating of the Bronze Coffin Tomb at Dabona in Xiangyun County, Yunnan]. Kaogu [Archaeology] (9):478-479.

Li, Kunsheng, 1980, Yunnan Niugeng de Qiyuan [The Origin of Ox-Plough in Yunnan]. Kaogu [Archaeology] (3):266-270.

Li, Kunsheng, 1983, Yunnan Yuanshi Wenhua Zuxi Shitan [Discussion on Promitive Ethnic Groups in Yunnan]. Yunnan Shehui Kexue [Social Sciences in Yunnan] (4):75-83.

Li, Kunsheng, 1985, Lun Yunnan yu Huanghe Liuyu Xinshiqi Shidai Wenhua de Guanxi [On Relations among Neolithic Cultures of Yunnan and the Yellow River Valley]. Shiqian Yanjiu [Prehistory] (1):36-42.

Li, Kunsheng, 1988, Cong Yunnan Kaogu Cailiao Kan Diqiang Wenhua [On *Di* and *Qiang* Cultures Based from Archaeological Materials of Yunnan]. Sixiang Zhanxian [Thinking] (1):56-62.

Li, Kunsheng, and Rui Min, 2011, Yunnan Zaoqi Qingtong Shidai Yanjiu [Study of the Early Bronze Age in Yunnan]. Sixiang Zhanxian [Thinking] 37(4):100-105.

Li, Kunsheng, and Qiu Xiao, 1980, Shilun Yunnan Xinshiqi Shidai Wenhua [On the Neolithic Cultures in Yunnan]. Wenwu Jikan [Collected Works on Cultural Relics] (2):133-142+188.

Li, Minchang, 1991, Jiangsu Shuyang Wanbei Xinshiqi Shidai Yizhi Dongwu Guge Jianding Baogao [Report on the Faunal Remains from Wanbei Neolithic Site in Shuyang, Jiangsu]. Dongnan Wenhua [Southeast Culture] (3-4):183-189.

Li, Shiping, Hong Chang, Guolong Ma, and Hongyu Chen, 2008, Molecular Phylogeny of the Gayal in Yunnan China Inferred from the Analysis of Cytochrome *b* Gene Entire Sequences. Asian-Aust. J. Anim. Sci. 21(6):789-793.

Li, Shiping, Hong Chang, GuoLong Ma, Hongyu Chen, Dejun Ji, and Rongqing Geng, 2008, Cong Xibao Sesu *b* Jinyin Quanxulie Tantao Daeniu de Fenzi Xitong Fasheng [Molecular Phylogeny of the Gayal Inferred from the Analysis of Cytochrome *b* Gene Entire Sequences]. Yichuan [Hereditas (Beijing)] 30(1):65-70.

Li, Tianyuan, 1992, Dongwu Guge Tupu [Atlas of Animal Bones: For Prehistorians, Archaeologists and Quaternary Geologists]. Wuhan: China University of Geosciences Press.

Li, Tianyuan, 2001, Fulu Yi: Yidu Chengbeixi Yizhi Nanqu Chutu de Dongwu Guge Jiandingbiao [Appendix I: List of Animal Taxa recovered from Chengbeixi in Yidu City]. *In* Yidu Chengbeixi [Excavation at Chengbeixi in Yidu City]. Hubei Provincial Institute of Cultural Relics, ed. Pp. 291. Beijing: Cultural Relics Press.

Li, Xiaoqiang, Xinying Zhou, Hongbin Zhang, Jie Zhou, Xue Shang, and John Dodson, 2007, The Record of Cultivated Rice from Archaeobiological Evidence in Northwestern China 5,000 Years Ago. Chinese Science Bulletin 52(10):1372-1378.

Li, Youheng, 1963, Fulu Er: Jiangxi Wannian Dayuan Xianrendong Dongxue Yizhi Chutu Dongwu Guge Qingdan [Appendix II: List of Animal Bones from the Xianrendong Cave in Wannian County, Jiangxi Province]. Kaogu Xuebao [Acta Archaeologia Sinica] (1):14-15.

Li, Youheng, and Defen Han, 1959, Shaanxi Xi'an Banpo Xinshiqi Shidai Yizhi zhong zhi Shoulei Guge [Mammal Remains from the Neolithic Site at Banpo, in Xi'an, Shaanxi]. Gu Jizhui Dongwu yu Gu Renlei [Certebrata Palasiatica] 1(4):173-185.

Li, Youheng, and Defen Han, 1978, Guangxi Guilin Zengpiyan Yizhi Dongwuqun [Fauna from the Zengpiyan Site in Guilin, Guangxi]. Gu Jizhui Dongwu yu Gu Renlei [Certebrata Palasiatica] 16(4):244-254.

Li, Zhixiang, 1963, Yunnan Yeniu [Gaur in Yunnan]. Zhongguo Shouyi Zazhi [Chinese Journal of Veterinary Medicine] (2):18-19.

Li, Zhixiang, Lan Yang, Yingxiang Wang, Rulun Xian, Qiming Xu, and Weitang Chen, 1980, Yunnan Yeniu (*Bos gaur*) Shengzhang Fayu de Chubu Guancha [A Preliminary Observation on the Growth and Development of Yunnan Gaur]. Dongwuxue Yanjiu [Zoological Research] 1(3):425-428.

Lin, Sheng, 1963, Ji Yi, Qiang, Naxi Zu de 'Yanggubu' [On Sheep Oracle Bones of *Yi*, *Qiang*, and *Naxi* Ethnic Minority Groups]. Kaogu [Archaeology] (3):162-164+166.

Liu, Guoxiang, 2004, Xinglonggou Juluo Yizhi Fajue Shouhuo ji Yiyi [Outcomes of Excavations at Xinglonggou Settlement Site]. *In* Dongbei Wenwu Kaogu Lunji [Collective Articles of Northeast Archaeology in China]. Pp. 58-86. Beijing: Science Press.

Liu, Guoxiang, Xiaobing Jia, Minghui Zhao, Guangling Tian, and Guotian Shao, 2004, Neimenggu Chifengshi Xinglonggou Juluo Yizhi 2002–2003 Nian de Fajue [2002–2003 Excavation on the Settlement-Site at Xinglonggou in Chifeng City, Inner Mongolia]. Kaogu [Archaeology] (7):579-584.

Liu, Li, and Xingcan Chen, 2001, China. *In* Encyclopedia of Archaeology: History and Discoveries. T. Murray, ed. Pp. 315-333. Santa Barbara, Denver, Oxford: ABC-CLIO.

Liu, Li, Xingcan Chen, and Leping Jiang, 2004, A Study of Neolithic Water Buffalo Remains from Zhejiang, China. *In* Indo-Pacific Prehistory: The Tapei Papers. P. Bellwood, D. Bowdery, S. Keates, L. Liu, and H. Martinsson-Wallin, eds. Pp. 113-120. Canberra: Australian National University.

Liu, Li, Gyoung-Ah Lee, Leping Jiang, and Juzhong Zhang, 2007, Evidence for the Early Beginning (c. 9000 cal. B.P.) of Rice Domestication in China: A Response. Holocene 17(8):1059-1068.

Liu, Li, Dongya Y. Yang, and Xingcan Chen, 2006, Zhongguo Jiayang Shuiniu Qiyuan Chutan [On the Origin of the *Bubalus Bubalis* in China]. Kaogu Xuebao [Acta Archaeologica Sinica] (2):141-178.

Liu, Yunhai, Chao Zhang, and Shaoqin He, 2010, Zhengui Chuzhong—Daeniu [Rare Livestock—Gayal]. Zhongguo Chuqin Zhongye [The Chinese Livestock and Poultry Breeding] (2):59-60.

Liu, Zhiyi, 1997, Yi Zu Jingdian zhong de Yuanshi Nonglinmuye yu Zhucha'niangjiu de Qiyuan [The Origins of Agriculture, Forestry, Faming, Tea-Making and Wine-Making in Classic Literature of the *Yi* Ethnic Minority Group]. Nongye Kaogu [Agricultural Archaeology] (1):215-219.

Lowe, V. P. W., 1967, Teeth as Indicators of Age with Special Reference to Red deer (*Cervus elaphus*) of Known Age from Rhum. Journal of Zoology 152(2):137-153.

Lu, Changkun, 1962, Xiongke [Ursidae]. *In* Zhongguo Jingji Dongwu Zhi: Shoulei [Economic Fauna of China: Mammals]. A. S. Editorial Committee of Fauna Sinica, ed. Pp. 325-331. Beijing: Science Press.

Lu, Haoquan, and Caiwu Zhou, 1990, Fulu Yi: Shandong Sishui Xian Yinjiacheng Yizhi Chutu Dongzhiwu Biaoben Jianding Baogao [Appendix I: Identification of Animal and Plant Remains from the Yinjiacheng Site in Sishui County, Shandong Province]. *In* Sishui Yinjiacheng [Yinjiacheng Site in Sishui County]. Historical Department of the Shandong University, ed. Pp. 350-352. Beijing: Cultural Relics Press.

Luo, Yunbing, and Juzhong Zhang, 2008, Henan Wuyang Xian Jiahu Yizhi Chutu Zhugu de Zaiyanjiu [Re-Analysis of the Pigs Bones from the Jiahu Site in Wuyang County, Henan]. Kaogu [Archaeology] (1):90-96.

Lydekker, Richard, 1888–1890, The New Natural History. Volume 2. New York: Merrill and Baker.

Lyman, R. Lee, 1984, Bone Density and Differential Survivorship of Fossil Classes. Journal of Anthropological Archaeology 3(4):259-299.

Lyman, R. Lee, 1985, Bone Frequencies: Differential Transport, *in situ* Destruction, and the MGUI. Journal of Archaeological Science 12(3):221-236.

Lyman, R. Lee, 1994a, Quantitative Units and Terminology in Zooarchaeology. American Antiquity 59(1):36-71.

Lyman, R. Lee, 1994b, Relative Abundances of Skeletal Specimens and Taphonomic Analysis of Vertebrate Remains. PALAIOS 9(3):288-298.

Lyman, R. Lee, 1994c, Vertebrate Taphonomy. Cambridge: Cambridge University Press.

Lyman, R. Lee, 2008, Quantitative Paleozoology. Cambridge: Cambridge University Press.

Lyman, R. Lee, and Gregory L. Fox, 1989, A Critical Evaluation of Bone Weathering as an Indication of Bone Assemblage Formation. Journal of Archaeological Science 16(3):293-317.

Ma, Guolong, Hong Chang, Shiping Li, Hongyu Chen, Dejun Ji, Rongqing Geng, Chunfang Chang, and Yonghong Li, 2007, Phylogenetic Relationships and Status Quo of Colonies for Gayal Based on Analysis of Cytochrome *b* Gene Partial Sequences. Journal of Genetics and Genomics 34(5):413-419.

Ma, Xiaolin, 2007, Lingbao Xipo Yizhi Jiazhu de Nianling Jiegou ji Xiangguan Wenti [Age Composition of the Domestic Pigs of the Xipo Site in Lingbao and Related Problems]. Huaxia Kaogu [Huaxia Archaeology] (1):55-74.

Mao, Huaming, Weidong Deng, and Jikun Wen, 2005, Daeniu de Shengwuxue Tezheng ji Yanjiu Kaifa Liyong Qianli [The Biological Characteristics of Gayal (*Bos frontalis*) and Potential Exploitation and Utilization]. Yunnan Nongye Daxue Xuebao [Journal of Yunnan Agricultural University] 20(2):258-261.

Marshall, Fiona, and Tom Pilgram, 1993, NISP vs. MNI in Quantification of Body-Part Representation. American Antiquity 58(2):261-269.

Marshall, L. G., 1989, Bone Modification and 'the Laws of Burial'. *In* Bone Modification. R. Bonnichsen and M. H. Sorg, eds. Pp. 7-24. Orono, Me.: Centre for the Study of the First Americans, Institute for Quaternary Studies, University of Maine.

Mason, I. L., ed., 1984a, Evolution of Domesticated Animals. London: Longman.

Mason, I. L., 1984b, Goat. *In* Evolution of Domesticated Animals. I. L. Mason, ed. Pp. 85-99. London: Longman.

Matschke, George H., 1967, Aging European Wild Hogs by Dentition. The Journal of Wildlife Management 31(1):109-113.

Mayer, John J., James M. Novak, and I. Lehr Brisbin, Jr., 1998, Evaluation of Molar Size as a Basis for Distinguishing Wild Boar from Domestic Swine: Employing the Present to Decipher the Past. *In* Ancestors for the Pigs: Pigs in Prehistory. S. M. Nelson, ed. Pp. 39-54. MASCA Research Papers in Science and Archaeology, Vol. 15. Philadelphia: University of Pennsylvania Museum of Archaeology and Anthropology.

Mayhew, Edward, 1849, The Horse's Mouth, Showing the Age by the Teeth. London: Messrs. Fores.

Meadow, Richard H., 1978, BONECODE. *In* Approaches to Faunal Analysis in the Middle East (Peabody Museum Bulletin 2), Cambridge, Mass, 1978, pp. 169-186. Peabody Museum of Archaeology and Ethnology, Harvard University.

Meadow, Richard H., 1980, Animal Bones: Problems for the Archaeologist Together with Some Possible Solutions. Paléorient 6:65-77.

Meadow, Richard H., 1989, Osteological Evidence for the Process of Animal Domestication. *In* The Walking Larder: Patterns of Domestication, Pastoralism, and Predation. J. Clutton-Brock, ed. Pp. 80-90. London: Unwin Hyman.

Metcalfe, Duncan, and Kevin T. Jones, 1988, A Reconsideration of Animal Body-Part Utility Indices. American Antiquity 53(3):486-504.

Moran, N. C., and T. P. O'Connor, 1994, Age Attribution in Domestic Sheep by Skeletal and Dental Maturation: A Pilot Study of Available Sources. International Journal of Osteoarchaeology 4(4):267-285.

Morris, P., 1972, A Review of Mammalian Age Determination Methods. Mammal Review 2(3):69-104.

Munson, Patrick J., 2000, Age-Correlated Differential Destruction of Bones and its Effect on Archaeological Mortality Profiles of Domestic Sheep and Goats. Journal of Archaeological Science 27(5):391-407.

Nagaoka, Lisa, 1994, Differential Recovery of Pacific Island Fish Remains: Evidence from the Moturakau Rockshelter, Aitutaki, Cook Islands. Asian Perspectives 33(1):1-17.

Nobis, G., 1979, Der älteste Haushunde lebte vor 14,000 Jahren [The Oldest Domestic Dog Lived More than 14,000 Years Ago]. Umschau 79:610.

Nobis, G., 1981, Aus Bonn: das älteste Haustier des Menschen. Unterkiefer eines Hundes aus dem Magdaleniengrab von Bonn-Oberkassel. Pp. 49-50. Bonn: Das Reinische Landesmuseum.

Noddle, Barbara, 1974, Ages of Epiphyseal Closure in Feral and Domestic Goats and Ages of Dental Eruption. Journal of Archaeological Science 1(2):195-204.

Noe-Nygaard, N., 1989, Man-Made Trace Fossils on Bones. Human Evolution 4(6):461-491.

Norbert, Benecke, 1987, Studies on Early Dog Remains from Northern Europe. Journal of Archaeological Science 14(1):31-49.

Nowak, Ronald M., 1991, Walker's Mammals of the World. Baltimore: Johns Hopkins University Press.

O'Connor, Terry Patrick, 2000, The Archaeology of Animal Bones. Stroud: Sutton Publishing.

Okumura, Naohiko, Naotaka Ishiguro, Masuo Nakano, Akira Matsui, Nobuo Shigehara, Toyohiko Nishimoto, and Makoto Sahara, 1999, Variations in Mitochondrial DNA of Dogs Isolated from Archaeological Sites in Japan and Neighbouring Islands. Anthropological Science 107(3):213-228.

Olsen, Sandra L., 1989, On Distinguishing Natural from Cultural Damage on a Archaeological Antler. Journal of Archaeological Science 16(2):125-135.

Ovodov, Nikolai D., Susan J. Crockford, Yaroslav V. Kuzmin, Thomas F. G. Higham, Gregory W. L. Hodgins, and Johannes van der Plicht, 2011, A 33,000-Year-Old Incipient Dog from the Altai Mountains of Siberia: Evidence of the Earliest Domestication Disrupted by the Last Glacial Maximum. PLoS ONE 6(7):e22821.

Payne, Sebastian, 1972a, On the Interpretation of Bone Samples from Archaeological Sites. *In* Papers in Economic Prehistory. E. S. Higgs, ed. Pp. 65-81. Cambridge: Cambridge University Press.

Payne, Sebastian, 1972b, Partial Recovery and Sample Bias: The Results of Some Sieving Experiments. *In* Papers in Economic Prehistory. E. S. Higgs, ed. Pp. 49-64. Cambridge: Cambridge University Press.

Payne, Sebastian, 1973, Kill-off Patterns in Sheep and Goats: The Mandibles from Aşvan Kale. Anatolian Studies 23:281-303.

Payne, Sebastian, 1975, Partial Recovery and Sample Bias. *In* Archaeozoological Studies: Papers of the Archaeozoological Conference 1974. A. T. Clason, ed. Pp. 7-17. Amsterdam, New York: North-Holland Pub. Co.

Payne, Sebastian, 1985, Morphological Distinctions between the Mandibular Teeth of Young Sheep, *Ovis*, and Goats, *Capra*. Journal of Archaeological Science 12(2):139-147.

Payne, Sebastian, 1987, Reference Codes for Wear States in the Mandibular Cheek Teeth of Sheep and Goats. Journal of Archaeological Science 14(6):609-614.

Perkins, Dexter, 1973, A Critique on the Methods of Quantifying Faunal Remains from Archaeological Sites. *In* Domestikationsforschung und Geschichte der Haustiere. J. Matolcsi, ed. Pp. 367-369. Budapest: Akademiai Kiado.

Perkins, Dexter. Jr., and Patricia Daly, 1968, A Hunters' Village in Neolithic Turkey. Scientific American 219(5):96-106.

Preparatory Office of Yunnan Provincial Museum, 1958, Jianchuan Haimenkou Guwenhua Yizhi Qingli Jianbao [Clearing on the Haimenkou Site, Jianchuan]. Kaogu Tongxun [Archaeological Communication] (6):5-12.

Prummel, Wietske, 1987a, Atlas for Identification of Foetal Skeletal Elements of Cattle, Horse, Sheep and Pig. Part 1. ArchaeoZoologia 1(1):23-30.

Prummel, Wietske, 1987b, Atlas for Identification of Foetal Skeletal Elements of Cattle, Horse, Sheep and Pig. Part 2. ArchaeoZoologia 1(2):11-42.

Prummel, Wietske, 1988, Atlas for Identification of Foetal Skeletal Elements of Cattle, Horse, Sheep and Pig. Part 3. ArchaeoZoologia 2(1/2):13-26.

Prummel, Wietske, 1989, Atlas for Identification of Foetal Skeletal Elements of Cattle, Horse, Sheep and Pig. Part 4. ArchaeoZoologia 3(1/2):71-78.

Prummel, Wietske, and Hans-Jörg Frisch, 1986, A Guide for the Distinction of Species, Sex and Body Side in Bones of Sheep and Goat. Journal of Archaeological Science 13(6):567-577.

Purdue, James R., 1983, Epiphyseal Closure in White-Tailed Deer. The Journal of Wildlife Management 47(4):1207-1213.

Qi, Guoqin, 1977, Fujian Minhou Tanshishan Xinshiqi Shidai Yizhizhong Chutu de Shougu [Animal Bones from Tanshishan Neolithic Site, Minhou, Fujian]. Gu Jizhui Dongwu yu Gu Renlei [Certebrata Palasiatica] 15(4):301-306.

Qi, Guoqin, 1983, Dongwu Kaoguxue Suoyao Yanjiu he Jiejue de Wenti [On the Issues of Zooarchaeological Research]. Renleixue Xuebao [Acta Anthropologica Sinica] 2(3):293-300.

Qi, Guoqin, 1988, Jiangzhai Xinshiqi Shidai Yizhi Dongwuqun de Fenxi [On Analysis of the Jiangzhai Neolithic Site]. *In* Jiangzhai. Xi'an Banpo Museum, Shanxi Provincial Institute of Archaeology, and Lintong County Museum, eds. Pp. 504-538. Beijing: Cultural Relics Press.

Qi, Guoqin, and Jing Yuan, 1997, Oumei Dongwu Kaoguxue Jianshi [A Brief History of Zooarchaeoloy in the Western World]. Huaxia Kaogu [Huaxia Archaeology] (3):91-99.

Reed, C. A., 1969, The Pattern of Animal Domestication in the Prehistoric Near East. *In* The Domestication and Exploitation of Plants and Animals. P. J. Ucko and G. W. Dimbleby, eds. Pp. 361-380. London: Duckworth.

Reitz, Elizabeth J., and Elizabeth S. Wing, 2008, Zooarchaeology. Cambridge: Cambridge University Press.

Ringrose, T. J., 1993, Bone Counts and Statistics: A Critique. Journal of Archaeological Science 20(2):121-157.

Rolett, Barry V., and Min-yung Chiu, 1994, Age Estimation of Prehistoric Pigs (*Sus scrofa*) by Molar Eruption and Attrition. Journal of Archaeological Science 21(3):377-386.

Russell, Nerissa, 1995, The Bone Tool Industry at Mehrgarh and Sibri. *In* Mehrgarh Field Reports 1974–1985. J.-F. J. Catherin Jarrige, Richard H. Meadow, Gonzague Quivron, ed. Pp. 583-613. Karachi: The Department of Culture and Tourism, Government of Sindh, Pakistan.

Ryder, M. L., 1984, Sheep. *In* Evolution of Domesticated Animals. I. L. Mason, ed. Pp. 63-85. London: Longman.

Savolainen, Peter, Ya-ping Zhang, Jing Luo, Joakim Lundeberg, and Thomas Leitner, 2002, Genetic Evidence for an East Asian Origin of Domestic Dogs. Science 298(5598):1610-1613.

Schmid, Elisabeth, 1972, Atlas of Animal Bones: For Prehistorians, Archaeologists and Quaternary Geologists. Knochenatlas. Für Prähistoriker, Archäologen und Quartärgeologen. Amsterdam: Elsevier Pub. Co.

Shaffer, Brian S., 1992, Quarter-Inch Screening: Understanding Biases in Recovery of Vertebrate Faunal Remains. American Antiquity 57(1):129-136.

Shaffer, Brian S., and Julia L. J. Sanchez, 1994, Comparison of 1/8′ '- and 1/4′ '-Mesh Recovery of Controlled Samples of Small-to-Medium-Sized Mammals. American Antiquity 59(3):525-530.

Shan, Xiangnian, Yifeng Chen, and Lihua Luo, 1980, Daeniu Hexing Fenxi [Karyotypic Studies on Gayal]. Yichuan [Hereditas (Beijing)] 2(5):25-27.

Sheng, Heling, 1992a, Chiji [Red Muntjac *Muntiacus muntjak*]. *In* Zhongguo Lulei Dongwu [The Deer in China]. H. Sheng, ed. Pp. 164-170. Shanghai: East China Normal University Press.

Sheng, Heling, 1992b, Huangji [Reeve's Muntjac *Muntiacus reevesi*]. *In* Zhongguo Lulei Dongwu [The Deer in China]. H. Sheng, ed. Pp. 126-144. Shanghai: East China Normal University Press.

Sheng, Heling, 1992c, Tunlu [Hog Deer *Axis porcinus*]. *In* Zhongguo Lulei Dongwu [The Deer in China]. H. Sheng, ed. Pp. 171-173. Shanghai: East China Normal University Press.

Shipman, Pat, 1981, Application of Scanning Electron Microscopy to Taphonomic Problems. *In* The Research Potential of Anthropological Museum Collections. A.-M. E. Cantwell, J. B. Griffin, and N. A. Rothschild, eds. Pp. 357-385. New York: New York Academy of Sciences.

Shipman, Pat, Giraud Foster, and Margaret Schoeninger, 1984, Burnt Bones and Teeth: An Experimental Study of Color, Morphology, Crystal Structure and Shrinkage. Journal of Archaeological Science 11(4):307-325.

Shipman, Pat, and Jennie Rose, 1983, Early Hominid Hunting, Butchering, and Carcass-Processing Behaviors: Approaches to the Fossil Record. Journal of Anthropological Archaeology 2(1):57-98.

Shotwell, J. Arnold, 1955, An Approach to the Paleoecology of Mammals. Ecology 36(2):327-337.

Shotwell, J. Arnold, 1958, Inter-Community Relationships in Hemphillian (Mid-Pliocene) Mammals. Ecology 39(2):271-282.

Shou, Zhenghuang, and Xitao Cai, 1958, Yunnan Xishuangbanna Faxian de Yeniu [The Gaur Discovered in Xishuangbanna, Yunnan]. Kexue Tongbao [Cinese Science Bulletin] (4):112-113.

Silver, I.A., 1970, The Aging of Domestic Animals. *In* Science in Archaeology: A Survey of Progress and Research. D. Brothwell and E. Higgs, eds. Pp. 283-302. New York: Praeger.

Simoons, F. J., 1984, Gayal or Mithan. *In* Evolution of Domesticated Animals. I. L. Mason, ed. Pp. 34-38. London: Longman.

Sino-Japanese Archaeological Expedition, 1998, Sichuan Xinjin Xian Baodun Yizhi 1996 Nian Fajue Jianbao [Excavation at the Baodun Site in Xinjin County, Sichuan, 1996]. Kaogu [Archaeology] (1):29-50.

Smith, Andrew T., and Yan Xie, eds., 2008, A Guide to the Mammals of China. Princeton, N.J.: Princeton University Press.

Smith, R. N., 1956, Fusion of the Epiphyses of the Limb Bones of the Sheep. Veterinary Record 68:257-259.

Sparks, B. W., 1961, The Ecological Interpretation of Quaternary Non-Marine Mollusca. Proceedings of the Linnean Society of London 172:71-80.

Stahl, Peter W., 1996, The Recovery and Interpretation of Microvertebrate Bone Assemblages from Archaeological Contexts. Journal of Archaeological Method and Theory 3(1):31-75.

Stampfli, H. R., 1983, The Fauna of Jarmo, with Notes on Animal Bones from Matarrah, the 'Amuq, and Karim Shahir'. *In* Prehistoric Archeology along the Zagros Flanks. L. S. Braidwood, R. J. Braidwood, B. Howe, C. A. Reed, and P. J. Watson, eds. Pp. 431-483. Oriental Institute Publications, Vol. 105. Chicago: The Oriental Institute of the University of Chicago.

Stock, Chester, 1929, A Census of the Pleistocene Mammals of Rancho La Brea, Based on the Collections of the Los Angeles Museum. Journal of Mammalogy 10:281-289.

Sun, Shuyun, and Rubin Han, 1997, Gansu Zaoqi Tongqi de Faxian yu Yelian, Zhizao Jishu de Yanjiu [Studies of Early Bronze Objects from Gansu in Terms of Their Casting and Manufacturing Techniques]. Wenwu [Cultural Relics] (7):75-84.

Teilhard de Chardin, Pierre, and Chung Chien Young, 1936, On the Mammalian Remains from the Archaeological Site of Anyang. Palaeontologia Sinica Series C, VoL XII (Fase. 1).

Thomas, David Hurst, 1969, Great Basin Hunting Patterns: A Quantitative Method for Treating Faunal Remains. American Antiquity 34(4):392-401.

Uerpmann, Hans-Peter, 1973, Animal Bone Finds and Economic Archaeology: A Critical Study of 'Osteo-Archaeological' Method. World Archaeology 4(3):307-322.

Vermeij, Geerat J., and Gregory S. Herbert, 2004, Measuring Relative Abundance in Fossil and Living Assemblages. Paleobiology 30(1):1-4.

Vilà, Carles, Peter Savolainen, Jesús E. Maldonado, Isabel R. Amorim, John E. Rice, Rodney L. Honeycutt, Keith A. Crandall, Joakim Lundeberg, and Robert K. Wayne, 1997, Multiple and Ancient Origins of the Domestic Dog. Science 276(5319):1687-1689.

von den Driesch, Angela, 1976, A Guide to the Measurement of Animal Bones from Archaeological Sites. Cambridge, Mass: Peabody Museum of Archaeology and Ethnology, Harvard University.

von den Driesch, Angela, and J. Boessneck, 1974, Kritische Anmerkungen zur Widerristhöherberechnung aus Längermassen vor- und frühgeschichtlicher Tierknochen. Säugetierkundliche Mitteilungen 22:325-348.

Walker, Ernest P., 1968, Mammals of the World. Baltimore: Johns Hopkins Press.

Wang, Dadao, 1994, Yunnan Qingtong Wenhua yu Xinshiqi Wanqi Wenhua de Guanxi [Relations between Cultures of the Bronze Age and the Late Neolithic Period in Yunnan]. *In* Nan Zhongguo ji Linjin Diqu Guwenhua Yanjiu [Ancient Cultures of South China and Neighbouring Regions]. C. Tang, ed. Pp. 269-279. Hong Kong: Chinese University Press.

Wang, Dadao, 1998, Zailun Yunnan Xinshiqi Shidai Wenhua de Leixing [A Further Typological Study on the Neolithic Culture in Yunnan Province]. *In* Yunnan Kaogu Wenji [Collected Works on Yunnan Cultural Relics and Archaeology]. D. Wang, M. Xiao, and F. Yang, eds. Pp. 41-61. Kunming: The Nationalities Publishing House of Yunnan.

Wang, Lanping, Rongqing Geng, Hong Chang, Dejun Ji, Yonghong Li, and Chunfang Chang, 2009, Daeniu Qiyuan yu Xitong Diwei de Yichuanxue Fenxi [Genetic Analysis on Origin and Phylogenetic Status of Gayal (*Bos frontalis*)]. Yunnan Nongye Daxue Xuebao [Journal of Yunnan Agricultural University] 24(2):231-234.

Wang, Ningsheng, 1992, Yunnan Kaogu [Archaeology of Yunnan]. Kunming: Yunnan People's Press.

Wang, Wenguang, and Xiaoyan Long, 2009, Yunnan Minzu de Lishi yu Wenhua Gaiyao [History and Culture of Minority Ethnic Groups of Yunnan]. Kunming: Yunnan University Press.

Watson, J. P. N., 1979, The Estimation of the Relative Frequencies of Mammalian Species: Khirokitia 1972. Journal of Archaeological Science 6(2):127-137.

Wei, Feng, Weitang Wu, Minghua Zhang, and Defen Han, 1989, Zhejiang Yuyao Hemudu Xinshiqi Shidai Yizhi Dongwuqun [Fauna from the Neolithic Site at Hemudu, Zhejiang]. Beijing: Ocean Press.

West, Barbara, and Benxiong Zhou, 1988, Did Chickens Go North? New Evidence for Domestication. Journal of Archaeological Science 15(5):515-533.

White, Theodore E., 1952, Observations on the Butchering Technique of Some Aboriginal Peoples: I. American Antiquity 17(4):337-338.

White, Theodore E., 1953a, A Method of Calculating the Dietary Percentage of Various Food Animals Utilized by Aboriginal Peoples. American Antiquity 18(4):396-398.

White, Theodore E., 1953b, Observations on the Butchering Technique of Some Aboriginal Peoples No. 2. American Antiquity 19(2):160-164.

White, Theodore E., 1954, Observations on the Butchering Technique of Some Aboriginal Peoples Nos. 3, 4, 5, and 6. American Antiquity 19(3):254-264.

White, Theodore E., 1955, Observations on the Butchering Technics of Some Aboriginal Peoples Numbers 7, 8, and 9. American Antiquity 21(2):170-178.

White, Theodore E., 1956, The Study of Osteological Materials in the Plains. American Antiquity 21(4):401-404.

White, Tim D., 1992, Prehistoric Cannibalism at Mancos 5MTUMR-2346. Princeton, N.J.: Princeton University Press.

Wilson, Bob, Caroline Grigson, and Sebastian Payne, eds., 1982, Ageing and Sexing Animal Bones from Archaeological Sites. Oxford: British Archaeological Reports.

Winter, H., B. Mayr, W. Schleger, E. Dworak, J. Krutzler, and B. Burger, 1984, Karyotyping, Red Blood Cell and Haemoglobin Typing of the Mithum (*Bos frontalis*), Its Wild Ancestor and Its Hybrids. Research in Veterinary Science 36(3):276-283.

Xie, Chong'an, 2005, Luelun Xi'nan Diqu Zaoqi Pingdi Shuang'erguan de Yuanliu jiqi Zushu Wenti [A Brief Research on the Origin of the Early Flat-Bottomed Amphoras from Southwest China and Their Ethnic Attribute]. Kaogu [Archaeology] (2):127-160.

Xie, Duanju, 2002, Gan Qing Diqu Shiqian Kaogu [Archaeology of Gansu and Qinghai Region]. Beijing: Cultural Relics Press.

Xu, Zhifan, and Minghua Xiao, 2004, Nanfang Wenhua yu Baiyue Dianyue Wenming [Southern Culture and Civilisations of Baiyue and Dianyue]. Nanjing: Jiangsu Education Press.

Yan, Wenming, 1978, Gansu Caitao de Yuanliu [Origin and Development of Painted Pottery in Gansu Region]. Wenwu [Cultural Relics] (10):62-76.

Yan, Wenming, 2005, The Beginning of Farming. *In* The Formation of Chinese Civilization: An Archaeological Perspective. S. Allan, ed. Pp. 27-41. New Haven and London: Yale University Press.

Yang, Dehua, Jiayin Zhang, and Chun Li, 1988, Yunnan Yeniu de Shuliang Fenbu [Number Distribution of Gaur in Yunnan]. Dongwuxue Zazhi [Chinese Journal of Zoology] 23(1):36-38, 54.

Yang, Dongya Y., Li Liu, Xingcan Chen, and Camilla F. Speller, 2008, Wild or Domesticated: DNA Analysis of Ancient Water Buffalo Remains from North China. Journal of Archaeological Science 35(10):2778-2785.

Yang, Fan, Yang Wang, and Changcheng Hu, 2010, Yunnan Kaogu: 1979–2009 [Archaeology of Yunnan: 1979–2009]. Kunming: Yunnan People's Press.

Yang, Zhengchun, Yongfu Xiao, Xingguo Li, and Fengchao Li, 1985, Qilu Hupan Zhongquanxinshi Renlei he Jizhui Dongwu Huashiqun [Fossil Group of Homo Sapiens and Vertebrates of Middle Holocene Epoch at Qilu Lake, Tonghai County]. Yunnan Dizhi [Yunnan Geology] 4(3):266-280.

Ye, Maolin, and Jinyong Luo, 1999, Sichuan Wenchuan Xian Zhaodian Cun Faxian de Shiguan Zang [Report of Stone Graves in Zhaodian Village, Wenchuan County, Sichuan]. Kaogu [Archaeology] (7):660-661.

Yuan, Jing, 1992, Genju Dongwu Yachi Zhuangkuang Panduan Buru Dongwu de Nianling Jiegou [Estimation of Animal Age File Based on Tooth Eruption and Ware]. Beifang Wenwu [Northern Cultural Relics] (3):104-106.

Yuan, Jing, 1993, Lun Misheng Shidai de Jiazhu [On Domestic Pigs in the Yayoi Period, Japan]. Nongye Kaogu [Agricultural Archaeology] (3):282-289.

Yuan, Jing, 1995a, Shilun Zhongguo Dongwu Kaoguxue de Xingcheng he Fazhan [The Formation and Development of Chinese Zooarchaeology: A Preliminary Discussion]. Jianghan Kaogu [Jianghan Archaeology] (2):84-88.

Yuan, Jing, 1995b, Yanjiu Dongwu Kaoguxue de Mubiao, Lilun he Fangfa [Aims, Theories, and Methods of Zooarchaeology]. Zhongguo Lishi Bowuguan Guankan [Journal of National Museum of China] (1):59-68.

Yuan, Jing, 1999, Lun Zhongguo Xinshiqi Shidai Jumin Huoqu Roushi Ziyuan de Fangshi [Man's Meat-acquiring Patterns in Neolithic China]. Kaogu Xuebao [Acta Archaeologica Sinica] (1):1-22.

Yuan, Jing, 2001, Zhongguo Xinshiqi Shidai Jiachu Qiyuan de Wenti [The Problems of the Origin of Domestic Animals in Neolithic China]. Wenwu [Cultural Relics] (5):51-58.

Yuan, Jing, 2002, Zhongguo Xinshiqi Shidai Jiachu Qiyuan de Jige Wenti [Several Problems of the Origin of Domestic Animals in the Chinese Neolithic]. Nongye Kaogu [Agricultural Archaeology] (3):26-28.

Yuan, Jing, and Rowan K. Flad, 2002, Pig Domestication in Ancient China. Antiquity 76(293):724-732.

Yuan, Jing, Yunping Huang, Mengfei Yang, Peng Lv, Yang Tao, and Jie Yang, 2007, Gongyuanqian 2500 Nian–Qian 1500 Nian Zhongyuan Diqu Dongwu Kaoguxue Yanjiu [Zooarchaeological Studies of Central Plain Region from 2500–1500 B.C.]. *In* Keji Kaogu [Archaeometry]. J. Yuan, S. Li, Z. Zhao, Z. Wang, J. Liu, and Z. Gu, eds. Pp. 12-34. Beijing: Science Press.

Yuan, Jing, and Nanfeng Jiao, 1993, Riben Dongwu Kaoguxue de Xianzhuang yu Keti [On Issues of Zooarchaeology in Japan]. Kaogu Yu Wenwu [Archaeology and Cultural Relics] (4):104-110.

Yuan, Jing, and Jun Li, 2010, 1997 Nian Hebei Xushui Nanzhuangtou Yizhi Fajue Baogao [Excavation of the Nanzhuangtou Site in Xushui County, Hebei in 1997], Fulu: Hebei Xushui Nanzhuangtou Yizhi Chutu Dongwu Yicun Yanjiu Baogao [Appendix: Report of the Analysis of the Faunal Assemblage from the Nanzhuangtou Site in Xushui County, Hebei]. Kaogu Xuebao [Acta Archaeologia Sinica] (3):385-391.

Yuan, Jing, and Xiaoli Qin, 1994, Dongwu Kaoguxue Yanjiu de Jinzhan [Progress of Zooarchaeology in Western Europe and North America]. Kaogu Yu Wenwu [Archaeology and Cultural Relics] (1):92-112.

Yunnan Provincial Archaeological Team, 1964, Yunnan Xiangyun Dabona Muguotongguan Qingli Baogao [Excavation of the Wooden and Bronze Coffins at Dabona in Xiangyun County, Yunnan]. Kaogu [Archaeology] (12):607-614.

Yunnan Provincial Archaeological Team, 1983, Chuxiong Wanjiaba Gumuqun Fajue Baogao [Excavation of Ancient Tombs at Wanjiaba in Chuxiong, Yunnan Province]. Kaogu Xuebao [Acta Archaeologia Sinica] (3):347-382.

Yunnan Provincial Institute of Cultural Relics and Archaeology, 1990, Jianchuan Aofengshan Gumu Fajue Baogao [Excavations of Ancient Tombs on Aofengshan Hill in Jianchuan County, Yunnan]. Kaogu Xuebao [Acta Archaeologia Sinica] (2):239-265.

Yunnan Provincial Institute of Cultural Relics and Archaeology, 2005, Yunnan Zhongdian Xian de Shiguan Mu [Stone Tombs in Zhongdian County, Yunnan]. Kaogu [Archaeology] (4):316-327.

Yunnan Provincial Institute of Cultural Relics and Archaeology, 2007, Jiangchuan Lijiashan [Lijiashan in Jiangchuan]. Beijing: Cultural Relics Press.

Yunnan Provincial Institute of Cultural Relics and Archaeology, Cultural Relics Management, Dali Prefecture, and Cultural Relics Management, Jianchuan County, 2009a, Yunnan Sheng Jianchuan Xian Haimenkou Yizhi [The Haimenkou Site in Jianchuan County, Yunnan]. Kaogu [Archaeology] (7): 594-599.

Yunnan Provincial Institute of Cultural Relics and Archaeology, Cultural Relics Management, Dali Prefecture, and Cultural Relics Management, Jianchuan County, 2009b, Yunnan Sheng Jianchuan Xian Haimenkou Yizhi Disanci Fajue [The Third Excavation on the Haimenkou Site in Jianchuan County, Yunnan]. Kaogu [Archaeology] (8):675-694.

Yunnan Provincial Institute of Cultural Relics and Archaeology, Dali Municipal Museum, Dali Municipal Cultural Relics Management, and Dali Prefecture Cultural Relics Management, 2009c, Yunnan Dali Shi Haidong Yinsuodao Yizhi Fajue Jianbao [Excavation on the Yinsuodao Site at Haidong in Dali City, Yunnan]. Kaogu [Archaeology] (8):695-713.

Yunnan Provincial Institute of Cultural Relics and Archaeology, Chinese Academy of Social Science Yunnan Archaeological Team of the Institute of Archaeology, Chengdu Municipal Institute of Cultural Relics and Archaeology, Museum of Chuxiong Autonomous Prefecture, and Cultural Centre of Yongren County, 2003, Yunnan Yongren Caiyuanzi, Mopandi Yizhi 2001 Nian Fajue Baogao [Excavations of the Caiyuanzi and Mopandi Sites in Yongren, Yunnan, in 2001]. Kaogu Xuebao [Acta Archaeologia Sinica] (2):263-296.

Yunnan Provincial Museum, 1977, Yuanmou Dadunzi Xinshiqi Shidai Yizhi [The Neolithic Site at Dadunzi in Yuanmou County, Yunnan Province]. Kaogu Xuebao [Acta Archaeologia Sinica] (1):43-72.

Yunnan Provincial Museum, 1981, Yunnan Binchuan Baiyangcun Yizhi [The Baiyangcun Site in Binchuan County, Yunnan]. Kaogu Xuebao [Acta Archaeologia Sinica] (9):349-368.

Yunnan Provincial Museum, 1995, Yunnan Jianchuan Haimenkou Qingtong Shidai Zaoqi Yizhi [Haimenkou Early Bronze Age Site in Jianchuan County, Yunnan]. Kaogu [Archaeology] (9):775-787.

Zeder, Melinda A., and Susan R. Arter, 2008, Meat Consumption and Bone Use in a Mississippian Village. *In* Case Studies In Environmental Archaeology. 2 edition. E. J. Reitz, C. M. Scarry, and S. J. Scudder, eds. Pp. 337-356. New York: Springer.

Zeder, Melinda A., and Brian Hesse, 2000, The Initial Domestication of Goats (*Capra hircus*) in the Zagros Mountains 10,000 Years Ago. Science 287(5461):2254-2257.

Zeder, Melinda A., and Suzanne E. Pilaar, 2010, Assessing the Reliability of Criteria Used to Identify Mandibles and Mandibular Teeth in Sheep, *Ovis*, and Goats, *Capra*. Journal of Archaeological Science 37(2):225-242.

Zeng, Yangzhi, Xiangnian Shan, Yifeng Chen, Lihua Luo, and Xiaomei Cao, 1979, Woguo Daeniu de Miaoshu jiqi Ranseti de Yanjiu [Observation of Gayal in China and the Study of Its Chromosomes]. Ziran Zazhi [Chinese Journal of Nature] 2(11):662-663.

Zeuner, Frederick Everard, 1963, A History of Domesticated Animals. New York: Harper & Row.

Zhang, Cizu, Xiang Zhu, and Bingzhang Pang, 1997, Birds of China. Beijing: China Forestry Publishing House.

Zhang, Taiyan, 1985, Xi'nan Shu Yi Xiaoji [Notes on the Barbarians in Southwestern China]. *In* Taiyan Wenlu Xubian [Continued Writings of Taiyan Zhang]. Pp. 331-337. Zhang Taiyan Quanji [Complete Works of Taiyan Zhang], Vol. 5. Shanghai: Shanghai People's Publishing House.

Zhang, Xingyong, 1985, Yuanmou Dadunzi Xinshiqi Shidai Yizhi Chutu de Dongwu Yigu [Animal Remains from the Neolithic Dadunzi Site, Yuanmou]. Yunnan Wenwu [Yunnan Cultural Relics] (17):29-34.

Zhang, Xingyong, 1987, Yunnan Xinshiqi Shidai de Jiachu [Domestic Animals in Neolithic Yunnan]. Nongye Kaogu [Agricultural Archaeology] (1):370-377.

Zhang, Xingyong, Deming Geng, and Hui Liu, 1992, Tangzigou Zaoquanxinshi Buru Dongwu [Early Holocene Mammal Fauna from Tangzigou]. *In* Baoshan Shiqian Kaogu [Prehistoric Archaeology in Baoshan]. X. Zhang, ed. Pp. 49-62. Kunming: Yunnan Science and Technology Publishing House.

Zhang, Zengqi, 1990, Zhongguo Xi'nan Minzu Kaogu [Ethnoarchaeology of Southwest China]. Kunming: Yunnan People's Press.

Zhang, Zengqi, 1998, Jinning Shizhaishan [Shizhaishan in Jinning]. Kunming: Yunnan Fine Arts Publishing House.

Zhang, Zhiheng, ed., 1995, Zhongguo Kaoguxue Tonglun [Archaeology of China]. Nanjing: Nanjing University Press.

Zhao, Zhijun, 2005, Zhiwu Kaoguxue jiqi Xinjinzhan [Archaeobotany and Its Recent Advances]. Kaogu [Archaeology] (7):522-529.

Zhao, Zhijun, 2011, New Archaeobotanic Data for the Study of the Origins of Agriculture in China. Current Anthropology 52(S4):S295-S306.

Zhao, Zhijun, and Jian Chen, 2011, Sichuan Mao Xian Yingpanshan Yizhi Fuxuan Jieguo ji Fenxi [Outcomes and Analysis of the Flotation Test at the Yingpanshan Site in Mao County, Sichuan]. Nanfang Wenwu [Cultural Relics in Southern China] (3):60-67.

Zhou, Benxiong, 1981, Hebei Wuan Cishan Yizhi de Dongwu Guhai [The Animal Remains Discovered at Cishan Village, Wu'an, Hebei Province]. Kaogu Xuebao [Acta Archaeologia Sinica] (3):339-347.

Zhou, Benxiong, 1994, Baijiacun Yizih Dongwu Yihai Jianding Baogao [Identification of Animal Remains from the Baijiacun site]. *In* Lintong Baijiacun [Baijiacun Site in Lintong]. Institue of Archaeology, Chinese Academy of Social Science, ed. Beijing: Cultural Relics Press.

Zhu, Caifa, 1987, Chengdu Zhihuijie Zhoudai Yizhi Fajue Baogao [Excavation at a Bronze-Age Site on Zhi Hui Street, Chengdu], Fulu Er: Dongwu Guge Jianding [Appendix II: Identification of Animal Remains]. *In* Nanfang Minzu Kaogu [Southern Ethnology and Archaeology]. Sichuan University Museum and Chinese Association for the Study of Ancient Bronze Drums, eds. Pp. 205-206, Vol. 1. Chengdu: Sichuan University Press.

Zong, Guanfu, and Xueshi Huang, 1985, Yunnan Baoshan Pupiao Quanxinshi Zaoqi Wenhua Yiwu ji Buru Dongwu De Yicun [Cultural Remains and Mammal Fossils from the Early Holocene Epoch Discovered at Pupiao in Baoshan city, Yunnan]. Shiqian Yanjiu [Prehistory] (4):46-50.

Appendix 1

Recording System for the Faunal Assemblage from Haimenkou

Based on BONECODE (Meadow 1978)

SIT: site

LOC: Locus, i.e., Grid Number

LAY: Excavation Layer

BON: Bone Identification Number

ART: Articulation

1 (unless below)
2 see succeeding specimen (s)
3 fused with preceding
4 articulated with preceding, weighed together
5 articulated with preceding, weighed separately
6 probably articulated
7 possibly articulated

ELM: Element Identification

N.A.	not applicable
????	indeterminate
SKULL	skull or more than two skull bones present
BASIO	basioccipital
OCCIP	occipital
SPHEN	sphenoid
PTERY	pterygoid
VOMER	vomer
PALAT	palatine
INPAR	interparietal
PARIE	parietal
FRONT	frontal
PETRS	petrous
TEMPR	temporal
ZYGOM	zygomatic
LACRI	lacrimal
NASAL	nasal
PMX/T	premaxillary with teeth
PMX	premaxillary without teeth
MAX/T	maxillary with teeth
MAX	maxillary without teeth
MAN/T	mandible with teeth
MAN	mandible without teeth
HORN	horn core
ANTLR	antler
HYOID	hyoid
G.VER	general vertebra(e)
ATLAS	atlas
AXIS	axis
CERV	cervical vertebra
THOR	thoracic vertebra
LUMB	lumbar vertebra
SACR	sacrum
CAUD	caudal vertebra
RIB	rib
COST	costal cartilage
STERN	sternebra
SCAP	scapula
CLAV	clavicle
HUMER	humerus
RAD	radius
ULNA	ulna
CARPL	carpal
C-RAD	radial carpal (except carnivores)
C-INT	intermediate carpal (except carnivores)
C-ULN	ulnar carpal
C-ACC	accessory carpal
C-1	first carpal (inconstant in equids. e.g., suids, canids, felids)
C-2	second carpal (e.g., equids, suids, canids, felids)
C-3	third carpal (same as C-2)
C-4	fourth carpal
C-R+I	radial + intermediate carpal (e.g., carnivores)
C-2+3	second + third carpal (e.g., ruminants)
MC1	metacarpal I
MC2	metacarpal II
MC3	metacarpal III
MC4	metacarpal IV
MC5	metacarpal V
MC3+4	metacarpal III + IV (ruminants)
MC?	indeterminate metacarpal
SESAP	anterior proximal sesamoid
SESAD	anterior distal sesamoid
SESAN	anterior sesamoid
PH1AN	anterior phalanx 1
PH2AN	anterior phalanx 2
PH3AN	anterior phalanx 3
PH?AN	anterior phalanx
PELV	pelvis (ilium + ischium + pubis)
IL/IS	ilium + ischium
IL/PB	ilium + pubis
IS/PB	ischium + pubis
ILIUM	ilium
ISCH	ischium
PUBIS	pubis
PENIS	penis bone (e.g., carnivores)
FEMUR	femur
PATEL	patella
TIBIA	tibia
FIBUL	fibula
MALL	lateral malleolus
TALUS	astragalus (talus)
CALC	calcaneus

TARSL	tarsal
T-CEN	central tarsal (e.g., equids, suids, canids, felids)
T-1	first tarsal (occ. sep. in equids)
T-2	second tarsal (occ. sep. in equids; as with T-CEN)
T-3	third tarsal (as with T-CEN)
T-4	fourth tarsal (as with T-CEN)
T-C+4	central + fourth tarsal (e.g., ruminants)
T-1+2	first + second tarsal (e.g., equids, usually)
T-2+3	second + third tarsal (e.g., ruminants)
MT1	metatarsal I
MT2	metatarsal II
MT3	metatarsal III
MT4	metatarsal IV
MT5	metatarsal V
MT3+4	metatarsal III + IV (ruminants)
MT?	indeterminate metatarsal
SESPP	posterior proximal sesamoid
SESPD	posterior distal sesamoid
SESPS	posterior sesamoid
PH1PS	posterior phalanx 1
PH2PS	posterior phalanx 2
PH3PS	posterior phalanx 3
PH?PS	posterior phalanx
MP3+4	metapodial III + IV
MP?	indeterminate metapodial
SES.P	proximal sesamoid
SES.D	distal sesamoid
SES.	sesamoid
PH1	phalanx 1
PH2	phalanx 2
PH3	phalanx 3
PH?	phalanx
TTHUP	loose tooth: upper
TTHLW	loose tooth: lower
HM/FR	humerus or femur fragment
LG.BN	long bone fragment
????	indeterminate fragment
TTHND	tooth fragment

(The following are for birds)

CORAC	coracoid
CARPO	carpometacarpus
TIBIO	tibiotarsus
TARSO	tarsometatarsus

ANM: Animal Identification

???	indeterminate
SMALL	smaller than rabbit-sized
SM-MD	rabbit-sized to medium dog-sized
MED.1	medium dog-sized to medium sheep-sized
MED.2	medium dog-sized to wild boar-sized
LG-MD	wild boar-sized to small cattle-sized
MD-LG	medium sheep-sized to medium cattle-sized
LARGE	domestic cattle/red deer-sized
V.LG.	larger than medium cattle-sized
S.ART	small artiodactyl
M.ART	medium artiodactyl
L.ART	large artiodactyl
S.BOV	small bovid
L.BOV	large bovid
OV/CP	*Ovis/Capra* (sheep/goat)
OVIS	*Ovis* sp. (sheep)
CAPRA	*Capra* sp. (goat)
BOS	*Bos* sp. (cattle)
B.GAU	*Bos gaurus* (gaur)
S.CER	small cervid
M.CER	medium cervid
L.CER	large cervid
R.UNI	*Rusa unicolor* (sambar)
A.POR	*Axis porcinus* (hog deer)
C.NIP	*Cervus nippon* (sika deer)
M.MUN	*Muntiacus muntjak* (red muntjac)
M.REE	*Muntiacus reevesi* (Reeve's muntjac)
M.BER	*Moschus berezovskii* (forest mask deer)
SUS	*Sus scrofa* (boar/pig)
S.CAR	small carnivore
M.CAR	medium carnivore
L.CAR	large carnivore
S.FEL	small felid
M.FEL	medium felid
L.FEL	large felid
PRNLR	*Prionailurus bengalensis* (leopard cat)
PARDO	*Pardofelis marmorata* (marbled cat)
PAGUM	*Paguma larvata* (masked palm civet)
S.CAN	small canid
M.CAN	medium canid
L.CAN	large canid
CANIS	*Canis* sp.
C.FAM	*Canis familiaris* (dog)
C.LUP	*Canis lupus* (wolf)
VULPS	*Vulpes* sp.
V.VUL	*Vulpes vulpes* (red fox)
URSUS	*Ursus* sp. (bear)
MACAC	*Macaca* sp. (macaque)
LEPUS	*Lepus* sp. (hare)
RHIZO	*Rhizomys* sp. (bamboo rat)
HYSTR	*Hystrix brachyuran* (porcupine)
RODEN	rodent
MURID	murid (rat/mouse)
MUS	*Mus musculus* (house mouse)
RATTU	*Rattus* sp. (rat)
HOMO	*Homo sapiens* (human)
BIRD	bird
ANATI	Anatidae (waterfowl)
ANAS	*Anas* sp. (dabbling duck)
CYGNU	*Cygnus* sp. (swan)
PHASI	Phasianidae (pheasant)
GALLUS	*Gallus gallus* (chicken)
GRUID	Gruidae (crane)

DOM: Domestication

ND	not determinable
W	wild
?W	probably wild
D	domestic
?D	probably domestic

CER: Certainty

1 standard identification
2 preliminary identification
3 possible identification
4 probable identification
5 definite identification of potentially questionable item

SYM: Symmetry (Right/Left)

— not applicable
? not determinable
R right
L left
M medial
4 right medial/left lateral
5 left medial/right lateral
B both left and right

SEX: Sex Determination

? not determinable
M male
?M possibly male
F female
?F possibly female
C castrate
?C possibly castrate
MC male or castrate
FC female or castrate

AGE: Relative Age

— not applicable
? not determinable
Y young (not grown up at all)
2 foetal or infantile
I infantile
4 infantile or juvenile
J juvenile
6 juvenile or sub-adult
S sub-adult
8 sub-adult or adult
A adult

SIZ: Relative Size (within age groups)

— not applicable
? not determinable
M medium
L large
S small

DIS: Disease

— not applicable
1 not defined below, see comment
2 typical alteration for that element
3 broken without dislocation
4 broken with dislocation
5 exostosis at articular surface
6 exostosis far from articular surfaces
7 not inflamed hyperplasia
8 holes and grinding in articular surface
9 holes far from articular surface

BUR: Burning

— not applicable
? indeterminate
1 completely calcined (white)
2 partly calcined
3 partly calcined, partly carbonised
4 carbonised (black)
5 partly carbonised
6 burnt and partly carbonised
7 burnt (reddish)
8 partly burnt
9 slightly burnt

MOD: Modification

— not applicable
?? possibly modified, see comment
R1 lightly rodent gnawed
R2 heavily rodent gnawed
C1 lightly carnivore gnawed
C2 heavily carnivore gnawed
BL blow marks present
RT retouched
PL polished
PR pierced
TL tool (shaped and used?)

FRG: Origin of Fragmentation

— not applicable
??? not determinable
WHL complete and unbroken
PRE pre-depositional
?PR ? pre-depositional
PST post-depositional but pre-modern
?PT ? post-depositional
MOD modern
?MD ? modern
P+R pre-depositional and minor recent
R+P pre-depositional and major recent

PCS: Number of Pieces

— not applicable (e.g., fused elements)
0 0
… …
99 99

ELS: Number of Elements

— (e.g., fragments not identifiable as to element)
0 0
… …
99 99

WGT: Weight (to nearest 0.1 gram)

—	not applicable (e.g., fused elements)
0.0	0.0
…	…
999.9	999.9

Cranial Format Only

RHC: Horn Core [Antler]

–right side or side not determinable

Horn Core

— not applicable
? not determinable
1 base
2 corpus
3 tip
4 base + corpus
5 corpus + tip
6 some of base missing
7 some of corpus missing
8 some of tip missing
9 complete

[Antler]

1 [base to beginning of first tine]
2 [first tine]
3 [base + first tine]
4 [fragment of column]
5 [column with tines]
6 [crown]
7 [single tine or fragments]
8 [base, first tine, and column with tines]
9 [complete or nearly complete brace of antlers]

LHC: Horn Core [Antler]

–left side
Same as for right side

FRN: Frontal

— not applicable
? side indeterminate
1 2/3 or more of right side present
2 2/3 or more of left side present
3 2/3 or more of both sides present
4 less than 2/3 of right side present
5 less than 2/3 of left side present
6 less than 2/3 of right side present, 2/3 or more of left side present
7 less than 2/3 of left side present, 2/3 or more of right side present
8 less than 2/3 of both right and left sides present
9 [for MAX and PMX, teeth only; for BAS, foramen magnum area]

The following are the same as for frontal (FRN)

PAR	Parietal
INP	Interparietal
OCC	Occipital
BAS	Basioccipital
SPH	Sphenoid
PTR	Pterygoid
PET	Petrous
TEM	Temporal
VOM	Vomer
PAL	Palatine
ZYG	Zygomatic
LAC	Lacrimal
NAS	Nasal
MAX	Maxillary
PMX	Premaxillary

AMN: Mandible–anterior

— not applicable
? fragment (or portion not determinable)
1 incisor area
2 diastema area
3 alveolus area
4 horizontal ramus area (including base)
5 incisor + diastema areas
6 incisor + diastema + alveolus
7 all anterior areas represented
8 diastema + alveolus + horizontal ramus area
9 alveolus + horizontal ramus

PMN: Mandible–posterior

— not applicable
? fragment (or portion not determinable)
1 heel area
2 ascending ramus area
3 area of articular process
4 coronoid process
5 heel + ascending ramus area
6 heel + ascending ramus + articular process areas
7 all posterior areas represented
8 ascending ramus + articular process + coronoid process
9 ascending ramus + articular or coronoid process areas

Tooth Wear

WI1: 1st Incisor

— not present/not applicable
? presence of tooth not determinable (that portion of jaw missing)
1 deciduous tooth lightly worn
2 deciduous tooth more than only lightly worn
3 permanent tooth bud present/deciduous tooth heavily worn or missing
P permanent tooth present but wear not determinable
U permanent tooth unworn/erupting
S permanent tooth slightly worn

M permanent tooth moderately worn
H permanent tooth heavily worn
V permanent tooth very heavily worn

The following are the same as for 1st Incisor (WI1)

WI2	2nd Incisor
WI3	3rd Incisor
WCN	Canine (='4th incisor')
WP1	1st Premolar
WP2	2nd Premolar

WP3: 3rd Premolar

[See Figure A1.1 for *Ovis/Capra* wear patterns]

— not applicable
?? presence of tooth not determinable (that portion of jaw missing)
CM comment on P3
P? P3 wear not determinable
U1 P3 visible in jaw
U2 P3 erupting
U3 P3 half-erupted
U4 P3 erupted but unworn
U5 P3 coming into wear
S+ P3 slight wear
M+ P3 moderate wear
H+ P3 heavy wear
V+ P3 very heavy wear

WP4: 4th Premolar

[See Figure A1.1 for *Ovis/Capra* wear patterns]

— not applicable
?? presence of tooth not determinable (that portion of jaw missing)
D? dP4 wear not determinable
DU dP4 unworn
D1 dP4 slightly worn 1
D2 dP4 slightly worn 2
D3 dP4 slightly worn 3
D4 dP4 moderately worn 1
D5 dP4 moderately worn 2
D6 dP4 moderately worn 3
D7 dP4 heavily worn 1
D8 dP4 heavily worn 2
D9 dP4 heavily worn 3
DV dP4 very heavily worn 1
DX dP4 very heavily worn 2
CM comment on dP4/P4
P? P4 wear not determinable
U1 P4 visible in jaw
U2 P4 erupting
U3 P4 half-erupted
U4 P4 erupted but unworn
U5 P4 coming into wear
S1 P4 slightly worn 1
S2 P4 slightly worn 2
S3 P4 slightly worn 3
S4 P4 slightly worn 4
M1 P4 moderately worn 1
M2 P4 moderately worn 2
H+ P4 heavily worn
V+ P4 very heavily worn

WMI: 1st Molar

[See Figure A1.1 for *Ovis/Capra* wear patterns]

— not applicable
?? presence of tooth not determinable (that portion of jaw missing)
P? wear not determinable
U0 perforation in crypt
U1 tooth visible in crypt
U2 tooth erupting
U3 tooth half erupted
U4 tooth erupted but not worn
U5 tooth coming into wear
UN (loose) tooth unworn
S1 tooth slightly worn 1
S2 tooth slightly worn 2
S3 tooth slightly worn 3
S4 tooth slightly worn 4
S5 tooth slightly worn 5
S6 tooth slightly worn 6
S7 tooth slightly worn 7
S8 tooth slightly worn 8
M1 tooth moderately worn 1
M2 tooth moderately worn 2
M3 tooth moderately worn 3
H1 tooth heavily worn 1
H2 tooth heavily worn 2
H3 tooth heavily worn 3
V1 tooth very heavily worn 1
V2 tooth very heavily worn 2

The following are the same as for 1st Molar (WM1)

WM2	2nd Molar
WM3	3rd Molar

Loose Teeth

TTY: Tooth Type

— not applicable
?? not determinable
UD upper deciduous (milk)
UA upper adult (permanent)
LD lower deciduous (milk)
LA lower adult (permanent)
U? upper tooth
L? lower tooth
?D deciduous (milk) tooth
?A adult (permanent) tooth

TCL: Tooth Class

—	not applicable
?	not determinable
I	incisor
C	canine
P	premolar
M	molar
C	cheek tooth (P or M)

TNO: Tooth Number

—	not applicable
?	not determinable
1	first
2	second
3	third
4	fourth
5	first or second
6	second or third
7	third or fourth
8	first, second, or third
9	second, third, or fourth

TRT: Tooth Root

—	not applicable
?	not determinable
1	open
2	separate but not closed
3	closed
4	swollen

TWR: Tooth Wear

[See CRANIAL FORMAT for tooth wear categories]

Post-cranial Format Only

FSZ: Fragment Size

—	not applicable
??	not determinable
EP	epiphysis
L1	less than 1/4 estimated length (or size)
L2	more than 1/4 but less than 2/4 (1/2) estimated length
L3	more than 1/2 but less than 3/4 estimated length
L4	more than 3/4 but less than 4/4 (whole) estimated length
WH	whole length (size) but still broken
UN	unbroken
VS	various sizes grouped together
MD	modern break, but fragment size greater than 1/2

END: Proximal/Distal

—	not applicable
???	not determinable
P-A	proximal articulation
P-E	proximal end
PES	proximal end and shaft
P-S	proximal shaft
SHF	shaft
D-S	distal shaft
DES	distal end and shaft
D-E	distal end
D-A	distal articulation
WHL	complete (whole)

FUS: State of Fusion

—	not applicable
???	not determinable
FUS	fused
LIN	epiphyseal line
UNF	unfused
F/F	proximal fused / distal fused
F/L	proximal fused / distal epiphyseal line
F/U	proximal fused / distal unfused
L/F	proximal epiphyseal line / distal fused
L/L	proximal epiphyseal line / distal epiphyseal line
L/U	proximal epiphyseal line / distal unfused
U/F	proximal unfused / distal fused
U/L	proximal unfused / distal epiphyseal line
U/U	proximal unfused / distal unfused

COM: comment

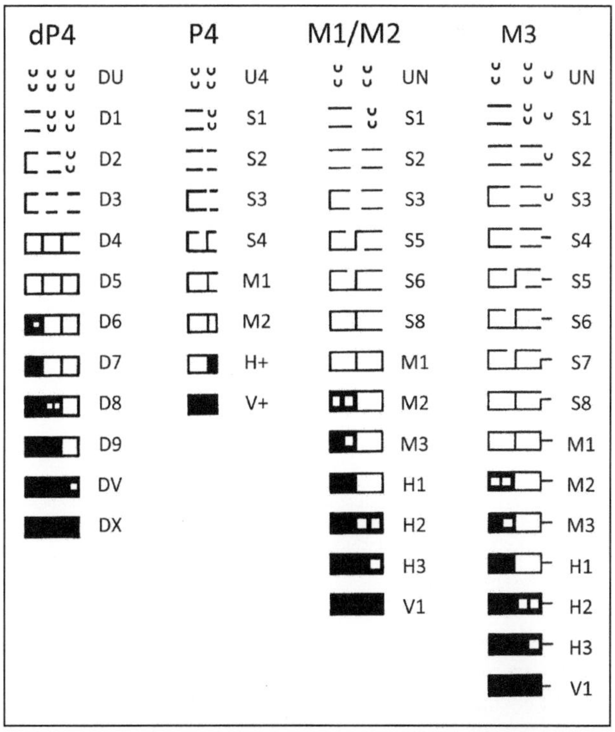

Figure A1.1 Codes for Mandibular Tooth Wear of Sheep and Goat (after Meadow 1978: figure 1; used with his permission)

Appendix 2

Measurements of Animal Bones from Haimenkou

Introduction:

1 All specimens were measured in millimeters. Measurements of teeth from gaurs, sheep & goats and cervids were after Klein and colleagues (1981: figure 1). Other measurements, if without special explanation, followed von den Driesch's (1976) directions. See detailed definitions and corresponding figures in von den Driesch (1976).

2 The numbers in parentheses refer to approximate estimates, as some specimens were slightly damaged and thus precise measurement could not be obtained.

3 See abbreviations not explained in tables in Appendix 1.

4 Measurements of modern sample are provided for comparison if possible.

5 08JH = Haimenkou in Jianchuan County, year 2008.

6 MS = measurement.

7 — = no data available in this category.

Contents

Table A1 Measurements of Gaur

A1.1 Horn Core

LOC	LAY	BON	SYM	MS 44	MS 45	MS 46
08JHAT2002	(5)	224	R	198.0	76.5	51.5

MS 44–46: See von den Driesch (1976:27-30).

A1.2 Cranium Bones except Maxilla

LOC	LAY	BON	ELM	SYM	MS 26	MS 28	MS 29	MS 39
08JHAT2002	(5)	271	TEMPR	L	—	—	—	(45.0)
08JHDT1304	(5)	276	OCCIP	M	(78.1)	40.3	30.5	—
Modern Domestic Cattle		No. 94	SKULL	B	90.1	41.7	34.9	31.4

MS 26, 28, 29, 39: See von den Driesch (1976:27-30).

A1.3 Maxilla with Teeth

LOC	LAY	BON	SYM	M1-L	M1-B	M1-H	M2-L	M2-B	M2-H	M3-L	M3-B	M3-H
08JHAT2605	(6)	125	L	—	—	—	26.2	26.6	11.7	29.4	26.0	(13.6)
Modern Domestic Cattle		No. 94	L	24.0	18.7	—	28.0	22.1	—	28.8	19.1	—

M = molar.
B = breadth.
L = length.
H = height.
See Klein and colleagues (1981: figure 1).

A1.4 Mandible with Teeth

a. Mandible (MS 3–9)

LOC	LAY	BON	SYM	MS 3	MS 4	MS 5	MS 6	MS 7	MS 8	MS 9
08JHAT2003	(6)	279	L	74.6	—	212.3	270.8	127.9	64.8	67.9
Modern Domestic Cattle		No. 94	L	99.8	257.4	247.2	300.9	139.9	90.2	48.0

MS 3–9: See von den Driesch (1976:56-57).

b. Mandible (MS 11–15c)

LOC	LAY	BON	SYM	MS 11	MS 12	MS 13	MS 14	MS 15a	MS 15b	MS 15c
08JHAT2003	(6)	1368	R	(88.0)	—	—	—	—	—	38.8
08JHAT2003	(6)	279	L	—	134.3	132.0	192.2	75.2	51.1	36.4
Modern Domestic Cattle		No. 94	L	92.6	165.6	—	229.0	73.1	52.0	35.5

MS 11–15: See von den Driesch (1976:56-57).

c. Teeth in Mandible (dP4)

LOC	LAY	BON	SYM	dP4-L	dP4-B	dP4-H
08JHAT2003	(6)	1368	R	—	(16.0)	(15.0)
08JHAT2003	(6)	279	L	35.3	16.0	16.0

dP4 = deciduous fourth premolar.
See other abbreviations in Table A1.3, see also Klein and colleagues (1981: figure 1).

A1.5 Loose Teeth

LOC	LAY	BON	ELM	SYM	TTY	L	B	R-B	H
08JHAT2002	(8)	1253	TTHUP	R	M3	30.7	20.8	(28+)	(47+)
08JHDT1304	(8)	138	TTHLW	R	M1/2	30.8	16.1	(19+)	(45+)
08JHAT2002	(7)	132	TTHUP	R	M3	33.3	21.7	27.3	59.7
08JHCT0304	(7)	2035	TTHUP	L	dP4	26.1	17.8	22.8	17.8
08JHAT2001	(6)	131	TTHUP	R	M1	29.7	21.3	26.9	36.5
08JHAT2001	(6)	133	TTHUP	R	M2	30.8	21.9	28.9	50.6
08JHAT2002	(6)	1254	TTHUP	R	P4	19.5	20.5	21.5	35.4
08JHAT2002	(6)	130	TTHUP	R	M2	30.5	21.4	25.4	49.2
08JHAT2002	(6)	135	TTHUP	R	M2	31.4	23.2	28.4	51.4
08JHAT2002	(6)	136	TTHUP	R	M1	26.5	21.7	26.0	31.0
08JHAT2004	(6)	1462	TTHLW	L	M1/2	33.0	14.6	18.9	(54+)
08JHAT2102	(6)	128	TTHUP	L	M2	30.9	23.3	28.0	53.8
08JHAT2102	(6)	134	TTHUP	R	M2	—	—	26.6	(47+)
08JHAT2505	(6)	1821	TTHUP	R	M1/2	—	24.0	24.0	9.7
08JHAT2605	(6)	1900	TTHUP	L	P4	17.6	18.4	20.5	24.8
08JHAT2002	(5)	1248	TTHUP	L	M3	30.8	20.9	28.0	41.8
08JHAT2002	(5)	1249	TTHUP	L	M3	33.4	22.8	(25+)	(53+)
08JHAT2002	(5)	1250	TTHUP	L	dP4	21.9	20.5	22.8	13.3
08JHAT2505	(5)	1822	TTHUP	L	M1/2	27.7	20.6	(23+)	—
08JHDT1104	(5)	611	TTHUP	R	M3	(32.0)	(25.2)	27.5	30.6
08JHDT1105	(5)	424	TTHLW	L	P4	22.3	13.3	13.9	17.7
08JHDT1204	(5)	129	TTHUP	L	M1/2	30.3	22.0	25.8	42.0
08JHDT1304	(5)	980	TTHUP	L	dP4	22.3	24.1	24.6	16.8
08JHDT1304	(5)	137	TTHLW	R	M1/2	31.6	13.9	18.8	(63+)
08JHDT1304	(5)	139	TTHLW	L	M1/2	37.1	15.5	(18+)	(49+)
08JHDT1104	(4)	610	TTHLW	R	M1/2	36.8	19.9	(24+)	(53+)
08JHDT1104	(4)	612	TTHUP	L	M1/2	25.8	26.1	26.1	15.7
08JHDT1105	(4)	422	TTHUP	R	M2	23.8	25.3	26.7	13.7
08JHDT1801	(4)	140	TTHLW	R	M3	39.5	16.0	19.8	(59+)
08JHDT1104	G1(2)	613	TTHLW	L	dP4	—	(15.1)	—	11.0
08JHDT1205	(1)	895	TTHLW	R	M1/2	28.8	17.1	19.9	40.0
Modern Domestic Cattle		No. 94	TTHLW	L	P4	21.2	12.4	—	—
Modern Domestic Cattle		No. 94	TTHLW	L	M1	24.1	—	—	—

R-B = breadth of root.
See other abbreviations in Table A1.3.

A1.6 Atlas

LOC	LAY	BON	SYM	GB	GL	BFcr	BFcd	GLF	H
08JHDT1304	(5)	155	M	—	—	—	—	89.3	—
08JHDT1005	(4)	755	M	—	—	—	—	(96.0)	—
Modern Domestic Cattle		No. 94	M	138.5	94.4	92.2	91.1	79.3	66.8

GB = greatest breadth over the wings.
GL = greatest length.
BFcr = (greatest) breadth of the cranial articular surface.
BFcd = (greatest) breadth of the caudal articular surface.
H = height.
See details in von den Driesch (1976:67).

A1.7 Scapula

LOC	LAY	BON	SYM	DHA	SLC	GLP	LG	BG
08JHDT1104	(8)	243	R	—	34.4	(51.3)	(44.9)	(33.0)
08JHAT2002	(6)	237	R	—	55.5	69.1	59.4	51.2
08JHDT1801	(6)	282	R	—	(63.0)	76.1	70.8	58.2
08JHAT2002	(4)	236	R	—	63.0	80.4	66.3	58.0
08JHDT1205	(4)	242	R	—	39.3	—	(55.0)	(39.8)
08JHAT2505	(3)	1842	R	—	—	—	52.6	45.0
Modern Domestic Cattle		No. 94	L	342.0	51.4	67.8	54.8	44.2

SLC = smallest length of the neck of the scapula.
GLP = greatest length of the glenoid process.
LG = length of the glenoid cavity.
BG = breadth of the glenoid cavity.
See details in von den Driesch (1976:74-75).

A1.8 Humerus

LOC	LAY	BON	SYM	Bd	BT	GLFd	LFd	Dd
08JHAT2002	(6)	53	L	105.0	93.0	52.0	40.4	85.0
08JHAT2003	(6)	212	L	101.7	85.1	51.2	37.6	89.0
08JHAT2003	(6)	220	R	(90.2)	(79.2)	(47.8)	36.7	79.4
08JHDT1104	(5)	55	L	(83.5)	(73.8)	(44.3)	(32.7)	—
08JHAT2002	(4)	56	L	(86.0)	(77.2)	(45.8)	(35.5)	—
Modern Domestic Cattle		No. 94	L	77.1	66.8	42.8	35.4	72.0

Bd = (greatest) breadth of the distal end.
BT = (greatest) breadth of the trochlea.
GLFd = grates length of the distal articular surface.
LFd =(mid) length of the distal articular surface.
Dd = (greatest) depth of the distal end.
See details of Bd and BT in von den Driesch (1976:76-77).
See GLFd, LFd, and Dd in Figure A2.1.

A1.9 Radius

LOC	LAY	BON	SYM	Dp	DFp	Bd	BFd	Dd	DFd
08JHAT2505	(8)	141	L	—	—	76.6	68.9	45.4	34.8
08JHDT1802	(7)	142	L	—	—	74.7	68.1	50.7	36.5
08JHAT2003	(6)	145	L	50.8	41.3	—	—	—	—
08JHAT2605	(6)	147	R	—	—	82.4	76.0	50.3	37.0
08JHAT2605	(6)	149	R	—	—	—	—	—	(33.4)
08JHDT1105	(5)	143	L	—	—	70.0	67.3	43.0	35.2
08JHDT1005	(5)	223	R	—	—	—	—	(48.5)	(34.1)
08JHDT1204	(4)	150	R	—	—	74.8	—	—	—
08JHDT1304	(4)	148	R	—	—	99.9	91.3	61.1	45.6
Modern Domestic Cattle	No. 94		L	40.3	37.2	66.3	61.4	42.0	31.8

Dp = (greatest) depth of the proximal end.
DFp = (greatest) depth of the proximal articular surface.
Bd = (greatest) breadth of the distal end.
BFd = (greatest) breadth of the distal articular surface.
Dd = (greatest) depth of the distal end.
DFd = (greatest) depth of the distal articular surface.
See details of Bd and BFd in von den Driesch (1976:79-81).
See Dp, DFp, Dd, and DFd in Figure A2.2.

A1.10 Ulna

LOC	LAY	BON	SYM	LO	DPA	SDO	BPC
08JHAT2605	(6)	1913	L	—	(54.6)	—	(32.5)
08JHAT2002	(4)	153	L	—	(70.7)	59.7	—
08JHDT1801	(4)	348	L	—	—	—	(51.6)
Modern Domestic Cattle		No. 94	L	90.5	59.5	48.9	42.1

LO = length of the olecranon.
DPA = depth across the anconeal process.
SDO = smallest depth of the olecranon.
BPC = (greatest) breadth across the coronoid process.
See details in von den Driesch (1976:79-81).

A1.11 Pelvis

LOC	LAY	BON	ELM	SYM	LA	LAR	SH	SB
08JHAT2004	(8)	246	PELV	L	80.4	67.0	—	—
08JHAT2004	(7)	244	ILIUM	L	—	—	48.3	20.8
08JHAT2605	(6)	255	IL/PB	R	—	—	40.8	30.4
08JHAT2002	(5)	248	IL/IS	L	79.1	68.5	—	—
08JHAT2002	(5)	250	IL/IS	L	86.1	75.4	—	—
Modern Domestic Cattle		No. 94	PELV	L	70.2	60.5	40.7	22.7

LA = length of the acetabulum including the lip.
LAR = length of the acetabulum on the rim.
SH = smallest height of the shaft of ilium.
SB = smallest breadth of the shaft of ilium.
See details in von den Driesch (1976:82-83).

A1.12 Femur

LOC	LAY	BON	SYM	DC	SD	Bd	Btpat
08JHDT1104	(8)	203	R	—	—	93.7	37.6
08JHAT2003	(6)	197	L	61.4	—	—	—
08JHAT2102	(6)	202	R	—	—	88.7	31.9
08JHAT2505	(6)	2066	R	53.0	—	—	—
08JHAT2605	(6)	199	R	—	—	—	(32.5)
08JHDT1104	(4)	200	R	56.0	—	—	—
Modern Domestic Cattle		No. 94	L	45.9	34.7	89.2	36.1

DC = (greatest) depth of the femoral head.
SD = smallest breadth of the diaphysis.
Bd = (greatest) breadth of the distal end.
Btpat = breadth of the patellar trochlea.
See details of DC, SD, and Bd in von den Driesch (1976:84-85).
See Btpat in Figure A2.3.

A1.13 Patella

LOC	LAY	BON	SYM	GL	GB
08JHDT1105	(4)	120	R	78.0	—

GL = greatest length.
GB = greatest breadth.
See details in von den Driesch (1976:85).

A1.14 Tibia

LOC	LAY	BON	SYM	Bp	BFp	Bd	Dd
08JHAT2505	(6)	1863	R	—	—	54.6	40.7
08JHAT2002	(5)	204	L	103.3	102.7	—	—
08JHAT2102	(4)	49	R	—	—	73.0	52.1
08JHAT2002	(4)	50	L	—	—	66.3	51.6
Modern Domestic Cattle		No. 94	L	94.6	90.8	61.1	46.2

Bp = (greatest) breadth of the proximal end.
BFp = (greatest) breadth of the proximal articular surface.
Bd = (greatest) breadth of the distal end.
Dd = (greatest) depth of the distal end.
See details of Bp, Bd, and Dd in von den Driesch (1976:86-87).
See BFp in Figure A2.4.

A1.15 Carpals

LOC	LAY	BON	ELM	SYM	GB	GD
08JHDT1803	(8)	116	C-ULN	R	25.4	37.4
08JHDT1803	(8)	119	C-INT	R	29.1	36.2
08JHCT0304	(7)	2040	C-2+3	L	(39.5)	(40.8)
08JHAT2103	(6)	115	C-2+3	L	34.1	35.8
08JHAT2605	(6)	1920	C-RAD	R	27.9	47.0
08JHAT2605	(6)	1921	C-INT	R	27.5	44.5
08JHAT2605	(6)	1922	C-ULN	R	26.0	37.7
08JHAT2605	(6)	112	C-RAD	L	28.2	45.8
08JHDT1205	(6)	113	C-RAD	L	29.3	42.6
08JHAT1901	(5)	114	C-RAD	R	31.1	50.4
08JHDT1105	(5)	501	C-ULN	R	26.4	37.9
08JHAT2102	(4)	117	C-INT	L	29.0	43.4
08JHDT1205	(4)	118	C-INT	R	28.0	42.8
Modern Domestic Buffalo		No. 63	C-RAD	R	26.0	44.2
Modern Domestic Buffalo		No. 63	C-INT	R	28.3	35.8
Modern Domestic Buffalo		No. 63	C-ULN	R	19.1	36.2
Modern Domestic Buffalo		No. 63	C-2+3	R	32.2	34.2

GB = greatest breadth.
GD = greatest depth.
See details in von den Driesch (1976:89).

A1.16 Calcaneus

LOC	LAY	BON	SYM	GL	GL w/o Ep	Dsustali	SLtuber
08JHAT0303	(6)	3	R	150.6	—	51.3	82.5
08JHAT2002	(6)	6	R	—	—	52.2	—
08JHAT2102	(6)	9	L	—	122.6	(46.0)	—
08JHDT1802	(6)	5	R	—	134.7	48.5	—
08JHDT1304	(5)	4	R	157.4	—	(47.0)	82.3
08JHDT1305	(4)	101	R	—	—	51.6	—
08JHDT1801	(4)	373	L	—	—	31.4	—
08JHDT1105	G1(2)	502	L	—	—	48.7	—
Modern Domestic Buffalo		No. 63	L	129.1	—	42.6	67.0

GL = greatest length.
GL w/o E = greatest length without epiphysis.
Dsustali = depth of the sustentaculum tali.
SLtuber = smallest length of the calcaneal tuber.
See details of GL in von den Driesch (1976:90-91).
See Dsustali and SLtuber in Figure A2.5.

A1.17 Astragalus

LOC	LAY	BON	SYM	GLl	GLm	Dl	Dm	Bd	BFp
08JHDT1802	(9)	48	L	66.3	58.2	36.6	35.1	42.8	40.5
08JHAT2004	(6)	1552	R	72.1	—	38.6	—	—	—
08JHDT1004	(6)	41	R	66.2	61.3	36.6	(38.3)	47.7	37.3
08JHDT1802	(6)	45	R	73.4	67.5	40.0	(41.6)	50.9	45.7
08JHAT2002	(5)	46	R	74.4	69.3	41.5	41.9	54.6	45.9
08JHAT2105	(5)	43	R	66.5	63.9	37.3	38.7	41.0	36.3
08JHDT1004	(5)	44	L	63.0	60.3	33.6	36.6	40.1	36.7
08JHDT1205	(4)	47	L	75.8	68.1	40.7	40.8	49.8	(44.0)
08JHDT1305	(4)	42	R	64.1	61.7	34.6	35.2	(40.0)	34.9

GLl: greatest length of the lateral half.
GLm: greatest length of the medial half
Dl: (greatest) depth of the lateral half.
Dm: (greatest) depth of the medial half.
Bd: (greatest) breadth of the distal end.
BFp: (greatest) breadth of the distal articular surface.
See details of GLl, GLm, Dl, Dm, and Bd in von den Driesch (1976:88-89).
See BFp in Figure A2.6.

A1.18 Tarsals except Calcaneus and Astragalus

LOC	LAY	BON	ELM	SYM	GB	GD
08JHAT2001	(6)	57	T-C+4	R	69.4	63.4
08JHAT2002	(5)	58	T-C+4	R	65.3	58.3
08JHAT2002	(5)	59	T-C+4	R	60.0	56.2
08JHDT1304	(5)	1043	T-C+4	R	(64.9)	—
08JHDT1801	(4)	372	T-2+3	L	26.2	41.6
Modern Domestic Buffalo		No. 63	T-C+4	L	55.0	47.3
Modern Domestic Cattle		No. 94	T-2+3	L	21.4	31.0

GB = greatest breadth.
GD = greatest depth.
See details in von den Driesch (1976:91).

A1.19 Metacarpal III+IV

LOC	LAY	BON	SYM	Bp	BFp	Dp	DFp	Lfm	Bfm	Lfl	Bfl
08JHDT1304	(8)	190	L	62.0	60.5	40.2	36.5	45.3	33.5	37.7	26.5
08JHAT2004	(6)	1562	L	—	—	—	—	—	—	(34.4)	(23.5)
08JHAT2102	(6)	191	L	57.5	56.2	35.6	33.0	39.4	31.6	36.9	24.1
08JHDT1801	(4)	376	L	76.0	74.1	44.0	40.5	50.3	39.4	44.9	35.6
Modern Domestic Buffalo		No. 63	L	58.4	54.9	33.8	32.3	39.3	27.3	34.3	24.0
Modern Domestic Cattle		No. 94	L	58.2	54.8	37.0	32.0	41.6	27.0	31.4	20.6

Bp = (greatest) breadth of the proximal end.
BFp = (greatest) breadth of the proximal articular surface.

Dp = (greatest) depth of the proximal end.
DFp = (greatest) depth of the proximal articular surface.
Lfm: length of the medial articular surface.
Bfm: breadth the medial articular surface.
Lfl: length of the lateral articular surface.
Bfl: breadth the lateral articular surface.
See details of Bp and Dp in von den Driesch (1976:92-93).
See BFp, DFp, Lfm, Bfm, Lfl, and Bfl in Figure A2.7.

A1.20 Metatarsal III + IV

a. Proximal End

LOC	LAY	BON	SYM	Bp	BFp	Dp	DFp	Lfm	Bfm	Lfl	Bfl	Bfpl	SD
08JHAT2003	(6)	193	L	—	—	—	—	49.6	30.4	—	—	—	—
08JHAT2605	(6)	1923	R	51.3	48.7	49.0	47.5	37.6	23.2	34.5	20.7	14.7	30.4
Modern Domestic Buffalo		No. 63	L	49.2	47.8	42.1	40.9	33.4	15.4	33.7	16.8	22.8	30.0
Modern Domestic Cattle		No. 94	L	49.8	—	49.0	—	—	—	—	—	—	27.3

Bp = (greatest) breadth of the proximal end.
BFp = (greatest) breadth of the proximal articular surface.
Dp = (greatest) depth of the proximal end.
DFp = (greatest) depth of the proximal articular surface.
Lfm: length of the medial articular surface.
Bfm: breadth the medial articular surface.
Lfl: length of the lateral articular surface.
Bfl: breadth the lateral articular surface.
Bfpl: breadth the lateral plantar articular surface.
SD = smallest breadth of the diaphysis.
See details of Bp, Dp and SD in von den Driesch (1976:92-93).
See BFp, DFp, Lfm, Bfm, Lfl, Bfl, and Bfpl in Figure A2.8.

b. Distal End

LOC	LAY	BON	SYM	Bd	MS (2)	MS (3)	MS (4)	MS (5)	MS (6)	MS (7)	MS (8)
08JHDT1304	(7)	195	?	57.1	—	—	—	—	—	—	—
08JHDT1304	(7)	195	?	57.0	—	—	—	—	—	—	—
08JHAT2001	(6)	1213	L	—	62.1	28.5	26.6	28.5	37.9	25.4	36.6
08JHAT2003	(6)	51	R	58.9	57.3	27.0	26.4	28.4	35.0	25.8	(35.0)
08JHAT2102	(6)	194	L	60.4	—	—	—	—	—	—	—
08JHAT2605	(6)	1923	R	—	—	—	—	—	—	—	—
Modern Domestic Buffalo		No. 63	L	55.0	57.2	27.2	27.0	26.8	35.1	22.0	33.7
Modern Domestic Cattle		No. 94	L	51.4	51.7	24.4	23.9	25.5	34.5	24.3	34.6

Bd = (greatest) breadth of the distal end.
MS (2)–(8): See Figure A2.9.
See details of Bd in von den Driesch (1976:92-93).

A1.21 Metapodial III+IV

LOC	LAY	BON	SYM	MS (3)	MS (4)	MS (5)	MS (6)	MS (7)	MS (8)
08JHDT1802	(7)	590	L	27.4	—	25.2	36.7	—	—
08JHAT2102	(6)	1696	?	24.2	—	26.0	33.5	—	—
08JHAT2605	(5)	1924	?	30.9	—	27.7	34.9	—	—
08JHDT1802	(5)	591	L	—	28.1	—	—	27.8	(37.5)
08JHAT2001	(4)	1214	?	21.7	—	21.6	30.5	—	—
08JHDT1801	(4)	378	?	26.9	—	28.6	36.3	—	—

See Table A1.20b.

A1.22 First Phalanx

LOC	LAY	BON	ELM	SYM	GLpe	Bp	Dp	SD	Bd	Dd
08JHDT1005	(8)	11	PH1	4	64.6	28.9	37.5	28.6	29.3	23.4
08JHAT2103	(7)	22	PH1	5	(67.5)	28.6	(38.8)	27.3	28.8	25.3
08JHAT2002	(6)	21	PH1	5	67.2	28.6	38.7	27.7	30.0	24.8
08JHAT2003	(6)	1423	PH1	5	—	—	—	—	32.8	27.2
08JHAT2102	(6)	1697	PH1	5	61.7	28.3	33.7	26.9	29.5	25.2
08JHAT2002	(5)	12	PH1	4	67.0	30.0	38.6	29.6	31.5	26.2
08JHAT2002	(5)	14	PH1	4	69.0	28.0	39.1	30.2	30.7	26.3
08JHAT2002	(5)	18	PH1	5	62.1	28.8	35.0	27.8	29.4	23.7
08JHAT2002	(5)	25	PH1	4	66.0	36.0	41.5	32.8	37.0	31.1
08JHDT1105	(5)	510	PH1	4	—	—	—	—	(33.2)	26.9
08JHDT1105	(5)	103	PH1	5	—	—	—	24.4	26.7	22.4
08JHDT1304	(5)	13	PH1	4	71.8	32.2	38.1	27.0	31.3	27.6
08JHDT1801	(5)	386	PH1	5	70.3	—	—	—	—	—
08JHAT2002	(4)	19	PH1	5	66.1	(28.3)	(37.8)	(27.0)	(29.5)	(22.1)
08JHAT2102	(4)	1698	PH1	4	—	25.5	—	—	—	—
08JHDT1005	(4)	23	PH1	5	72.4	29.7	39.3	27.3	29.6	25.6
08JHDT1305	(4)	16	PH1	5	66.9	31.7	38.0	31.5	34.6	26.1
08JHDT1801	(4)	385	PH1	4	(69.6)	31.7	38.2	—	—	—
08JHDT1801	(4)	15	PH1	4	74.0	34.1	41.3	30.6	34.0	29.6
08JHDT1801	(4)	20	PH1	5	66.0	(28.6)	37.2	(27.4)	30.2	24.3
08JHDT1105G1	(2)	17	PH1	5	67.3	31.9	36.7	28.3	32.3	26.1
Modern Domestic Buffalo		No. 63	PH1AN	4	55.6	33.7	33.0	28.6	31.4	22.7
Modern Domestic Buffalo		No. 63	PH1PS	4	58.6	30.1	34.5	24.9	27.3	21.3

GLpe = greatest length of the peripheral half.
Bp = (greatest) breadth of the proximal end.
Dp = (greatest) depth of the proximal end.
SD = smallest breadth of the diaphysis.
Bd = (greatest) breadth of the distal end.
Dd = (greatest) depth of the distal end.
See details in von den Driesch (1976:96-97).

A1.23 Second Phalanx

LOC	LAY	BON	ELM	SYM	GL	Bp	SD	Bd	Dd
08JHAT1901	(6)	33	PH2AN	5	52.7	31.2	25.0	26.5	35.9
08JHAT2002	(6)	27	PH2AN	5	48.3	33.9	29.7	31.0	37.3
08JHAT2002	(6)	30	PH2PS	4	50.6	34.2	30.4	31.0	36.6
08JHAT2002	(6)	37	PH2PS	4	52.6	32.5	25.3	28.0	36.7
08JHAT2102	(6)	40	PH2PS	4	48.3	28.7	23.7	24.5	28.6
08JHAT2605	(6)	35	PH2PS	5	45.0	31.5	24.3	25.6	30.3
08JHAT1901	(5)	26	PH2AN	5	47.8	34.5	31.8	32.6	35.5
08JHAT2002	(5)	31	PH2AN	4	48.0	35.8	28.5	31.1	38.5
08JHDT1105	(5)	28	PH2AN	5	(46.5)	(35.3)	(28.7)	(28.0)	(34.4)
08JHAT2002	(4)	36	PH2AN	5	50.4	30.2	25.0	24.9	31.4
08JHAT2102	(4)	39	PH2AN	4	48.4	31.7	23.7	24.3	28.9
08JHAT2103	(4)	32	PH2PS	4	(46.5)	30.0	25.4	26.5	32.2
08JHDT1204	(4)	29	PH2PS	4	49.3	37.6	30.0	31.5	39.3
08JHDT1305	(4)	34	PH2PS	5	54.5	32.5	26.0	27.3	34.8
08JHDT1801	(4)	38	PH2PS	4	50.5	(26.5)	(22.5)	(21.6)	(32.4)
Modern Domestic Buffalo		No. 63	PH2AN	4	39.9	33.4	24.8	26.1	30.0
Modern Domestic Buffalo		No. 63	PH2PS	4	41.5	30.2	21.4	24.1	29.5

GL = greatest length.
Bp = (greatest) breadth of the proximal end.
SD = smallest breadth of the diaphysis.
Bd = (greatest) breadth of the distal end.
Dd = (greatest) depth of the distal end.
See details in von den Driesch (1976:98-99).

A1.24 Third Phalanx

LOC	LAY	BON	ELM	SYM	DLS	MBS	Ld	HP
08JHAT1901	(6)	107	PH3	5	77.4	21.3	56.7	37.9
08JHAT2002	(5)	104	PH3	4	(62.5)	18.7	(52.4)	(34.2)
08JHAT2002	(5)	105	PH3	4	80.2	25.3	60.2	41.2
08JHDT1801	(5)	108	PH3	5	71.3	23.0	58.3	39.3
08JHAT2102	(4)	106	PH3	4	—	21.3	50.1	32.3
08JHDT1105	(4)	511	PH3	4	80.3	22.8	59.4	35.4
08JHDT1105	G1(2)	514	PH3	5	86.4	28.4	67.4	40.0
Modern Domestic Buffalo		No. 63	PH3AN	4	85.9	28.5	64.4	29.8
Modern Domestic Buffalo		No. 63	PH3PS	4	85.5	26.2	68.6	30.3

See details in von den Driesch (1976:101).

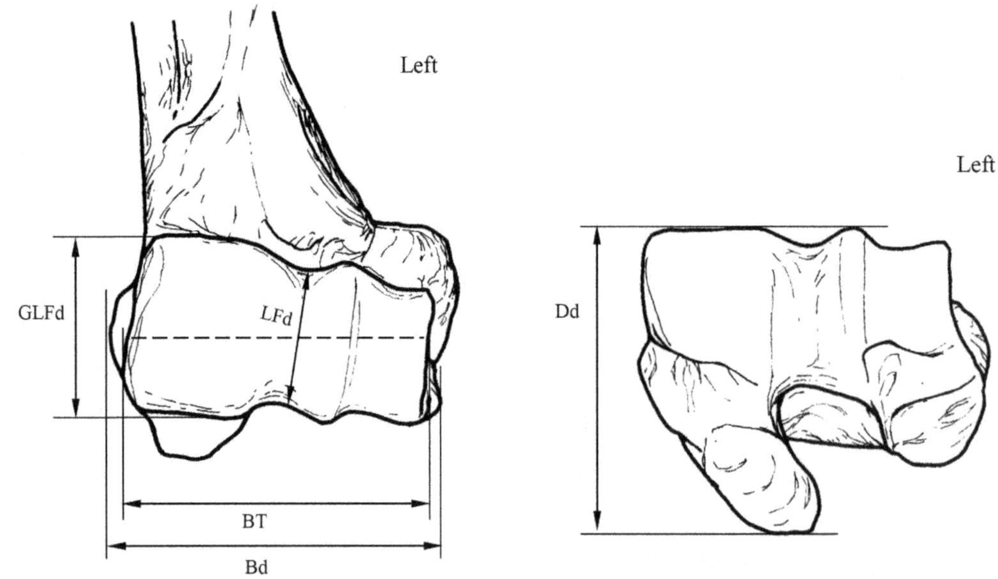

a. *Bos* humerus, distal end, cranial view b. *Bos* humerus, distal view

Figure A2.1 Measurements of Distal Humerus (Drawn by Juan Wang)

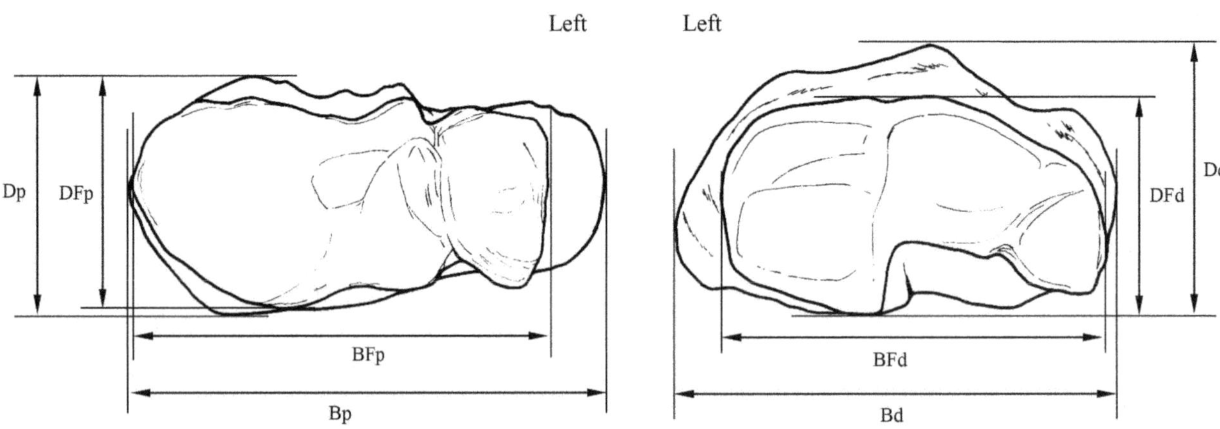

a. *Bos* radius, proximal view b. *Equus* radius, distal view

Figure A2.2 Measurements of Radius (Drawn by Juan Wang)

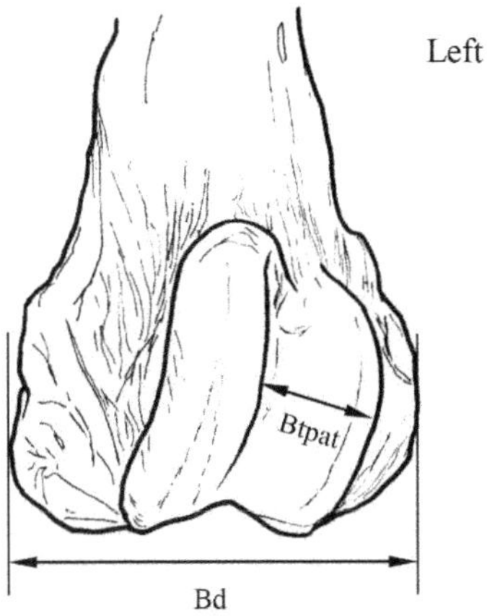

Ovis femur, distal end, cranial view

Figure A2.3 Measurements of Distal Femur (Drawn by Juan Wang)

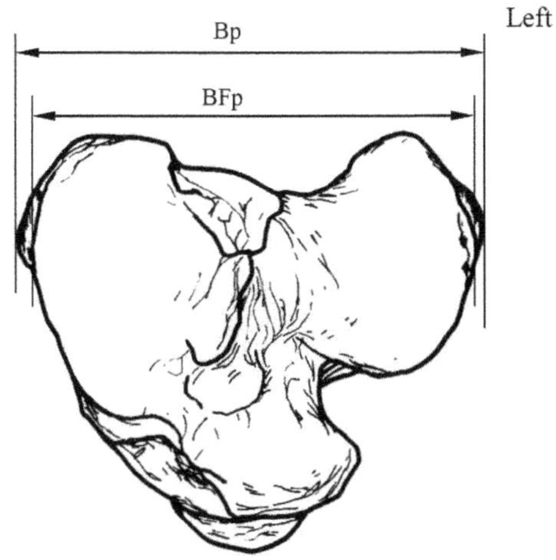

Bos tibia, proximal view

Figure A2.4 Measurements of Proximal Tibia (Drawn by Juan Wang)

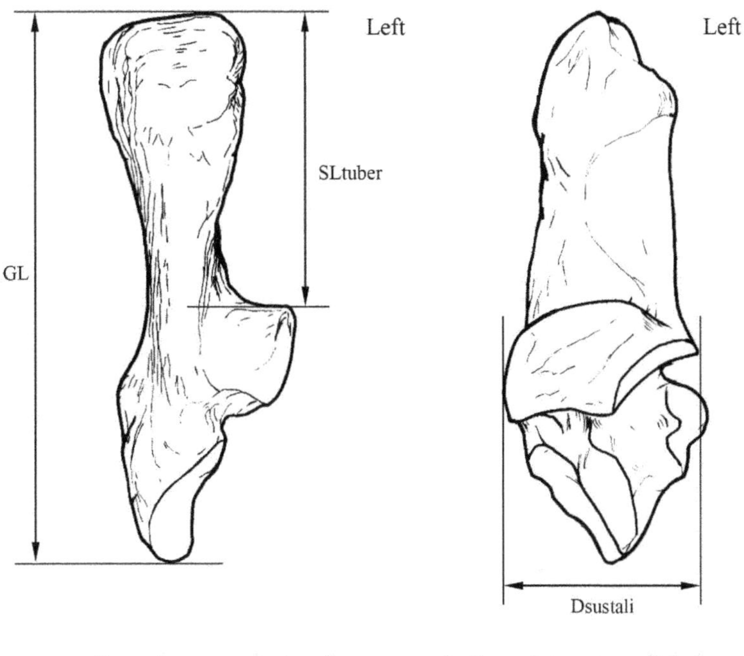

a. *Bos* calcaneus, plantar view b. *Bos* calcaneus, medial view

Figure A2.5 Measurements of Calcaneus (Drawn by Juan Wang)

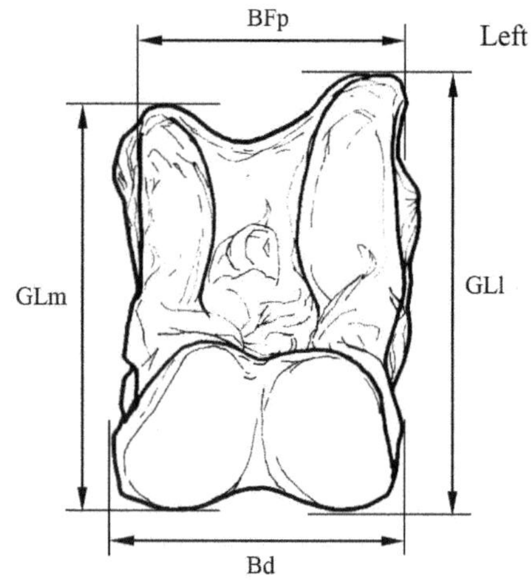

Bos astragalus, dorsal view

Figure A2.6 Measurements of Astragalus (Drawn by Juan Wang)

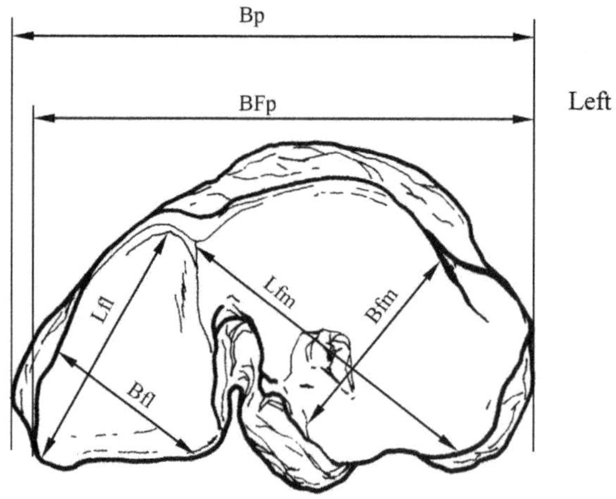

Bos metacarpal III + IV, proximal view.

Figure A2.7 Measurements of Proximal Metacarpal III + IV (Drawn by Juan Wang)

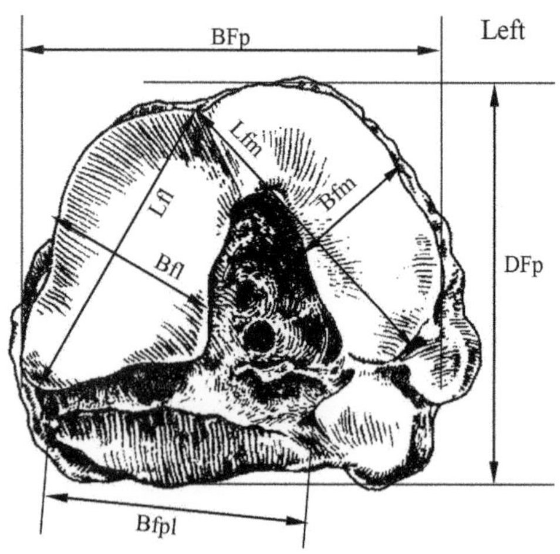

Bubalus metatarsal III + IV, proximal view.

Figure A2.8 Measurements of Proximal Metatarsal III + IV (Modified from Higham 1975: figure 7, used with his permission)

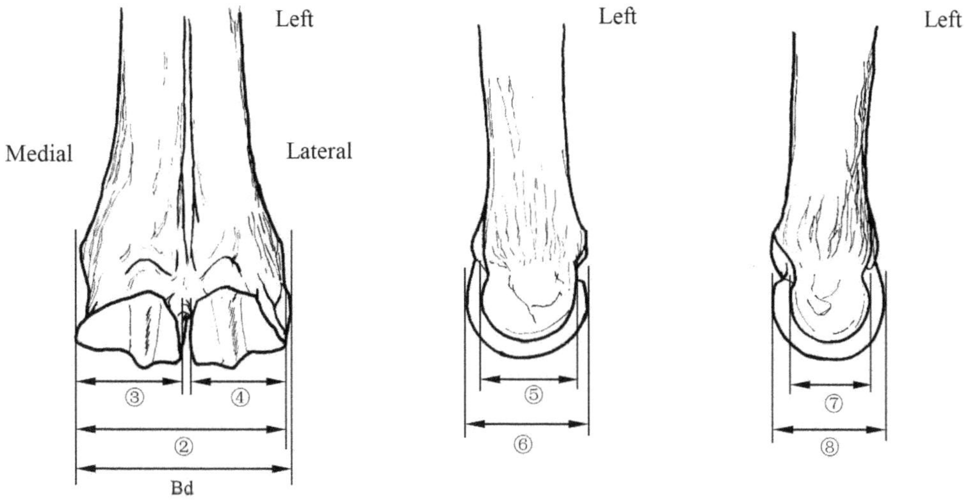

a. *Bos* metatarsal III + IV, dorsal view

b. *Bos* metatarsal III + IV, medial view

c. *Bos* metatarsal III + IV, lateral view

Figure A2.9 Measurements of Distal Metatarsal III + IV (Drawn by Juan Wang)

Table A2 Measurements of Pig

A2.1 Cranium Bones except Maxilla

LOC	LAY	BON	ELM	SYM	MS 13	MS 38	MS 39	MS 40
08JHAT1901	(6)	(5)1944	PARIE	B	(39.0)	—	(43.0)	34.1
08JHAT2003	(6)	1358	PARIE	L	(37.0)	—	—	—
08JHAT2003	(6)	1359	PARIE	L	(38.0)	—	—	—
08JHAT2004	(6)	1443	SKULL	B	49.0	49.4	44.0	29.6
08JHAT2004	(6)	1444	PARIE	L	(40.7)	—	—	—
08JHAT2004	(6)	1445	SKULL	R	(36.0)	—	—	—
08JHDT1204	(6)	796	OCCIP	M	—	44.6	40.7	—
08JHDT1304	(6)	948	SKULL	B	47.1	45.6	44.2	35.3
08JHDT1304	(6)	950	PARIE	B	(46.8)	—	—	—
08JHDT1304	(6)	951	OCCIP	B	—	54.5	47.0	—
08JHAT2004	(5)	1442	SKULL	B	47.4	50.8	44.0	30.3
08JHDT1005	(5)	728	SKULL	B	(42.9)	—	—	25.7
08JHDT1205	(5)	868	PARIE	R	(36.0)	—	—	—
08JHDT1205	(5)	870	OCCIP	M	—	40.3	34.1	—
08JHAT2001	(4)	1187	PARIE	L	(30.0)	—	—	—
08JHAT2001	(4)	1188	PARIE	R	(30.0)	—	—	—
08JHAT2105	(4)	1766	SKULL	B	(38.1)	50.0	49.7	29.7
08JHAT2505	(4)	1803	PARIE	L	(32.0)	—	—	—
08JHDT1205	(4)	867	SKULL	L	42.6	42.7	38.3	—
Modern Wild Boar		No. 99	SKULL	B	47.1	82.7	61.8	44.0

MS 13, 38–40: See von den Driesch (1976:38-41)

A2.2 Maxilla with Teeth

a. Maxilla

LOC	LAY	BON	SYM	MS 27	MS 27a	MS 28	MS 29	MS 29a	MS 32
08JHDT1802	(7)	534	L	—	—	—	45.9	35.9	—
08JHDT1802	(7)	535	L	—	—	23.8	—	—	—
08JHDT1802	(7)	536	R	—	—	—	—	—	—
08JHAT1901	(6)	1958	L	81.4	—	34.6	46.3	35.8	12.7
08JHAT1901	(6)	1959	R	78.6	—	33.9	44.5	35.4	12.4
08JHAT2004	(6)	1452	R	—	—	33.9	—	—	—
08JHAT2505	(6)	1812	L	—	—	32.4	—	—	—
08JHAT2505	(5)	1813	R	—	—	34.6	—	—	—
08JHAT2103	(4)	1718	R	65.2	55.7	24.7	40.3	31.9	—
Modern Wild Boar		No. 99	L	118.8	109.8	73.0	46.8	35.9	15.5

MS 27–29a, 32: See von den Driesch (1976:38-41)

A2.2 Maxilla with Teeth (continued)

b. Teeth in Maxilla (dP4 and P4)

LOC	LAY	BON	SYM	dP4-L	dP4-B	P4-L	P4-B
08JHAT2002	(8)	1243	L	—	—	—	—
08JHAT2002	(7)	1234	L	—	—	—	—
08JHAT2002	(7)	1235	R	—	—	—	—
08JHCT0304	(7)	2034	R	13.8	12.1	—	—
08JHDT1104	(7)	605	R	—	—	—	—
08JHDT1205	(7)	887	L	12.0	11.8	(15.9)	13.1
08JHDT1205	(7)	889	R	12.1	11.9	(15.6)	13.0
08JHDT1802	(7)	534	L	—	—	10.6	12.5
08JHDT1802	(7)	535	L	12.5	12.1	—	—
08JHDT1802	(7)	536	R	—	—	—	—
08JHAT1901	(6)	1958	L	11.7	11.9	—	—
08JHAT1901	(6)	1959	R	12.1	12.0	—	—
08JHAT2002	(6)	1236	R	—	—	—	—
08JHAT2004	(6)	1452	R	12.2	11.2	—	—
08JHAT2102	(6)	1658	R	—	—	—	—
08JHAT2505	(6)	1812	L	11.9	10.5	—	—
08JHDT1304	(6)	958	R	—	—	—	—
08JHDT1305	(6)	1075	L	12.1	10.9	—	—
08JHDT1305	(6)	1076	L	—	—	10.8	12.4
08JHDT1305	(6)	1077	R	—	—	10.5	12.4
08JHDT1801	(6)	302	L	—	—	—	—
08JHAT2505	(5)	1813	R	12.6	10.4	—	—
08JHDT1104	(5)	604	R	—	—	—	—
08JHDT1105	(5)	411	L	—	—	—	—
08JHDT1205	(5)	888	L	—	—	15.3	13.1
08JHDT1205	(5)	890	R	12.3	12.2	—	—
08JHAT2102	(4)	1657	L	11.8	11.2	—	—
08JHAT2103	(4)	1718	R	10.5	9.9	—	—
08JHDT1801	(4)	301	L	—	—	—	—
Modern Wild Boar		No. 99	L	—	—	12.2	14.9

See Table A1.3

A2.2 Maxilla with Teeth (continued)

c. Teeth in Maxilla (M1–M3)

LOC	LAY	BON	SYM	M1-L	M1-B	M2-L	M2-B	M3-L	M3-B
08JHAT2002	(8)	1243	L	—	—	17.0	15.3	—	—
08JHAT2002	(7)	1234	L	14.5	12.3	—	—	—	—
08JHAT2002	(7)	1235	R	14.0	12.4	16.5	14.9	—	—
08JHCT0304	(7)	2034	R	14.7	14.8	—	—	—	—
08JHDT1104	(8)	605	R	13.6	13.3	(18.8)	(15.2)	—	—
08JHDT1205	(7)	887	L	—	—	—	—	—	—
08JHDT1205	(7)	889	R	—	—	—	—	—	—
08JHDT1802	(7)	534	L	—	—	—	—	—	—
08JHDT1802	(7)	535	L	—	—	—	—	—	—
08JHDT1802	(7)	536	R	13.7	14.0	17.9	16.4	—	—
08JHAT1901	(6)	1958	L	14.1	14.7	—	—	—	—
08JHAT1901	(6)	1959	R	13.6	14.4	—	—	—	—
08JHAT2002	(6)	1236	R	—	—	16.4	15.5	26.9	16.3
08JHAT2004	(6)	1452	R	14.5	12.3	(20.5)	14.8	—	—
08JHAT2102	(6)	1658	R	13.6	12.8	16.4	15.7	—	—
08JHAT2505	(6)	1812	L	13.0	12.6	—	—	—	—
08JHDT1304	(6)	958	R	—	—	17.5	16.4	—	—
08JHDT1305	(6)	1075	L	13.8	13.2	—	—	—	—
08JHDT1305	(6)	1076	L	13.0	14.4	17.5	17.3	—	—
08JHDT1305	(6)	1077	R	14.1	14.2	19.3	17.6	—	—
08JHDT1801	(6)	302	L	14.6	14.1	18.4	16.5	—	—
08JHAT2505	(5)	1813	R	13.3	13.0	—	—	—	—
08JHDT1104	(5)	604	R	—	—	18.4	15.9	—	—
08JHDT1105	(5)	411	L	15.4	13.6	—	—	—	—
08JHDT1205	(5)	888	L	—	—	—	—	—	—
08JHDT1205	(5)	890	R	15.8	13.3	—	—	—	—
08JHAT2102	(4)	1657	L	14.2	13.0	—	—	—	—
08JHAT2103	(4)	1718	R	13.8	12.3	—	—	—	—
08JHDT1801	(4)	301	L	13.0	13.4	(17.5)	16.6	—	—
Modern Wild Boar		No. 99	L	15.0	15.2	19.3	18.1	36.8	20.7

See Table A1.3

A2.3 Mandible without Teeth

LOC	LAY	BON	SYM	MS 12	MS 21
08JHAT2003	(6)	1361	L	37.6	—
08JHDT1801	(4)	310	R	—	14.5

MS 12, 21: See von den Driesch (1976:58-59)

A2.4 Mandible with Teeth

a. Mandible (MS 1–8)

LOC	LAY	BON	SYM	MS 1	MS 2	MS 3	MS 4	MS 5	MS 6	MS 7	MS 7a	MS 8
08JHDT1801	(7)	305	L	—	—	—	—	—	73.3	72.4	63.3	31.9
08JHDT1801	(7)	306	R	151.1	—	—	98.6	—	75.0	70.6	64.2	30.8
08JHAT2003	(6)	1364	L	—	—	—	—	—	—	—	—	60.0
08JHAT2003	(6)	1365	L	—	—	51.7	—	—	—	—	—	33.5
08JHAT2003	(6)	1366	R	—	—	—	—	—	—	—	—	34.2
08JHAT2004	(6)	1455	L	—	—	56.3	—	—	—	—	—	44.5
08JHAT2004	(6)	1456	L	—	—	56.9	—	—	—	—	—	47.7
08JHAT2004	(6)	1457	R	—	—	49.8	—	—	—	—	—	27.6
08JHAT2004	(6)	1459	R	—	—	47.6	—	—	—	—	—	25.0
08JHAT2004	(6)	1460	R	—	—	53.8	—	—	—	—	—	46.3
08JHAT2004	(6)	1461	R	—	—	48.2	—	—	—	—	—	33.1
08JHAT2102	(6)	121	L	(204.0)	(208.0)	57.0	134.1	155.0	100.3	99.0	91.4	54.5
08JHAT2102	(6)	281	L	183.0	188.0	55.2	119.0	142.2	90.8	—	79.7	46.7
			R	183.0	188.0	56.7	122.8	141.8	91.9	—	77.5	45.6
08JHAT2002	(5)	1242	L	—	—	—	—	—	—	—	—	35.7
08JHAT2003	(5)	1363	L	—	—	—	—	—	—	—	—	61.0
08JHAT2505	(5)	1816	L	—	—	59.0	—	—	—	—	—	60.2
08JHDT1105	(5)	413	L	50.0	—	—	—	—	—	—	—	—
08JHDT1802	(5)	122	R	214.7	—	65.7	146.6	—	112.3	103.6	94.3	57.7
08JHAT2002	(4)	1246	L	179.5	181.5	48.9	127.7	142.3	—	—	—	58.5
			R	180.0	182.0	48.7	127.3	142.0	—	—	—	58.8
08JHAT2005	(4)	280	L	210.5	—	64.4	135.1	157.7	98.3	97.2	82.9	47.0
			R	210.5	—	65.8	135.0	159.8	100.4	99.0	84.8	47.0
08JHAT2102	(4)	1663	L	—	—	49.4	—	—	—	—	—	33.0
08JHAT2102	(4)	1664	L	163.9	171.2	44.9	115.0	125.1	82.4	81.3	69.6	41.9
08JHAT2102	(4)	1667	R	—	—	48.8	—	—	—	—	—	—
08JHAT2103	(4)	1720	L	—	—	—	—	—	—	—	83.4	45.5
08JHAT2103	(4)	1721	L	—	—	—	—	—	—	—	—	—
08JHAT2103	(4)	1722	L	144.2	153.0	48.2	96.0	112.3	74.2	—	60.6	27.7
08JHAT2104	(4)	1760	R	—	—	37.0	—	—	—	—	—	—
Modern Wild Boar		No. 99	L	256.5	272.8	84.4	180.9	194.5	138.8	130.4	114.3	74.8

MS 1–8: See von den Driesch (1976:58-59)

A2.4 Mandible with Teeth (continued)

b. Mandible (MS 9, 9a, 11–16c, 21)

LOC	LAY	BON	SYM	MS 9	MS 9a	MS 11	MS 12	MS 13	MS 14	MS 15	MS 16a	MS 16b	MS 16c	MS 21
08JHAT2605	(7)	1899	L	—	—	—	50.5	—	—	—	—	—	—	—
08JHDT1801	(7)	305	L	41.0	34.3	—	—	—	—	—	39.8	28.3	30.7	6.6
08JHDT1801	(7)	306	R	40.9	34.1	19.9	40.4	—	—	—	37.3	28.7	30.5	7.2
08JHAT2003	(6)	1364	L	—	—	—	—	—	—	—	45.6	—	—	—
08JHAT2003	(6)	1365	L	—	—	—	—	68.6	—	—	35.6	26.8	—	—
08JHAT2003	(6)	1366	R	—	34.9	—	—	—	—	—	34.5	—	—	—
08JHAT2004	(6)	1453	L	—	—	—	37.8	—	—	—	—	—	31.2	—
08JHAT2004	(6)	1455	L	—	—	—	—	84.3	82.4	93.1	40.8	30.9	—	—
08JHAT2004	(6)	1456	L	—	—	—	—	—	—	—	43.1	—	—	—
08JHAT2004	(6)	1457	R	—	—	—	—	—	—	—	37.9	—	—	—
08JHAT2004	(6)	1459	R	—	—	—	—	67.9	64.2	71.6	33.4	26.1	—	—
08JHAT2004	(6)	1460	R	—	—	—	—	86.4	—	—	39.6	32.4	—	—
08JHAT2004	(6)	1461	R	—	—	—	—	78.2	72.9	(78.9)	33.2	28.3	—	—
08JHAT2005	(6)	1611	L	—	—	—	—	69.9	67.9	75.6	35.7	—	—	—
08JHAT2005	(6)	1612	R	—	—	—	18.6	—	—	—	—	—	—	—
08JHAT2102	(6)	1665	L	—	—	—	—	—	—	—	30.0	22.3	—	—
08JHAT2102	(6)	121	L	47.7	38.8	28.4	56.1	94.0	84.8	91.7	46.4	37.1	40.7	13.7
08JHAT2102	(6)	281	L	—	35.6	—	49.5	79.1	75.4	85.9	40.0	28.4	37.2	12.5
			R	—	34.2	—	49.5	82.6	77.5	87.3	38.6	28.7	33.5	10.6
08JHDT1305	(6)	1081	L	—	—	—	35.7	—	—	—	—	—	—	—
08JHAT2002	(5)	1242	L	—	—	—	—	—	—	—	33.1	23.2	—	—
08JHAT2003	(5)	1363	L	—	—	—	—	—	—	—	46.5	33.7	—	—
08JHAT2505	(5)	1816	L	—	—	—	—	117.6	105.8	115.3	59.8	—	—	—
08JHDT1004	(5)	697	L	—	—	—	—	45.9	—	—	—	—	35.5	9.3
08JHDT1005	(5)	734	L	—	—	—	—	—	—	—	26.1	—	—	—
08JHDT1304	(5)	966	L	—	—	—	—	—	—	—	39.0	33.3	—	—
08JHDT1802	(5)	539	L	—	—	—	—	—	—	—	43.0	36.1	—	—
08JHDT1802	(5)	122	R	45.7	35.2	35.1	50.7	—	—	—	40.7	34.4	38.6	12.3
08JHAT2001	(4)	1201	R	—	—	—	—	—	—	—	30.8	—	—	—
08JHAT2002	(4)	1246	L	—	—	—	38.0	84.9	—	—	36.9	25.8	—	—
			R	—	—	—	38.0	83.7	(73.6)	—	38.5	25.7	—	—
08JHAT2005	(4)	280	L	49.1	34.7	34.3	51.7	—	—	—	41.3	35.3	35.7	12.3
			R	49.5	35.4	34.9	51.7	—	—	—	41.7	34.0	35.7	12.2
08JHAT2102	(4)	1663	L	—	—	—	—	74.0	70.3	(79.0)	35.2	30.2	—	—
08JHAT2102	(4)	1664	L	38.4	30.0	26.6	42.6	66.8	61.3	71.3	33.5	27.3	31.9	10.6
08JHAT2102	(4)	1667	R	—	—	—	—	70.0	64.1	74.5	—	—	—	—
08JHAT2103	(4)	1720	L	—	35.0	26.8	47.7	—	—	—	—	35.5	32.5	10.8
08JHAT2103	(4)	1721	L	—	—	—	44.5	—	—	—	—	—	31.1	—
08JHAT2103	(4)	1722	L	—	34.6	19.5	33.0	66.7	64.6	71.0	29.9	25.9	29.2	5.0
08JHAT2104	(4)	1760	R	—	—	—	—	49.2	46.2	—	—	—	—	—
08JHDT1204	(4)	801	R	—	—	—	(28.3)	—	—	—	—	32.2	33.6	—
08JHDT1801	(4)	309	L	—	—	—	—	—	—	—	36.6	—	—	—
Modern Wild Boar		No. 99	L	53.5	36.8	41.4	74.0	122.7	110.5	118.6	52.3	45.8	47.0	13.0

MS 9, 9a, 11–16c, 21: See von den Driesch (1976:58-59)

A2.4 Mandible with Teeth (continued)

c. Teeth in Mandible (dP4 and P4)

LOC	LAY	BON	SYM	dP4-L	dP4-B	P4-L	P4-B
08JHDT1801	(7)	306	R	15.4	8.2	—	—
08JHAT2003	(6)	1364	L	—	—	12.0	8.6
08JHAT2003	(6)	1365	L	—	—	12.8	9.6
08JHAT2004	(6)	1455	L	—	—	11.6	8.4
08JHAT2004	(6)	1456	L	—	—	12.4	8.8
08JHAT2004	(6)	1459	R	16.7	8.7	—	—
08JHAT2004	(6)	1461	R	16.0	8.2	—	—
08JHAT2005	(6)	1612	R	16.3	8.0	—	—
08JHAT2005	(6)	1613	R	17.7	8.1	—	—
08JHAT2102	(6)	1665	L	16.3	8.1	—	—
08JHAT2102	(6)	281	L	—	—	12.0	8.2
			R	—	—	11.9	8.2
08JHAT2505	(5)	1816	L	—	—	13.9	9.6
08JHDT1005	(5)	734	L	17.0	8.6	—	—
08JHDT1304	(5)	966	L	—	—	(12.0)	—
08JHDT1802	(5)	122	R	—	—	12.2	8.9
08JHAT2002	(5)	1246	L	—	—	11.5	7.7
			R	—	—	11.6	7.9
08JHAT2005	(4)	280	L	—	—	12.2	8.9
			R	—	—	12.3	8.9
08JHAT2102	(4)	1664	L	—	—	10.4	8.1
08JHAT2102	(4)	1666	L	18.3	9.2	—	—
08JHAT2102	(4)	1667	R	—	—	—	(8.2)
08JHAT2103	(4)	1720	L	—	—	11.4	8.1
08JHAT2103	(4)	1722	L	16.8	8.8	—	—
08JHAT2103	(4)	1725	R	18.5	8.6	—	—
08JHAT2103	(4)	1726	R	(18.0)	—	—	—
08JHAT2104	(4)	1760	R	17.9	8.2	—	—
08JHDT1801	(4)	308	L	16.6	8.3	—	—
Modern Wild Boar		No. 99	L	—	—	12.6	10.3

dP4 = deciduous fourth premolar.
P4 = fourth premolar.
L = length.
B = breadth.
See details in von den Driesch (1976:58-59).

A2.4 Mandible with Teeth (continued)

d. Teeth in Mandible (M1–M3)

LOC	LAY	BON	SYM	M1-L	M1-B	M2-L	M2-B	M3-L	M3-B
08JHDT1801	(7)	305	L	12.9	10.1	—	—	—	—
08JHDT1801	(7)	306	R	13.4	10.5	—	—	—	—
08JHDT1802	(7)	541	R	15.1	11.2	—	—	—	—
08JHAT2002	(6)	1244	L	—	—	—	—	(28.4)	13.0
08JHAT2003	(6)	1364	L	13.0	9.9	17.3	13.1	28.8	13.8
08JHAT2003	(6)	1365	L	(16.5)	12.0	—	—	—	—
08JHAT2003	(6)	1366	R	13.4	10.6	17.2	12.9	—	—
08JHAT2004	(6)	1455	L	14.0	10.1	17.2	12.5	—	—
08JHAT2004	(6)	1456	L	14.2	9.7	18.4	12.5	—	—
08JHAT2004	(6)	1457	R	16.3	11.1	—	—	—	—
08JHAT2004	(6)	1458	R	14.4	10.5	—	—	—	—
08JHAT2004	(6)	1459	R	14.3	10.4	—	—	—	—
08JHAT2004	(6)	1460	R	15.0	10.8	18.0	13.3	—	—
08JHAT2004	(6)	1461	R	13.4	10.4	18.2	12.3	—	—
08JHAT2005	(6)	1611	L	15.3	10.2	—	—	—	—
08JHAT2102	(6)	1665	L	15.0	10.2	—	—	—	—
08JHAT2102	(6)	121	L	—	—	20.6	14.3	—	—
08JHAT2102	(6)	281	L	14.3	10.4	17.5	12.9	—	—
			R	14.0	10.2	18.2	12.9	—	—
08JHAT2106	(6)	1776	L	15.0	10.2	—	—	—	—
08JHAT2505	(6)	1817	L	—	—	18.7	12.4	—	—
08JHAT2505	(6)	1818	R	—	—	17.8	12.5	—	—
08JHDT1305	(6)	1081	L	14.8	10.5	—	—	—	—
08JHDT1802	(6)	540	L	—	—	—	—	30.6	15.1
08JHAT2002	(5)	1242	L	13.5	10.4	(20.0)	12.9	—	—
08JHAT2003	(5)	1363	L	(13.0)	10.1	17.6	12.7	29.9	14.2
08JHAT2505	(5)	1816	L	15.0	11.0	18.7	14.7	(34.4)	17.0
08JHDT1105	(5)	419	L	15.6	11.2	17.5	13.4	—	—
08JHDT1105	(5)	421	R	—	—	17.5	—	—	—
08JHDT1304	(5)	966	L	14.9	11.1	18.6	14.0	—	—
08JHDT1802	(5)	539	L	—	—	17.7	13.1	(25.8)	14.3
08JHDT1802	(5)	122	R	13.1	9.7	17.5	13.0	(25.7)	14.5
08JHAT2001	(4)	1201	R	(15.5)	(10.3)	—	—	—	—
08JHAT2001	(4)	1202	L	15.5	10.6	18.8	13.3	—	—
08JHAT2002	(4)	1246	L	13.4	9.7	17.0	12.3	27.3	13.3
			R	13.3	9.6	16.9	12.1	26.0	13.2
08JHAT2005	(4)	280	L	13.8	10.4	19.3	13.4	—	—
			R	14.1	10.5	18.8	13.7	—	—
08JHAT2102	(4)	1663	L	13.9	10.5	17.2	12.8	—	—
08JHAT2102	(4)	1664	L	13.4	10.2	16.4	12.0	—	—
08JHAT2102	(4)	1667	R	14.1	9.7	15.8	11.0	—	—
08JHAT2103	(4)	1720	L	13.8	9.7	15.9	12.3	—	—
08JHAT2103	(4)	1722	L	16.0	10.7	—	—	—	—
08JHAT2103	(4)	1723	L	16.0	10.2	—	—	—	—
08JHAT2103	(4)	1724	R	—	—	—	—	(28.0)	14.7
08JHDT1105	(4)	420	R	—	—	18.3	—	—	—
08JHDT1204	(4)	801	R	14.8	11.0	19.8	15.0	—	—
08JHDT1801	(4)	307	L	13.8	10.5	(18.9)	(12.8)	—	—
08JHDT1801	(4)	308	L	(13.1)	(9.3)	—	—	—	—
08JHDT1801	(4)	309	L	14.2	10.1	17.7	12.8	—	—
Modern Wild Boar		No. 99	L	15.4	12.2	19.5	14.7	38.9	17.2

See Table A1.3, see also von den Driesch (1976:58-59).

A2.5 Loose Teeth

LOC	LAY	BON	ELM	SYM	TTY	L	B
08JHAT2002	(8)	1260	TTHUP	L	M1	16.1	12.6
08JHAT2002	(8)	1262	TTHUP	L	M3	30.1	18.4
08JHDT1104	(8)	616	TTHUP	L	M1	16.4	14.9
08JHDT1104	(8)	2061	TTHUP	L	M1	15.2	14.1
08JHDT1104	(8)	2062	TTHUP	R	dP4	12.8	11.9
08JHDT1104	(8)	2063	TTHUP	R	M1	—	14.2
08JHDT1104	(8)	2065	TTHLW	R	M2	17.6	—
08JHDT1802	(8)	544	TTHUP	L	M1	15.3	13.8
08JHAT2002	(7)	1261	TTHUP	L	M2	17.7	15.3
08JHAT2003	(7)	1375	TTHUP	L	M1	16.5	15.0
08JHAT1901	(6)	1965	TTHUP	L	M2	20.6	16.7
08JHAT2005	(6)	1617	TTHLW	L	M1	14.3	9.4
08JHDT1304	(6)	983	TTHUP	L	M1	14.4	13.8
08JHDT1305	(6)	1089	TTHUP	L	M2	19.6	15.7
08JHDT1305	(6)	1092	TTHUP	L	P4	12.5	12.7
08JHDT1305	(6)	1093	TTHUP	L	M1	15.5	14.6
08JHDT1305	(6)	1094	TTHUP	L	M2	18.9	16.6
08JHDT1305	(6)	1095	TTHUP	L	M3	25.9	17.0
08JHDT1305	(6)	1096	TTHUP	R	M1	15.4	14.2
08JHDT1802	(6)	543	TTHLW	L	M2	19.1	13.4
08JHDT1004	(5)	700	TTHLW	L	M1	15.0	10.7
08JHDT1005	(5)	749	TTHLW	L	M1	15.5	10.5
08JHDT1005	(5)	750	TTHLW	R	M1	15.0	10.3
08JHDT1005	(5)	751	TTHLW	L	M2	18.2	12.0
08JHDT1005	(5)	752	TTHLW	L	dP4	16.7	8.5
08JHDT1104	(5)	2064	TTHUP	R	M1	14.2	—
08JHDT1105	(5)	429	TTHLW	L	M3	27.3	14.1
08JHAT1901	(4)	1966	TTHUP	R	M3	—	16.5
08JHAT1901	(4)	1967	TTHLW	L	M3	34.5	13.8
08JHAT2001	(4)	1206	TTHLW	L	M3	31.4	15.3
08JHAT2102	(4)	1669	TTHUP	R	M2	17.0	14.8
08JHAT2103	(4)	1733	TTHUP	L	M2	20.7	14.0
08JHAT2103	(4)	1734	TTHLW	R	M2	17.8	12.2
08JHAT2105	(4)	1770	TTHLW	L	dP4	16.6	8.3
08JHCT0304	(4)	2037	TTHUP	L	M2	20.1	16.7
08JHDT1003	(4)	679	TTHUP	L	M1	14.9	13.4
08JHDT1801	(4)	316	TTHUP	L	M3	25.8	17.5
08JHDT1801	(4)	317	TTHLW	R	M3	—	(14.5)
08JHDT1802	(4)	545	TTHLW	R	M2	18.0	(12.2)

See Table A1.5, see also von den Driesch (1976:58-59).

A2.6 Atlas

LOC	LAY	BON	SYM	GB	GL	BFcr	BFcd	GLF	H
08JHDT1801	(9)	332	M	—	—	52.6	(49.8)	40.2	(42.1)
08JHAT2003	(7)	1377	M	—	—	—	—	34.0	—
08JHAT1901	(6)	1974	M	—	—	(48.0)	—	—	35.2
08JHAT2003	(6)	1376	M	—	—	52.6	48.5	37.0	(40.0)
08JHAT2004	(6)	1470	M	55.6	28.8	43.1	43.7	28.4	34.0
08JHAT2102	(6)	1675	M	—	—	51.7	45.5	37.6	38.3
08JHDT1204	(6)	808	M	(66.1)	35.9	45.1	44.0	34.0	32.6
08JHDT1304	(6)	988	M	—	—	(47.0)	—	33.2	39.5
08JHDT1105	(5)	433	M	(74.1)	43.1	53.5	46.8	42.2	41.6
08JHDT1304	(5)	987	M	—	—	48.4	42.4	35.9	41.7
08JHDT1801	(5)	331	M	(77.5)	(44.6)	55.1	47.8	39.2	45.2
08JHAT2002	(4)	1265	M	—	—	44.8	46.0	30.6	33.4
08JHAT2004	(4)	1469	M	—	—	49.5	40.8	31.8	39.6
08JHAT2102	(4)	1673	M	—	—	45.8	—	(31.5)	36.5
08JHAT2102	(4)	1674	M	(69.6)	—	53.5	46.5	34.5	37.3
08JHAT2103	(4)	1728	M	(65.0)	—	48.9	47.1	34.4	36.2
08JHAT2103	(4)	1736	M	—	—	(47.4)	(46.1)	(33.3)	35.9
08JHDT1003	(4)	681	M	—	—	—	—	(33.0)	—
08JHDT1005	(4)	756	M	—	—	—	—	34.6	—
08JHDT1801	(4)	333	M	—	—	(48.7)	(46.5)	32.4	(37.3)
08JHDT1105	G1(2)	435	M	81.2	41.4	51.4	46.8	38.2	42.8
Modern Wild Boar		No. 99	M	94.4	52.6	62.8	56.0	48.4	51.0

See Table A1.6, see also von den Driesch (1976:67).

A2.7 Axis

LOC	LAY	BON	SYM	LCDe	LAPa	BFcr	BPacd	BPtr	SBV	BFcd	H
08JHAT2605	(7)	1901	M	28.2	25.8	42.2	29.2	(31.5)	22.5	22.2	46.3
08JHAT2003	(6)	1378	M	26.4	21.2	39.8	24.6	—	22.9	(22.5)	(41.7)
08JHAT2003	(5)	1379	M	—	—	—	—	—	—	(26.5)	—
08JHAT2505	(5)	1828	M	32.2	(23.0)	41.2	29.4	—	26.7	(26.6)	(52.0)
08JHDT1304	(5)	989	M	29.8	29.9	40.4	28.7	28.8	24.5	24.4	50.6
Modern Wild Boar		No. 99	M	41.5	43.1	52.8	38.5	40.9	30.8	(30.8)	74.8

LCDe = (greatest) length in the region of the corpus including the dens.
LAPa = (greatest) length of the arch including the caudal articular process.
BFcr = (greatest) breadth of the cranial articular surface.
BPacd = (greatest) breadth across the caudal articular process.
BPtr = (greatest) breadth across the transverse process.
SBV = smallest breadth of the vertebra.
BFcd = (greatest) breadth of the caudal articular surface.
H = (greatest) height.
See details in von den Driesch (1976:68-69).

A2.8 Scapula

LOC	LAY	BON	SYM	DHA	SLC	GLP	LG	BG
08JHDT1005	(8)	2033	L	—	19.3	—	—	—
08JHDT1204	(8)	822	R	—	18.0	—	—	—
08JHDT1304	(8)	996	L	—	13.6	—	—	—
08JHAT2002	(7)	1293	L	—	16.7	(26.1)	(22.8)	(18.0)
08JHAT2106	(7)	1779	L	—	11.1	—	—	—
08JHDT1005	(7)	765	L	—	19.1	29.4	24.6	20.1
08JHAT1901	(6)	1988	L	—	—	32.9	24.2	23.3
08JHAT1901	(6)	1992	L	—	18.5	—	—	—
08JHAT2003	(6)	1393	L	—	20.5	30.1	24.5	21.9
08JHAT2004	(6)	1497	L	—	18.5	—	—	—
08JHAT2004	(6)	1498	R	—	20.9	34.8	28.9	23.6
08JHAT2106	(6)	1780	R	—	17.0	(26.8)	(23.9)	(20.1)
08JHAT2505	(6)	1840	L	—	21.1	30.8	23.1	21.5
08JHDT1104	(6)	625	R	—	21.4	(29.0)	(24.5)	19.1
08JHDT1105	(6)	452	R	—	32.0	(40.2)	35.0	(30.0)
08JHDT1204	(6)	821	R	—	18.8	—	—	—
08JHDT1304	(6)	995	L	—	21.0	31.2	26.0	21.8
08JHDT1801	(6)	339	L	—	23.4	31.0	25.3	19.8
08JHDT1802	(6)	554	L	—	16.0	—	—	—
08JHAT1901	(5)	1989	L	—	—	—	23.6	21.2
08JHAT1901	(5)	1991	L	—	16.6	—	—	—
08JHAT2003	(5)	1394	R	—	17.0	—	—	—
08JHAT2505	(5)	1841	R	—	10.4	—	—	—
08JHDT1105	(5)	448	L	—	19.5	(31.0)	—	(20.6)
08JHDT1105	(5)	449	L	—	16.2	—	—	—
08JHDT1204	(5)	819	R	—	21.0	31.7	22.8	21.8
08JHDT1204	(5)	820	R	—	24.6	—	—	(21.7)
08JHDT1304	(5)	997	R	—	21.3	—	—	—
08JHDT1304	(5)	998	R	—	20.7	30.0	25.8	19.9
08JHDT1304	(5)	999	L	—	25.3	38.4	31.5	28.4
08JHDT1305	(5)	1108	R	—	15.0	(21.5)	—	(13.2)
08JHAT2102	(4)	1681	L	(161.1)	20.8	29.8	25.7	21.3
08JHAT2102	(4)	1682	R	—	20.8	(29.7)	—	(22.3)
08JHAT2103	(4)	1738	L	(149.0)	20.6	—	—	(18.5)
08JHDT1005	(4)	766	R	—	19.7	31.1	25.1	20.5
08JHDT1104	(4)	623	L	—	19.4	(26.2)	(21.3)	(16.0)
08JHDT1105	(4)	451	L	—	20.4	32.4	28.0	(18.8)
08JHDT1204	(4)	818	L	—	21.4	(31.4)	(22.5)	(20.7)
08JHDT1305	(4)	1105	L	—	17.1	26.5	23.5	18.2
08JHDT1801	(4)	340	L	—	18.9	27.4	22.0	18.8
08JHDT1205	(1)	904	L	—	31.8	45.9	37.0	32.4
Modern Wild Boar		No. 99	L	203.3	28.7	38.4	32.2	24.6

See Table A1.7, see also von den Driesch (1976:74-75).

A2.9 Humerus

LOC	LAY	BON	SYM	GL w/o PE	GL w/o BE	Bp	SD	Bd	BT	GLFd	LFd	Dd
08JHDT1802	(9)	557	L	—	—	—	—	36.3	27.0	26.2	17.5	34.6
08JHDT1104	(8)	630	L	—	92.4	—	10.8	—	—	—	—	—
08JHDT1204	(8)	827	L	—	—	—	—	34.3	25.6	22.4	15.9	29.5
08JHDT1204	(8)	828	R	—	—	—	13.6	34.1	25.8	24.1	15.5	30.0
08JHDT1304	(8)	1003	L	—	—	—	—	32.5	23.9	22.9	15.0	30.8
08JHDT1104	(8)	629	L	—	—	—	—	—	40.4	—	—	—
08JHAT0303	(7)	2049	L	—	—	—	—	34.5	27.8	25.0	15.8	33.1
08JHAT1901	(7)	1995	L	137.8	—	—	13.5	36.1	28.3	23.2	16.6	32.4
08JHAT2005	(7)	1625	L	—	—	—	—	(27.6)	22.7	21.8	15.0	(28.7)
08JHAT2106	(7)	1783	R	—	—	—	—	40.0	31.7	27.9	18.0	34.9
08JHAT2505	(7)	1843	L	—	—	—	10.8	—	—	—	—	—
08JHAT2505	(7)	1850	R	—	90.4	—	9.9	—	—	—	—	—
08JHAT1901	(6)	1996	L	—	—	—	—	35.5	26.7	23.5	16.4	—
08JHAT1901	(6)	1997	L	—	—	—	—	31.0	23.9	22.3	16.0	—
08JHAT2003	(6)	1397	R	—	—	—	—	(31.5)	(25.7)	(22.7)	14.8	—
08JHAT2003	(6)	1398	R	—	—	—	12.6	—	—	—	—	—
08JHAT2003	(6)	1399	R	—	121.7	—	13.3	—	—	—	—	—
08JHAT2004	(6)	1501	L	—	—	—	—	(34.0)	—	—	—	—
08JHAT2004	(6)	1502	R	—	115.0	—	11.2	—	—	—	—	—
08JHAT2102	(6)	215	L	—	—	—	15.3	37.5	(28.9)	(26.0)	18.0	(34.6)
08JHAT2605	(6)	1912	R	—	—	—	—	(36.3)	31.3	25.4	15.1	(35.0)
08JHDT1105	(6)	457	L	—	—	—	11.9	—	—	—	—	—
08JHDT1205	(6)	908	L	—	115.7	—	14.9	—	—	—	15.9	—
08JHDT1205	(6)	909	R	—	—	—	14.2	—	—	—	—	—
08JHDT1304	(6)	1001	L	—	—	42.7	—	—	—	—	—	—
08JHDT1304	(6)	1002	R	—	—	—	12.8	—	—	—	—	—
08JHDT1305	(6)	1113	R	—	—	—	13.0	—	—	—	—	—
08JHDT1305	(6)	1115	R	—	—	—	14.1	—	—	—	—	—
08JHDT1802	(6)	556	L	131.5	—	—	13.2	33.0	25.4	21.7	14.8	30.6
08JHDT1802	(6)	558	L	—	108.8	—	11.0	—	—	—	—	—
08JHDT1802	(6)	559	L	—	(94.0)	—	9.8	—	—	—	—	—
08JHAT2002	(5)	1298	L	—	—	—	—	33.1	25.7	22.4	14.9	30.7
08JHDT1305	(5)	1114	R	—	94.0	—	10.1	—	—	—	—	—
08JHDT1305	(5)	216	L	—	—	—	20.8	49.4	36.4	32.5	20.7	47.4
08JHAT2102	(4)	1687	L	—	—	—	12.3	—	—	—	—	—
08JHAT2505	(4)	1844	L	—	—	—	12.0	33.4	24.5	24.1	16.6	31.5
08JHDT1105	(4)	456	L	—	—	—	—	36.8	28.3	26.9	17.5	34.6
08JHDT1105	(4)	459	R	—	—	—	—	(32.4)	25.6	—	17.9	—
08JHDT1204	(4)	829	R	—	—	—	10.8	—	—	—	—	—
08JHDT1205	(4)	907	L	—	—	—	—	32.5	24.5	20.6	15.5	29.8
08JHDT1802	(4)	1605	L	—	—	—	—	32.7	24.9	23.3	14.8	—
08JHDT1204	(3)	826	L	—	—	—	—	36.9	26.7	24.5	17.3	31.1
08JHDT1105	G1(2)	455	L	—	—	—	—	36.6	27.1	25.9	16.9	34.7
Modern Wild Boar		No. 99	L	188.3	—	49.8	20.2	46.6	32.7	30.4	20.4	44.0

GL w/o PE = greatest length without proximal epiphysis.
GL w/o BE = greatest length without both epiphyses.
Bp = (greatest) breadth of the proximal end.
SD = smallest breadth of the diaphysis.
See other abbreviations in Table A1.8, see also von den Driesch (1976:76-77)

A2.10 Radius

LOC	LAY	BON	SYM	GL w/o DE	GL w/o BE	Bp	Dp	SD	Bd	Dd
08JHDT1105	(7)	466	L	—	—	21.7	15.5	—	—	—
08JHDT1802	(7)	568	L	—	—	25.8	17.8	15.4	—	—
08JHAT1901	(6)	2000	R	—	—	19.8	14.8	12.0	—	—
08JHAT2002	(6)	1300	R	—	—	—	—	—	(22.0)	(18.1)
08JHAT2003	(6)	1401	R	—	68.9	(18.4)	(12.8)	10.6	(22.9)	(16.3)
08JHAT2003	(6)	1402	L	—	77.8	(21.1)	(14.2)	12.5	(24.9)	(19.4)
08JHAT2004	(6)	1507	R	—	—	24.6	16.4	—	—	—
08JHAT2106	(6)	1784	L	—	—	23.5	16.4	13.0	—	—
08JHDT1004	(6)	707	L	—	67.6	(20.0)	(12.9)	12.6	(21.7)	(16.0)
08JHDT1204	(6)	832	R	—	79.6	(22.8)	(14.3)	13.6	(29.0)	(20.6)
08JHDT1204	(6)	834	R	105.1	—	27.1	17.9	16.5	(30.0)	(20.8)
08JHDT1304	(6)	1005	L	—	—	—	—	—	(24.9)	(19.9)
08JHDT1304	(6)	1006	R	102.8	—	24.8	16.8	14.0	(28.7)	(21.4)
08JHDT1304	(6)	1007	R	—	—	22.5	15.1	13.3	—	—
08JHDT1802	(6)	569	R	—	—	21.0	15.4	13.2	—	—
08JHDT1004	(5)	705	L	—	—	22.3	15.9	14.6	—	—
08JHDT1004	(5)	706	L	—	—	—	—	14.0	—	—
08JHDT1105	(5)	467	R	—	—	24.0	17.8	—	—	—
08JHDT1204	(5)	833	R	—	—	24.3	17.2	13.4	—	—
08JHDT1305	(5)	1120	L	—	—	—	—	—	(23.4)	(18.3)
08JHDT1305	(5)	1121	R	—	—	—	—	—	(16.0)	(12.6)
08JHAT2001	(4)	1208	R	—	—	23.6	15.8	14.4	—	—
08JHAT2102	(4)	1690	R	—	—	22.4	15.4	—	—	—
08JHAT2102	(4)	1691	L	—	48.3	(12.0)	(9.1)	6.9	(13.2)	(11.2)
08JHAT2103	(4)	1740	L	—	—	22.8	16.0	—	—	—
08JHDT1205	(4)	910	L	104.5	—	23.7	15.3	13.7	—	—
08JHDT1205	(4)	911	L	—	—	24.9	(15.9)	13.2	—	—
08JHDT1305	(4)	1119	L	—	88.4	(24.0)	(15.4)	13.9	(24.0)	(18.3)
08JHDT1801	(4)	347	R	—	—	24.2	16.5	12.9	—	—
08JHDT1205	(1)	912	L	—	—	23.4	17.3	—	—	—
08JHDT1205	(1)	913	R	—	—	21.8	14.3	—	—	—
Modern Wild Boar		No. 99	L	—	—	31.3	22.3	17.5	36.9	27.8

GL w/o DE = greatest length without distal epiphysis.
GL w/o BE = greatest length without both epiphyses.
Bp = (greatest) breadth of the proximal end.
SD = smallest breadth of the diaphysis.
See other abbreviations in Table A1.9, see also von den Driesch (1976:79-81).

A2.11 Ulna

LOC	LAY	BON	SYM	GL w/o BE	DPA	SDO	BPC
08JHAT2106	(7)	1788	L	83.4	22.2	18.5	14.9
08JHAT2106	(7)	1790	R	—	27.0	21.0	18.0
08JHAT1901	(6)	2002	R	—	—	—	(16.0)
08JHAT1901	(6)	2003	R	—	23.1	—	(15.2)
08JHAT2003	(6)	1403	L	109.3	28.3	20.3	17.9
08JHAT2106	(6)	1787	L	—	27.8	—	16.9
08JHAT2106	(6)	1789	R	—	29.0	22.2	19.7
08JHDT1103	(6)	667	L	91.9	24.3	19.4	16.4
08JHDT1204	(6)	835	L	123.3	30.2	20.5	17.6
08JHDT1204	(6)	836	L	—	—	—	(16.0)
08JHDT1204	(6)	837	R	—	26.9	—	17.4
08JHDT1304	(6)	1011	L	—	26.9	19.7	16.4
08JHDT1304	(6)	1012	R	—	30.5	—	18.5
08JHDT1304	(6)	1013	R	—	30.4	20.9	17.5
08JHDT1304	(6)	1015	R	—	31.3	—	18.1
08JHDT1802	(6)	572	R	—	31.9	23.9	19.2
08JHAT2605	(5)	1914	L	—	(30.6)	—	18.1
08JHDT1004	(5)	711	L	—	29.7	23.7	16.9
08JHDT1105	(5)	468	L	—	30.7	22.6	18.0
08JHDT1105	(5)	469	L	—	25.2	18.6	15.3
08JHDT1105	(5)	471	R	—	26.7	21.4	17.0
08JHDT1105	(5)	472	R	—	27.4	22.2	17.6
08JHDT1105	(5)	473	R	—	19.8	15.9	13.7
08JHDT1205	(5)	916	L	—	(31.6)	23.1	(17.1)
08JHDT1205	(5)	917	L	—	43.8	—	25.8
08JHDT1304	(5)	1014	R	—	24.9	19.1	17.6
08JHDT1304	(5)	1016	R	—	31.0	—	17.8
08JHDT1304	(5)	1017	R	—	28.7	22.3	—
08JHDT1305	(5)	1125	R	—	(24.0)	—	(15.5)
08JHDT1802	(5)	570	L	—	33.9	23.3	(19.0)
08JHDT1802	(5)	573	R	—	26.6	20.2	16.1
08JHAT2505	(4)	1851	R	—	32.9	24.9	18.7
08JHDT1205	(4)	915	L	(112.6)	26.5	20.8	(17.6)
08JHDT1305	(4)	1123	L	—	25.2	—	16.2
08JHDT1305	(4)	1124	L	—	28.1	—	—
08JHDT1801	(4)	350	L	—	29.6	21.0	18.8
08JHDT1801	(4)	351	L	—	29.2	21.8	16.8
08JHDT1801	(4)	352	L	—	25.1	—	(13.9)
08JHDT1801	(4)	353	R	—	—	—	16.5
08JHDT1801	(4)	354	R	—	26.6	—	15.1
08JHDT1802	(4)	571	L	—	(25.5)	—	—
Modern Wild Boar		No. 99	L	—	41.1	33.1	23.7

GL w/o BE = greatest length without both epiphyses.
See other abbreviations in Table A1.10, see also von den Driesch (1976:79-81).

A2.12 Pelvis

LOC	LAY	BON	ELM	SYM	LAR	SH	SB	LFo
08JHAT2002	(8)	1304	PELV	L	27.7	—	—	—
08JHAT2002	(7)	1303	PELV	L	27.4	20.6	9.8	—
08JHDT1304	(7)	1021	PELV	R	26.9	19.0	—	—
08JHDT1304	(7)	1023	PELV	R	27.2	—	—	—
08JHAT2002	(6)	1305	PELV	R	26.6	17.5	8.8	—
08JHAT2003	(6)	1405	PELV	L	25.8	18.1	10.1	—
08JHAT2003	(6)	1406	PELV	L	—	16.9	10.2	—
08JHAT2004	(6)	1517	PELV	R	29.7	17.3	9.8	—
08JHAT2106	(6)	1791	PELV	R	(27.0)	—	—	—
08JHAT2605	(6)	1915	ILIUM	R	—	(18.0)	(11.1)	—
08JHDT1004	(6)	712	PELV	R	27.1	18.3	9.8	—
08JHDT1103	(6)	668	PELV	R	28.2	19.7	10.8	—
08JHDT1105	(6)	479	ILIUM	L	—	15.8	9.1	—
08JHDT1304	(6)	1022	PELV	R	31.1	20.0	10.1	—
08JHDT1802	(6)	577	ILIUM	R	—	19.0	9.6	—
08JHAT2002	(5)	261	PELV	L	(28.3)	(16.5)	10.7	—
08JHAT2004	(5)	1516	ILIUM	L	—	17.4	8.5	—
08JHDT1005	(5)	770	IL/IS	R	—	18.7	11.4	—
08JHDT1005	(5)	771	PELV	L	26.7	—	—	—
08JHDT1005	(5)	772	ILIUM	L	—	19.2	11.3	—
08JHDT1104	(5)	638	IL/IS	L	(29.4)	—	—	—
08JHDT1105	(5)	478	PELV	R	27.0	—	—	—
08JHDT1205	(5)	920	PELV	R	27.6	19.3	9.2	—
08JHDT1304	(5)	1019	PELV	L	28.1	20.1	10.4	—
08JHDT1305	(5)	1130	PELV	R	28.8	—	—	—
08JHDT1305	(5)	1131	PELV	R	31.2	18.8	10.3	—
08JHDT1105	(4)	477	PELV	L	(29.8)	—	—	—
08JHDT1204	(4)	260	PELV	L	29.6	—	—	(36.2)
08JHDT1305	(4)	1129	PELV	R	28.6	21.2	12.4	—
08JHDT1801	(4)	355	ILIUM	L	—	19.5	10.5	—
08JHDT1801	(4)	356	PELV	L	27.3	—	—	—
08JHDT1801	(4)	357	ILIUM	R	—	20.6	8.5	—
08JHDT1801	(4)	402	ILIUM	L	—	16.2	10.2	—
08JHDT1205	(1)	921	PELV	R	30.4	—	—	—
Modern Wild Boar		No. 99	PELV	L	32.7	29.5	14.4	48.0

LFo = inner length of the foramen obturatum.
See other abbreviations in Table A1.11, see also von den Driesch (1976:82-83).

A2.13 Sacrum

LOC	LAY	BON	SYM	GL	PL	GB	BFcr	HFcr
08JHDT1305	(5)	1132	M	—	—	—	(23.0)	(8.3)
Modern Wild Boar		No. 99	M	112.0	100.0	93.6	36.7	17.5

GL = greatest length of the ventral side.
PL = physiological length.
GB = greatest breadth (across the wings).
BFcr = (greatest) breadth of the cranial articular surface.
HFcr = (greatest) height of the cranial articular surface.
See details in von den Driesch (1976:71).

A2.14 Femur

LOC	LAY	BON	SYM	GL w/o BE	Bp	DC	SD	Bd	Btpat
08JHDT1803	(8)	2045	L	—	—	22.0	—	—	—
08JHDT1005	(7)	775	L	—	—	—	11.4	—	—
08JHDT1005	(7)	776	R	142.7	—	—	14.8	—	—
08JHAT2002	(6)	1315	R	134.8	—	—	12.0	—	—
08JHAT2003	(6)	1411	L	(51.0)	—	—	6.3	—	—
08JHAT2003	(6)	1412	R	(142.5)	—	—	12.8	—	—
08JHAT2004	(6)	1525	L	—	—	—	—	36.5	15.0
08JHAT2004	(6)	1526	L	—	—	—	—	38.0	16.7
08JHAT2104	(6)	1764	R	147.9	—	—	13.9	—	—
08JHAT2106	(6)	1793	L	(92.0)	—	—	10.6	—	—
08JHDT1103	(6)	1943	L	—	—	—	12.5	—	—
08JHDT1105	(6)	486	L	—	—	—	16.1	—	—
08JHDT1105	(6)	490	R	—	—	—	16.2	—	—
08JHDT1204	(6)	843	L	—	—	29.5	—	—	—
08JHDT1304	(6)	1028	R	—	—	—	15.8	—	—
08JHDT1305	(6)	1136	L	—	—	—	17.8	—	—
08JHDT1305	(6)	1139	R	—	—	—	13.2	—	—
08JHDT1802	(6)	580	R	—	(49.3)	—	—	—	—
08JHAT2002	(5)	1313	L	—	—	20.7	—	—	—
08JHAT2002	(5)	1314	L	—	—	—	—	—	(14.8)
08JHAT2002	(5)	1316	R	—	—	—	—	(43.0)	(16.5)
08JHAT2002	(5)	198	L	—	—	—	12.6	—	—
08JHAT2004	(5)	1527	R	142.6	—	—	14.8	—	—
08JHAT2505	(5)	1857	R	—	—	—	—	38.7	18.4
08JHDT1005	(5)	777	R	—	—	—	15.6	—	—
08JHDT1204	(5)	844	R	—	—	—	18.2	—	—
08JHDT1205	(5)	923	R	—	—	—	15.5	—	—
08JHDT1304	(5)	1027	L	—	—	—	15.4	—	—
08JHDT1304	(5)	1031	R	—	—	—	—	41.0	16.1
08JHDT1305	(5)	1138	R	—	—	—	15.2	—	—
08JHDT1305	(5)	1144	L	—	—	—	8.3	—	—
08JHAT2105	(4)	1772	L	—	—	—	7.0	—	—
08JHDT1104	(4)	639	L	—	—	—	14.8	—	—
08JHDT1204	(4)	842	L	—	52.6	23.4	—	—	—
08JHDT1305	(4)	1134	L	—	—	—	—	43.7	19.6
08JHDT1305	(4)	1140	R	139.7	—	—	15.3	—	—
08JHDT1801	(4)	364	L	(88.6)	—	—	10.4	—	—
08JHDT1105	G1(2)	489	L	—	—	—	—	44.3	19.2
Modern Wild Boar		No. 99	L	—	68.7	28.7	21.0	52.1	23.5

GL w/o BE = greatest length without both epiphyses.
Bp = (greatest) breadth of the proximal end.
See other abbreviations in Table A1.12, see also von den Driesch (1976:84-85).

A2.15 Tibia

LOC	LAY	BON	SYM	GL	GL w/o PE	GL w/o BE	Bp	BFp	SD	Bd	Dd
08JHDT1802	(8)	588	R	—	—	(56.0)	—	—	6.9	—	—
08JHAT2004	(7)	1540	R	—	150.2	—	—	—	14.4	25.3	23.0
08JHDT1005	(7)	778	R	—	—	—	—	—	13.2	—	—
08JHAT2002	(6)	1322	L	—	—	—	—	—	—	25.9	23.4
08JHAT2003	(6)	1416	L	—	—	(69.0)	—	—	7.7	—	—
08JHAT2003	(6)	1417	R	—	—	—	—	—	10.2	—	—
08JHAT2003	(6)	1418	R	—	—	—	—	—	8.7	—	—
08JHAT2106	(6)	1794	L	—	—	—	—	—	14.0	—	—
08JHDT1103	(6)	669	R	—	—	99.8	—	—	12.3	—	—
08JHDT1104	(6)	648	L	—	—	—	—	—	11.3	—	—
08JHDT1105	(6)	499	R	—	—	—	48.8	48.3	—	—	—
08JHDT1304	(6)	1035	R	—	—	—	—	—	13.0	—	—
08JHDT1304	(6)	1036	R	—	—	—	—	—	14.8	—	—
08JHDT1305	(6)	1148	L	(158.5)	—	—	—	—	16.4	26.7	23.5
08JHDT1305	(6)	1149	L	—	—	—	—	—	16.0	26.9	21.2
08JHDT1305	(6)	1152	L	—	—	—	—	—	15.2	—	—
08JHDT1305	(6)	1155	R	—	—	—	—	—	12.7	—	—
08JHDT1305	(6)	1156	R	—	—	(105.0)	—	—	11.7	—	—
08JHDT1802	(6)	584	L	—	—	(95.0)	—	—	11.3	—	—
08JHAT2105	(5)	210	L	—	—	109.8	—	—	11.9	—	—
08JHDT1105	(5)	496	L	—	—	—	—	—	15.3	27.3	(21.6)
08JHDT1105	(5)	500	R	—	—	142.0	—	—	14.8	—	—
08JHDT1204	(5)	848	L	—	—	110.3	—	—	12.0	—	—
08JHDT1304	(5)	1033	L	—	—	—	—	—	15.7	—	—
08JHDT1304	(5)	1034	L	—	—	—	—	—	13.7	—	—
08JHDT1801	(5)	369	R	—	—	—	—	—	11.4	—	—
08JHDT1104	(4)	649	R	—	—	—	—	—	12.4	—	—
08JHDT1105	(4)	497	L	—	—	—	—	—	—	—	23.9
08JHDT1205	(4)	929	R	—	165.4	—	—	—	16.4	26.0	21.9
08JHDT1305	(4)	1147	L	—	—	—	—	—	18.0	26.5	24.2
08JHDT1305	(4)	1158	R	—	—	—	—	—	—	26.1	21.1
08JHDT1801	(4)	370	R	—	—	—	—	—	—	25.0	21.7
08JHDT1801	(4)	371	R	—	—	—	—	—	—	(27.5)	—
08JHDT1802	(4)	583	L	—	—	—	—	—	14.5	—	—
08JHDT1205	(1)	926	L	—	—	—	—	—	15.8	27.1	23.9
Modern Wild Boar		No. 99	L	218.0	—	—	55.3	54.5	22.0	32.8	28.6

GL = greatest length.
GL w/o PE = greatest length without proximal epiphysis.
GL w/o BE = greatest length without both epiphyses.
SD = smallest breadth of the diaphysis.
See other abbreviations in Table A1.14, see also von den Driesch (1976:86-87).

A2.16 Fibula

LOC	LAY	BON	SYM	GL w/o PE	GL w/o BE
08JHAT2004	(6)	1547	L	—	116.5
Modern Wild Boar		No. 99	L	190.5	—

GL w/o PE = greatest length without proximal epiphysis.
GL w/o BE = greatest length without both epiphyses.
See von den Driesch (1976:87).

A2.17 Calcaneus

LOC	LAY	BON	SYM	GL	GL w/o Ep	Dsustali	SLtuber
08JHAT1901	(6)	2017	L	—	67.4	15.5	—
08JHAT1901	(6)	2018	L	—	—	14.3	—
08JHAT2004	(6)	1551	R	—	61.9	14.5	—
08JHAT2002	(5)	1329	R	76.0	—	16.2	42.8
08JHDT1305	(5)	1164	R	—	60.0	16.1	—
08JHDT1104	(4)	653	L	—	64.0	14.8	—
08JHDT1204	(4)	853	R	—	63.5	14.8	—
08JHDT1205	(4)	931	R	—	55.9	12.1	—
Modern Wild Boar		No. 99	L	89.7	—	18.5	49.8

See Table A1.16, see also von den Driesch (1976:90-91)

A2.18 Astragalus

LOC	LAY	BON	SYM	GLl	GLm	Dl	Dm	Bd	BFp
08JHAT2002	(4)	1330	L	36.6	33.0	19.2	20.0	(23.4)	11.6
08JHAT2002	(7)	1331	L	(35.3)	(31.5)	19.8	19.1	20.9	(17.3)
08JHAT2003	(6)	1420	R	39.3	35.7	19.7	20.2	23.2	18.8
08JHAT2005	(7)	1639	L	—	—	—	—	—	16.9
08JHAT2005	(6)	1640	R	34.1	31.0	17.2	18.1	20.3	17.5
08JHAT2505	(4)	1869	R	43.6	40.1	22.4	25.2	25.7	20.3
08JHAT2505	(6)	1870	R	33.8	30.6	18.1	17.7	(17.7)	(16.2)
08JHDT1204	(5)	854	R	39.4	36.7	20.8	20.9	24.4	19.3
08JHDT1304	(5)	1041	L	—	33.7	(19.2)	21.2	—	17.9
08JHDT1304	(5)	1042	R	35.5	32.0	17.5	19.5	19.9	16.5
08JHDT1305	(4)	1165	L	37.9	—	19.3	20.4	—	17.8
Modern Wild Boar		No. 99	L	47.3	41.2	23.3	26.2	25.7	22.5

See Table A1.17, see also von den Driesch (1976:88-91).

A2.19 Metacarpal

a. Metacarpal III/IV

LOC	LAY	BON	ELM	SYM	GL	Bp	B	Bd
08JHAT2505	(7)	1874	MC 4	L	—	13.3	—	—
08JHAT2004	(6)	1560	MC 3	R	—	15.0	10.5	—
08JHAT2605	(6)	1925	MC 4	R	(64.0)	(12.0)	(12.2)	(14.8)
08JHDT1004	(6)	718	MC 3	L	—	—	9.5	—
08JHDT1304	(6)	1050	MC 4	L	—	14.1	11.3	—
08JHDT1801	(6)	381	MC 4	R	63.4	13.9	10.0	13.5
08JHAT2002	(5)	1334	MC 3	R	—	(13.8)	—	—
08JHDT1104	(5)	655	MC 3	L	—	15.1	12.5	—
Modern Wild Boar		No. 99	MC 3	L	80.4	18.9	14.9	17.9
Modern Wild Boar		No. 99	MC 4	L	81.3	18.4	14.4	17.8

GL = greatest length.
Bp = (greatest) breadth of the proximal end.
B = breadth in the middle of the diaphysis.
Bd = (greatest) breadth of the distal end.
See details in von den Driesch (1976:94-95)

b. Metacarpal II/V

LOC	LAY	BON	ELM	SYM	GL
08JHAT2505	(6)	1875	MC 5	R	53.4
Modern Wild Boar		No. 99	MC 2	L	61.2
Modern Wild Boar		No. 99	MC 5	L	65.6

GL = greatest length.
See details in von den Driesch (1976:95)

A2.20 Metatarsal

a. Metatarsal III/IV

LOC	LAY	BON	ELM	SYM	GL	LeP	Bp	B	Bd
08JHAT2004	(6)	1558	MT 3	R	—	—	18.3	14.5	—
08JHAT2004	(6)	1559	MT 4	R	(92.4)	90.8	19.5	16.5	20.5
08JHDT1304	(5)	1049	MT 3	L	62.8	62.1	13.0	12.7	—
08JHAT2103	(4)	1749	MT 3	R	—	—	(12.7)	11.3	—
08JHDT1105	(4)	508	MT 4	L	70.0	67.6	13.0	(11.6)	15.1
Modern Wild Boar		No. 99	MT 3	L	87.2	84.3	16.6	14.5	18.3
Modern Wild Boar		No. 99	MT 4	L	96.3	87.0	16.9	14.9	17.9

LeP = length excepting the plantar projection.
See other abbreviations in Table A2.19a, see also von den Driesch (1976:94-95).

A2.20 Metatarsal (continued)

b. Metatarsal II/V

LOC	LAY	BON	ELM	SYM	GL
08JHDT1305	(6)	1169	MT 2	L	67.8
Modern Wild Boar		No. 99	MT 2	L	57.4
Modern Wild Boar		No. 99	MT 5	L	57.7

See Table A2.19b, see also von den Driesch (1976:95)

A2.21 Metapodial

LOC	LAY	BON	ELM	SYM	B	Bd
08JHDT1801	(9)	382	MP ?	4	11.5	14.6

See Table A2.19a.

A2.22 First Phalanx

LOC	LAY	BON	SYM	Glpe	Bp	SD	Bd
08JHAT2003	(6)	1424	5	—	—	10.8	12.7
08JHAT2004	(5)	1568	5	43.6	19.4	16.7	18.2
08JHDT1305	(4)	1173	5	—	—	12.2	13.8
Modern Wild Boar		No. 99	4	40.3	17.6	14.9	15.7

See Table A1.22, see also von den Driesch (1976:96-97).

A2.23 Second Phalanx

LOC	LAY	BON	SYM	GL	Bp	SD	Bd	Dd
08JHDT1801	(9)	387	5	22.5	15.8	12.8	12.2	12.3
08JHDT1305	(4)	1174	5	20.4	13.8	11.6	11.1	(10.6)
Modern Wild Boar		No. 99	4	26.3	16.6	13.9	13.7	14.0

See Table A1.23, see also von den Driesch (1976:98-99).

Table A3 Measurements of Dog

A3.1 Mandible with Teeth

a. Mandible (MS 1–9)

LOC	LAY	BON	SYM	MS 1	MS 2	MS 3	MS 4	MS 5	MS 6	MS 7	MS 8	MS 9
08JHAT2002	(5)	1240	L	—	—	—	—	—	—	68.5	65.0	—
08JHAT2102	(4)	1661	L	—	—	(107.4)	—	94.2	—	68.2	64.8	59.3
08JHAT2102	(4)	1662	R	96.4	96.0	92.5	82.3	79.4	82.8	58.6	58.5	55.3
Modern Rural Dog		No. 46	L	134.1	135.1	129.8	120.6	112.9	118.8	75.1	69.3	63.4

MS 1–9: See von den Driesch (1976:60-61).

A3.1 Mandible with Teeth (continued)

b. Mandible (MS 10–12, 14, 17–20)

LOC	LAY	BON	SYM	MS 10	MS 11	MS 12	MS 14	MS 17	MS 18	MS 19	MS 20
08JHAT2002	(5)	1240	L	33.0	—	—	17.9	10.3	—	20.0	16.6
08JHAT2102	(4)	1661	L	32.5	32.4	27.3	17.6	9.8	41.2	18.6	16.3
08JHAT2102	(4)	1662	R	29.1	32.6	27.0	17.4	8.9	34.2	16.0	13.9
Modern Rural Dog		No. 46	L	33.1	38.0	32.0	18.6	11.3	54.3	22.3	19.0

MS 10–12, 14, 17–20: See von den Driesch (1976:60-61).

c. Teeth in Mandible (M1–M3)

LOC	LAY	BON	SYM	M1-L	M1-B	M2-L	M2-B	M3-L	M3-B
08JHAT2002	(5)	1240	L	19.9	8.1	8.2	6.6	—	—
08JHAT2102	(4)	1661	L	18.7	7.3	8.6	6.5	—	—
08JHAT2102	(4)	1662	R	18.1	7.5	7.8	5.6	—	—
Modern Rural Dog		No. 46	L	20.5	8.5	7.5	6.2	4.7	4.9

See Table A1.3, see also von den Driesch (1976:60-61).

A3.2 Atlas

LOC	LAY	BON	SYM	GB	GL	BFcr	BFcd	GLF	H	LAd
08JHAT2002	(4)	1279	M	—	(23.2)	32.5	25.6	20.7	22.6	(10.1)
Modern Rural Dog		No. 46	M	79.3	39.5	39.7	31.7	30.7	27.1	17.8

LAd = length of the dorsal arch.
See other abbreviations in Table A1.6, see also von den Driesch (1976:67).

A3.3 Humerus

LOC	LAY	BON	SYM	GLC	Bp	Dp	SD	Bd	BT	GLFd	LFd	Dd
08JHAT2004	(8)	1504	R	—	—	—	7.6	17.9	12.6	11.6	7.8	14.0
08JHAT2004	(6)	1503	L	123.3	—	—	8.4	19.7	14.0	12.0	8.3	14.3
Modern Rural Dog		No. 46	L	162.7	32.9	41.7	13.5	34.6	22.2	18.6	13.1	25.2

GLC = greatest length from the humeral head.
Dp = depth of the proximal end.
See other abbreviations in Tables A1.8 and A2.9, see also von den Driesch (1976:76-77).

A3.4 Radius

LOC	LAY	BON	SYM	Bp	Dp	SD
08JHDT1003	(8)	685	R	14.0	9.4	9.7
08JHDT1104	(8)	635	L	13.9	9.3	—
Modern Rural Dog		No. 46	L	18.8	12.2	13.5

See Tables A1.9 and A2.10, see also von den Driesch (1976:79-81).

A3.5 Ulna

LOC	LAY	BON	SYM	DPA	SDO	BPC
08JHDT1003	(8)	686	R	19.0	17.3	14.2
08JHDT1003	(7)	687	L	17.5	—	(10.4)
08JHDT1802	(6)	575	L	17.4	15.2	12.0
08JHDT1104	(5)	637	L	20.8	17.9	14.0
08JHDT1305	(4)	1126	L	17.0	15.4	11.1
Modern Rural Dog		No. 46	L	24.7	22.2	17.9

See Table A1.10, see also von den Driesch (1976:79-81).

A3.6 Femur

LOC	LAY	BON	SYM	DC	SD
08JHDT1104	(8)	644	L	—	8.5
08JHDT1104	(8)	645	R	—	8.4
08JHDT1305	(6)	1145	R	—	6.9
08JHDT1305	(4)	1146	R	14.1	—
Modern Rural Dog		No. 46	L	19.6	13.6

See Table A1.12, see also von den Driesch (1976:84-85).

A3.7 Tibia

LOC	LAY	BON	SYM	SD
08JHDT1104	(8)	650	L	8.0
08JHDT1104	(8)	651	R	8.0
Modern Rural Dog		No. 46	L	12.9

See Table A2.15, see also von den Driesch (1976:86-87).

A3.8 Metacarpal

LOC	LAY	BON	ELM	SYM	GL	Bd
08JHAT2004	(7)	1557	MC 4	R	48.6	5.2
08JHDT1802	(7)	593	MC 5	L	48.1	5.4
Modern Rural Dog		No. 46	MC 4	L	66.9	10.0
Modern Rural Dog		No. 46	MC 5	L	56.6	10.1

GL = greatest length.
Bd = (greatest) breadth of the distal end.
See details in von den Driesch (1976:95).

Table A4 Measurements of Sheep & Goat

A4.1 Maxilla with Teeth

a. Maxilla (MS 21–23)

LOC	LAY	BON	ANM	SYM	MS 21	MS 22	MS 23
08JHDT1305	(6)	1078	sheep/goat	R	64.0	41.2	23.2
Modern Sample		No. 66	goat	L	64.7	41.4	24.5
Modern Sample		No. 40	sheep	L	78.5	51.5	26.6

MS 21–23: See von den Driesch (1976:31-34).

b. Teeth in Maxilla (P4–M3)

LOC	LAY	BON	ANM	SYM	P4-L	P4-B	M1-L	M1-B	M2-L	M2-B	M3-L	M3-B
08JHDT1305	(6)	1078	sheep/goat	R	8.0	9.6	10.3	12.0	15.0	12.0	16.5	11.0
Modern Sample		No. 66	goat	L	7.8	11.1	8.9	12.8	13.8	14.0	18.0	13.5
Modern Sample		No. 40	sheep	L	9.6	10.0	12.6	12.4	18.9	12.8	20.2	12.9

P = premolar
M = molar.
B = breadth.
L = length.

A4.2 Mandible with Teeth

a. Mandible (MS 1–8)

LOC	LAY	BON	ANM	SYM	MS 1	MS 2	MS 3	MS 4	MS 5	MS 6	MS 7	MS 8
08JHAT2002	(7)	1241	sheep	R	(165.0)	(170.5)	49.0	(115.0)	(123.0)	139.6	73.5	45.8
08JHAT2003	(6)	1367	sheep/goat	R	—	—	—	—	—	—	—	50.5
08JHDT1205	(5)	123	sheep/goat	L	—	—	50.9	—	(126.0)	—	71.9	50.0
08JHAT2103	(4)	1727	sheep/goat	R	—	—	—	—	—	—	76.3	53.5
Modern Sample		No. 66	goat	L	179.5	176.8	55.1	121.4	125.7	142.2	70.7	47.1

MS 1–8: See von den Driesch (1976:56-57).

b. Mandible (MS 9, 11–15c)

LOC	LAY	BON	ANM	SYM	MS 9	MS 11	MS 12	MS 13	MS 14	MS 15a	MS 15b	MS 15c
08JHAT2002	(7)	1241	sheep	R	24.3	32.8	71.0	64.3	—	42.0	23.8	17.1
08JHAT2003	(6)	1367	sheep/goat	R	—	—	—	—	—	40.7	17.0	—
08JHAT2505	(6)	1819	sheep/goat	R	31.9	31.5	—	—	—	—	25.4	15.6
08JHDT1802	(6)	542	sheep/goat	L	32.7	—	—	—	—	—	22.4	12.4
08JHDT1205	(5)	123	sheep/goat	L	21.3	—	85.7	73.7	(113.5)	41.5	23.9	16.5
08JHAT2103	(4)	1727	sheep/goat	R	21.8	—	—	—	—	37.0	20.4	14.3
08JHAT2105	(4)	1768	sheep/goat	R	—	—	—	—	—	—	—	10.3
Modern Sample		No. 66	goat	L	22.5	42.2	69.2	69.7	104.4	36.1	24.1	16.4

MS 9, 11–15c: See von den Driesch (1976:56-57).

A4.2 Mandible with Teeth (continued)

c. Teeth in Mandible (dP4 and P4)

LOC	LAY	BON	ANM	SYM	dP4-L	dP4-B	dP4-H	P4-L	P4-B
08JHAT2002	(7)	1241	sheep	R	—	—	—	11.0	7.0
08JHAT2505	(6)	1819	sheep/goat	R	18.0	7.6	9.5	—	—
08JHDT1802	(6)	542	sheep/goat	L	19.8	6.5	14.0	—	—
08JHDT1205	(5)	123	sheep/goat	L	—	—	—	9.7	6.3
08JHAT2103	(4)	1727	sheep/goat	R	—	—	—	8.6	7.3
08JHAT2105	(4)	1768	sheep/goat	R	19.3	6.5	18.8	—	—
Modern Sample		No. 66	goat	L	—	—	—	9.4	7.0

See Tables A1.4c and A2.4c, see also von den Driesch (1976:56-57).

d. Teeth in Mandible (M1–M3)

LOC	LAY	BON	ANM	SYM	M1-L	M1-B	M2-L	M2-B	M3-L	M3-B
08JHAT2002	(7)	1241	sheep	R	13.4	7.6	17.4	9.1	—	—
08JHAT2003	(6)	1367	sheep/goat	R	—	—	13.0	8.7	25.5	9.3
08JHDT1205	(5)	123	sheep/goat	L	9.3	8.7	13.8	9.0	25.5	9.9
08JHAT2103	(4)	1727	sheep/goat	R	11.7	8.3	14.9	(9.9)	25.2	9.7
Modern Sample		No. 66	goat	L	9.8	7.3	13.1	8.6	23.1	9.2

See Table A1.3, see also von den Driesch (1976:56-57).

A4.3 Loose Teeth

LOC	LAY	BON	ELM	ANM	SYM	TTY	L	B	H
08JHAT1901	(6)	1968	TTHLW	sheep/goat	R	M2	13.0	8.6	6.6
08JHAT1901	(6)	1969	TTHUP	sheep/goat	R	M2	16.7	13.1	21.8
08JHDT1105	(4)	425	TTHUP	sheep/goat	R	M3	21.5	10.8	(24+)
08JHDT1802	(4)	546	TTHLW	sheep/goat	R	M2	14.6	8.5	17.3

See Table A1.5.

A4.4 Atlas

LOC	LAY	BON	ANM	SYM	GL	BFcr	GLF	H
08JHAT1901	(6)	1972	sheep	M	46.1	42.6	39.1	29.6

See Table A1.6, see also von den Driesch (1976:67).

A4.5 Axis

LOC	LAY	BON	ANM	SYM	LCDe	LAPa	BFcr	BPacd	SBV	BFcd	H
08JHAT1901	(6)	1973	sheep	M	51.8	42.5	37.9	26.3	20.9	20.2	(41.7)

See Table A2.7, see also von den Driesch (1976:68-69).

A4.6 Scapula

LOC	LAY	BON	ANM	SYM	DHA	SLC	GLP	LG	BG
08JHDT1304	(5)	1000	sheep/goat	L	—	14.8	—	—	—
08JHAT2104	(4)	1763	sheep	R	(163.8)	19.3	32.4	26.0	22.9
Modern Sample		LM	goat	R	158.4	21.4	36.0	27.1	22.2

LM = local market.
See other abbreviations in Table A1.7, see also von den Driesch (1976:74-75).

A4.7 Humerus

LOC	LAY	BON	ANM	SYM	GL	GL w/o PE	SD	Bd	BT	GLFd	LFd	Dd
08JHAT0303	(7)	2050	sheep	L	—	—	14.0	(31.4)	(30.2)	—	14.9	—
08JHAT2005	(6)	1624	sheep	L	—	125.1	14.7	32.0	29.7	21.5	14.9	25.8
08JHAT2115	(5)	2052	sheep	R	—	—	—	31.4	28.7	19.3	15.1	26.0
08JHDT1802	(5)	565	sheep	L	—	—	11.7	—	—	—	—	—
Modern Sample		No. 81	goat	L	158.0	—	15.6	36.1	35.4	20.5	15.6	29.9

GL = greatest length.
See other abbreviations in Tables A1.8 and A2.9, see also von den Driesch (1976:76-77).

A4.8 Radius

a. Proximal End

LOC	LAY	BON	ANM	SYM	GL	Bp	BFp	Dp	DFp
08JHAT1901	(7)	2001	sheep	L	161.2	32.6	32.4	17.2	15.8
08JHAT2005	(6)	1631	sheep	L	—	31.3	29.9	17.3	16.4
08JHDT1103	(5)	665	sheep	R	—	31.5	29.5	15.9	14.5
08JHAT2001	(4)	1209	sheep	R	—	25.6	24.5	14.9	(13.0)
08JHAT2102	(4)	1689	sheep	L	—	30.2	27.9	17.0	16.0
Modern Sample		No. 81	goat	L	152.2	34.3	33.8	17.9	17.6

GL = greatest length.
See other abbreviations in Tables A1.9 and A2.10, see also von den Driesch (1976:79-81).

A4.8 Radius (continued)

b. Distal End

LOC	LAY	BON	ANM	SYM	SD	Bd	BFd	Dd	DFd
08JHDT1205	(8)	914	sheep	L	—	(27.0)	24.5	17.8	14.5
08JHAT2005	(7)	1630	sheep/goat	L	10.2	—	—	—	—
08JHAT1901	(7)	2001	sheep	L	15.7	29.3	28.0	20.7	17.7
08JHAT2106	(6)	1785	sheep/goat	R	9.6	(21.6)	—	(12.9)	—
08JHAT2106	(6)	1786	sheep/goat	L	8.6	—	—	—	—
08JHAT2004	(6)	1509	sheep	R	—	29.4	28.6	20.9	16.0
08JHAT2005	(6)	1631	sheep	L	16.1	—	—	—	—
08JHDT1103	(5)	665	sheep	R	14.6	—	—	—	—
08JHAT2001	(4)	1209	sheep	R	12.9	—	—	—	—
Modern Sample		No. 81	goat	L	17.1	34.1	33.0	23.8	19.4

See Tables A1.9 and A2.10, see also von den Driesch (1976:79-81).

A4.9 Ulna

LOC	LAY	BON	ANM	SYM	GL	LO	DPA	SDO	BPC
08JHAT2505	(7)	1853	sheep	R	—	—	28.3	—	17.9
08JHAT2005	(6)	1632	sheep	L	—	38.0	27.0	23.2	18.6
Modern Sample		No. 81	goat	L	189.1	35.4	26.3	24.7	25.3

GL = greatest length.
See other abbreviations in Table A1.10, see also von den Driesch (1976:79-81).

A4.10 Pelvis

LOC	LAY	BON	ELM	ANM	SYM	LA	LAR	SH	SB	LFo
08JHAT2505	(6)	1854	PELV	sheep	L	26.0	23.1	11.7	6.9	—
08JHDT1003	(5)	688	PELV	sheep	R	30.2	25.3	—	—	—
08JHAT2102	(4)	1693	PELV	sheep	R	28.7	24.9	12.8	7.7	35.3
Modern Sample		LM	PELV	goat	L	29.3	24.8	16.1	11.0	36.0

LM = local market
See other abbreviations in Tables A1.11 and A2.12, see also von den Driesch (1976:82-83).

A4.11 Sacrum

LOC	LAY	BON	ANM	SYM	GB	BFcr	HFcr
08JHAT2005	(6)	1634	sheep/goat	M	(61.5)	32.0	12.8
08JHAT2002	(5)	1312	sheep/goat	M	—	31.0	10.8

See Table A2.13, see also von den Driesch (1976:71).

A4.12 Femur

LOC	LAY	BON	ANM	SYM	Bp	DC	SD	Bd	Btpat
08JHAT1901	(5)	2011	sheep/goat	R	—	—	—	37.3	(12.0)
08JHAT2004	(6)	1529	sheep/goat	L	—	—	—	36.9	(17.4)
08JHAT2505	(6)	1856	sheep/goat	L	—	—	14.5	—	15.4
08JHDT1304	(5)	1030	sheep/goat	R	—	—	—	(37.5)	(13.0)
08JHAT2002	(6)	1318	sheep	L	—	19.8	—	—	—
08JHAT2102	(4)	1694	sheep	R	45.2	20.6	—	—	—
Modern Sample		No. 81	goat	L	45.4	22.6	15.7	37.4	16.1

See Tables A1.12 and A2.14, see also von den Driesch (1976:84-85).

A4.13 Tibia

LOC	LAY	BON	ANM	SYM	SD	Bd	Dd
08JHDT1801	(7)	365	sheep	R	15.2	28.1	19.9
08JHDT1802	(7)	586	sheep	L	—	26.4	21.1
Modern Sample		No. 81	goat	L	12.4	26.0	19.4

See Tables A1.14 and A2.15, see also von den Driesch (1976:86-87).

A4.14 Calcaneus

LOC	LAY	BON	ANM	SYM	GL	Dsustali	SLtuber
08JHAT2103	(4)	1748	sheep	R	54.1	18.4	30.0
Modern Sample		No. 81	goat	L	55.6	17.0	30.6

See Table A1.16, see also von den Driesch (1976:90-91).

A4.15 Astragalus

LOC	LAY	BON	ANM	SYM	GLl	GLm	Dl	Dm	Bd	BFp
08JHAT2004	(6)	1553	sheep	L	28.4	28.5	16.0	16.6	19.1	17.6
Modern Sample		No. 81	goat	L	28.1	27.0	15.3	16.5	18.4	17.0

See Table A1.17, see also von den Driesch (1976:88-91).

A4.16 Metatarsal III+IV

LOC	LAY	BON	ELM	ANM	SYM	Bp	Dp	SD
08JHDT1204	(5)	858	MT3+4	sheep/goat	L	18.2	15.9	10.2
Modern Sample		No. 81	MT3+4	goat	L	19.5	19.4	12.3

See Table A1.20a, see also von den Driesch (1976:92-93).

Table A5 Measurements of Cervids

A5.1 Antler

LOC	LAY	BON	ANM	SYM	MS 39	MS 40	MS 41	PL	AL
08JHDT1205	(7)	892	C.UNI	L	—	146.0	—	45.8	—
08JHDT1205	(7)	893	C.UNI	R	—	145.0	—	32.8	—
08JHDT1105	(6)	430	C.UNI	L	—	—	190.0	—	—
08JHDT1304	(6)	2054	C.UNI	L	—	—	20.0	—	—
08JHAT2104	(4)	293	C.UNI	R	160.0	—	135.0	—	—
08JHDT1005	(6)	294	A.POR	R	—	—	132.0	—	—
08JHDT1103	(6)	287	M.MUN	L	—	54.0	—	119.0	—
08JHDT1802	(6)	284	M.MUN	L	128.0	64.0	115.0	98.5	120.0
08JHDT1802	(6)	285	M.MUN	R	130.0	60.0	113.0	94.7	117.0
08JHAT2102	(6)	2056	M.MUN	L	—	64.0	86.0	97.1	111.4
08JHDT1003	(6)	2057	M.MUN	L	—	73.0	85.0	79.2	—
08JHAT2605	(6)	2058	M.MUN	R	—	70.0	90.0	81.0	85.5
08JHAT2002	(6)	2059	M.MUN	R	—	61.0	80.0	79.3	—
08JHDT1005	(5)	744	M.MUN	R	130.0	70.0	110.0	—	—
08JHDT1802	(5)	286	M.MUN	R	114.0	—	99.0	—	111.2
08JHAT2102	(4)	1668	M.MUN	R	135.0	—	170.0	—	—
08JHDT1305	(1)	2055	M.MUN	L	—	72.0	99.0	98.3	124.6
08JHDT1104	(5)	283	M.REE	L	72.0	—	60.0	—	96.7

MS 39–40: See von den Driesch (1976:36-37).
PL = pedicle length.
AL = antler length.

A5.2 Cranium Bones

LOC	LAY	BON	ELM	ANM	SYM	MS 26	MS 27	MS 28	MS 29
08JHAT2003	(6)	1360	OCCIP	M.CER	M	40.6	(54.0)	24.3	21.0
Modern Sample		No. 87	SKULL	C.NIP	B	59.3	90.6	27.5	24.2
Modern Sample		No. 103	SKULL	M.REE	B	29.7	43.1	16.9	16.2

MS 26–29: See von den Driesch (1976:36-37).

A5.3 Mandible with Teeth

a. Mandible (MS 1–8)

LOC	LAY	BON	ANM	SYM	MS 1	MS 2	MS 3	MS 4	MS 5	MS 6	MS 7	MS 8
08JHAT1901	(7)	1961	M.MUN	R	(162.1)	168.1	33.8	118.8	104.1	140.5	65.0	38.8
08JHAT2003	(6)	1369	M.REE	R	—	—	—	—	—	—	—	26.0
Modern Sample		No. 103	M.REE	L	127.6	125.5	25.5	95.7	80.3	99.8	52.6	32.5

MS 1–8: See von den Driesch (1976:56-57).

A5.3 Mandible with Teeth (continued)

b. Mandible (MS 9–15c)

LOC	LAY	BON	ANM	SYM	MS 9	MS 11	MS 12	MS 13	MS 14	MS 15a	MS 15b	MS 15c
08JHAT1901	(7)	1961	M.MUN	R	26.4	53.7	47.1	48.3	80.0	24.4	20.2	17.2
08JHAT2003	(6)	1369	M.REE	R	—	—	—	—	—	14.0	—	—
Modern Sample		No. 103	M.REE	L	20.1	36.7	43.1	38.5	65.9	20.6	14.5	12.4

MS 9–15c: See von den Driesch (1976:56-57).

c. Teeth in Mandible (P4–M1)

LOC	LAY	BON	ANM	SYM	P4-L	P4-B	P4-H	M1-L	M1-B	M1-H
08JHAT1901	(7)	1961	M.MUN	R	10.0	(6.7)	6.7	10.1	7.8	5.2
08JHAT2003	(6)	1369	M.REE	R	—	—	—	7.2	—	4.5
Modern Sample		No. 103	M.REE	L	7.2	4.6	6.2	8.7	5.8	4.4

P = premolar
M = molar.
B = breadth.
L = length.
H = height.
See Klein and colleagues (1981: figure 1).

d. Teeth in Mandible (M2–M3)

LOC	LAY	BON	ANM	SYM	M2-L	M2-B	M2-H	M3-L	M3-B	M3-H
08JHAT1901	(7)	1961	M.MUN	R	12.3	8.3	6.7	17.4	8.6	6.4
08JHAT2003	(6)	1369	M.REE	R	8.0	5.1	5.9	10.4	4.7	6.1
Modern Sample		No. 103	M.REE	L	9.8	6.2	6.4	14.2	6.1	6.3

See Table A5.3c.

A5.4 Loose Teeth

LOC	LAY	BON	ELM	ANM	SYM	TTY	L	B	R-B	H
08JHAT2002	(8)	1258	TTHUP	C.UNI	L	M3	23.8	18.8	24.6	17.6
08JHDT1104	(8)	614	TTHUP	C.UNI	L	M3	26.0	20.6	25.9	18.0
08JHDT1304	(8)	978	TTHUP	C.UNI	L	M2	22.3	(19.5)	23.1	14.5
08JHAT2002	(6)	1259	TTHUP	C.UNI	L	M2	(24.0)	18.4	25.1	16.9
08JHDT1004	(6)	126	TTHUP	C.UNI	L	M3	23.1	22.1	25.6	13.1
08JHDT1304	(5)	979	TTHUP	C.UNI	L	M2	22.9	19.8	24.3	19.2
08JHDT1304	(5)	981	TTHUP	C.UNI	L	M1	18.9	17.2	22.4	16.1
08JHDT1801	(5)	314	TTHUP	C.UNI	L	M2	23.7	23.7	25.5	14.2
08JHAT2103	(4)	1735	TTHUP	C.UNI	R	M2	22.9	20.4	24.3	14.8
08JHDT1104	(4)	615	TTHUP	C.UNI	L	M2	24.0	22.4	—	13.7
08JHDT1801	(4)	127	TTHUP	C.UNI	R	M2	22.3	24.8	26.4	13.0
08JHDT1005	(4)	753	TTHLW	M.MUN	R	M3	17.2	8.9	9.9	9.5

See Table A1.5, see also Klein and colleagues (1981: figure 1).

A5.5 Axis

LOC	LAY	BON	ANM	SYM	LCDe	LAPa	BFcr	BPacd	SBV	BFcd	H
08JHDT1205	(5)	900	M.CER	M	49.3	45.2	39.1	31.4	22.1	19.7	(48.5)
08JHDT1801	(4)	328	M.CER	M	50.0	45.7	37.9	30.8	23.7	17.5	(42.4)

See Table A2.7, see also von den Driesch (1976:68-69).

A5.6 Scapula

LOC	LAY	BON	ANM	SYM	DHA	SLC	GLP	LG	BG
08JHAT2001	(7)	238	L.CER	R	—	—	—	—	(48.5)
08JHAT2003	(7)	1395	L.CER	L	—	—	—	(40.8)	(34.0)
08JHAT1901	(6)	234	L.CER	L	—	—	63.7	50.1	45.1
08JHAT1901	(6)	239	L.CER	R	—	34.5	68.2	55.8	46.8
08JHAT2004	(6)	1500	L.CER	L	—	25.1	—	—	(33.0)
08JHAT2505	(6)	240	L.CER	R	—	36.5	62.6	49.9	43.4
08JHDT1305	(6)	232	L.CER	L	—	38.5	65.7	50.6	43.1
08JHDT1305	(6)	233	L.CER	L	—	35.6	65.1	48.3	41.4
08JHDT1304	(5)	241	L.CER	R	—	36.9	66.9	52.8	44.8
08JHDT1305	(4)	1109	L.CER	R	—	36.8	—	(45.6)	43.5
08JHAT2102	(4)	1683	M.CER	L	—	(16.2)	(27.1)	(20.7)	(19.3)
Modern Sample		No. 87	C.NIP	L	243.4	28.4	48.8	37.7	36.4
Modern Sample		No. 103	M.REE	L	111.4	11.4	24.1	19.1	16.5

See Table A1.7, see also von den Driesch (1976:74-75).

A5.7 Humerus

LOC	LAY	BON	ANM	SYM	GL	GLC	Bp	SD	Bd	BT	GLFd	LFd	Dd
08JHAT2505	(7)	221	L.CER	R	—	—	—	—	73.2	59.1	43.6	33.4	57.8
08JHDT1104	(4)	628	L.CER	L	—	—	—	—	60.3	54.3	38.6	29.5	54.0
08JHDT1305	(4)	213	L.CER	L	—	—	—	—	71.8	61.1	46.9	32.8	62.3
08JHDT1801	(4)	214	L.CER	L	—	—	—	—	72.6	61.9	43.0	32.1	59.7
08JHDT1204	(8)	831	M.CER	R	—	—	—	—	(34.2)	(33.2)	20.9	17.3	(27.0)
08JHDT1105	(5)	463	M.CER	R	—	—	—	—	(29.6)	26.9	19.8	15.8	(26.1)
08JHDT1305	(5)	1116	M.CER	L	169.6	154.5	37.1	14.6	34.1	28.7	21.0	15.8	27.3
08JHDT1802	(5)	566	M.CER	R	—	—	—	—	30.3	27.1	20.5	16.0	26.9
08JHDT1305	(4)	1117	M.CER	R	—	—	36.6	—	—	—	—	—	—
08JHDT1801	(4)	343	M.CER	R	—	—	—	—	32.6	28.3	20.4	15.6	26.8
08JHDT1801	(4)	344	M.CER	L	—	—	—	—	(30.4)	—	21.2	—	(26.7)
08JHDT1801	(4)	346	M.CER	L	—	—	—	13.6	—	—	—	—	—
08JHAT2505	(7)	1846	S.CER	L	—	—	—	7.9	—	—	—	—	—
08JHDT1003	(7)	684	S.CER	R	—	—	—	9.4	—	—	—	—	—
08JHDT1105	(6)	462	S.CER	R	—	121.5	—	10.0	20.3	19.0	13.1	10.3	16.4
08JHAT2002	(5)	1299	S.CER	L	—	—	—	—	20.9	18.1	12.0	10.6	16.3
08JHDT1103	(5)	664	S.CER	L	—	—	—	—	19.2	18.1	12.8	9.8	16.7
08JHDT1104	(5)	634	S.CER	R	—	—	—	—	(27.2)	(24.8)	—	13.8	—
08JHAT2103	(4)	1739	S.CER	L	129.6	121.1	23.5	9.1	19.9	17.6	13.0	9.4	15.6
Modern Sample		No. 87	C.NIP	L	235.6	204.0	62.8	24.1	53.0	44.7	33.9	26.4	44.1
Modern Sample		No. 103	M.REE	L	116.7	105.3	25.6	9.6	21.7	18.8	14.7	11.0	18.1

GL = greatest length.
See other abbreviations in Tables A1.8, A2.9 and A3.3, see also von den Driesch (1976:76-77).

A5.8 Radius

a. Proximal End

LOC	LAY	BON	ANM	SYM	GL	GL w/o DE	Bp	BFp	Dp	DFp
08JHAT2004	(6)	152	L.CER	R	—	—	—	—	—	—
08JHDT1104	(6)	636	L.CER	R	—	—	45.8	45.1	24.3	23.6
08JHDT1304	(6)	144	L.CER	L	—	—	—	—	—	—
08JHDT1304	(5)	1008	L.CER	L	—	—	62.2	58.8	33.4	31.3
08JHDT1304	(5)	151	L.CER	R	—	—	—	—	—	—
08JHDT1205	(4)	211	L.CER	L	—	—	61.7	55.6	31.5	29.7
08JHDT1801	(4)	146	L.CER	L	—	—	61.5	58.9	31.6	30.0
08JHDT1304	(6)	1009	M.CER	L	128.9	—	25.7	24.6	14.4	13.5
08JHDT1304	(6)	1010	M.CER	L	—	—	28.7	26.8	16.1	15.8
08JHAT2004	(5)	1508	M.CER	R	—	123.6	23.7	23.5	14.3	13.2
08JHDT1305	(4)	1122	M.CER	R	—	—	29.2	27.9	17.3	15.5
08JHDT1103	(2)	666	M.CER	R	—	—	—	—	—	—
Modern Sample		No. 87	C.NIP	L	233.8	—	48.4	44.7	24.6	23.4
Modern Sample		No. 103	M.REE	L	99.8	—	18.4	17.9	11.0	10.0

See Tables A1.9, A2.10, and A4.8, see also von den Driesch (1976:79-81).

b. Distal End

LOC	LAY	BON	ANM	SYM	SD	Bd	BFd	Dd	DFd
08JHAT2004	(6)	152	L.CER	R	—	55.0	51.5	41.0	32.4
08JHDT1104	(6)	636	L.CER	R	—	—	—	—	—
08JHDT1304	(6)	144	L.CER	L	—	52.5	(50.3)	36.4	29.9
08JHDT1304	(5)	1008	L.CER	L	—	—	—	—	—
08JHDT1304	(5)	151	L.CER	R	—	(55.4)	(51.8)	40.7	27.1
08JHDT1205	(4)	211	L.CER	L	—	—	—	—	—
08JHDT1801	(4)	146	L.CER	L	—	—	—	—	—
08JHDT1304	(6)	1009	M.CER	L	15.5	23.4	20.9	18.4	14.0
08JHDT1304	(6)	1010	M.CER	L	15.8	—	—	—	—
08JHAT2004	(5)	1508	M.CER	R	14.0	(20.0)	—	(16.5)	—
08JHDT1305	(4)	1122	M.CER	R	15.7	—	—	—	—
08JHDT1103	(2)	666	M.CER	R	—	26.7	25.8	18.9	16.0
Modern Sample		No. 87	C.NIP	L	26.3	39.8	39.1	28.9	22.6
Modern Sample		No. 103	M.REE	L	10.7	17.6	16.7	13.5	9.5

See Tables A1.9 and A2.10, see also von den Driesch (1976:79-81).

A5.9 Ulna

LOC	LAY	BON	ANM	SYM	LO	DPA	SDO	BPC
08JHAT2505	(6)	154	L.CER	R	89.1	54.6	52.8	35.5
08JHDT1005	(6)	769	L.CER	L	—	57.7	—	35.9
08JHDT1304	(5)	1018	L.CER	R	—	52.1	—	(24.0)
08JHAT2004	(4)	1510	L.CER	R	—	—	—	33.4
08JHDT1801	(4)	349	L.CER	R	—	46.7	38.3	28.1
08JHAT2003	(6)	1404	M.CER	L	40.5	28.0	26.4	14.3
08JHAT2505	(5)	1852	M.CER	R	40.0	26.9	24.6	15.4
Modern Sample		No. 87	C.NIP	L	57.1	43.0	42.5	25.3
Modern Sample		No. 103	M.REE	L	29.5	18.4	16.7	11.8

See Table A1.10, see also von den Driesch (1976:79-81).

A5.10 Pelvis

LOC	LAY	BON	ELM	ANM	SYM	LA	LAR
08JHDT1304	(5)	1024	IL/IS	M.CER	R	29.9	26.7
08JHAT2103	(4)	1742	PELV	M.CER	L	31.8	28.0
Modern Sample		No. 103	PELV	M.REE	L	21.0	18.5

See Table A1.11, see also von den Driesch (1976:82-83).

A5.11 Sacrum

LOC	LAY	BON	ELM	ANM	SYM	GL	PL	GB	BFcr	HFcr
08JHDT1205	(4)	922	SACR	M.CER	M	(88.7)	(85.4)	68.0	29.1	11.7

See Table A2.13, see also von den Driesch (1976:71).

A5.12 Femur

LOC	LAY	BON	ANM	SYM	Bp	DC	SD	Bd	Btpat
08JHAT2002	(8)	1319	M.CER	L	—	—	—	(36.8)	(12.0)
08JHDT1802	(8)	581	M.CER	L	—	—	—	40.1	17.0
08JHAT2002	(7)	1320	M.CER	R	44.7	21.0	—	—	—
08JHDT1003	(7)	689	M.CER	L	43.9	21.3	—	—	—
08JHDT1204	(6)	846	M.CER	R	—	—	—	40.0	16.0
08JHDT1304	(5)	1032	M.CER	R	—	21.0	—	—	—
08JHDT1305	(5)	1141	M.CER	L	42.6	21.2	—	—	—
08JHDT1305	(5)	1143	M.CER	L	—	—	—	43.2	18.7
08JHAT1901	(4)	2007	M.CER	L	—	19.7	—	—	—
08JHDT1204	(4)	845	M.CER	L	—	—	—	40.2	15.8
08JHDT1205	(4)	925	M.CER	R	—	—	—	41.7	15.7
08JHDT1305	(4)	1142	M.CER	L	43.8	22.0	—	—	—
08JHDT1801	(4)	361	M.CER	R	—	—	—	41.9	17.8
08JHDT1801	(4)	363	M.CER	R	—	—	—	39.7	18.0
08JHAT2004	(4)	1528	S.CER	R	—	—	—	26.8	10.6
Modern Sample		No. 87	C.NIP	L	66.2	31.4	25.9	64.3	25.2
Modern Sample		No. 103	M.REE	L	29.1	13.9	10.4	27.8	12.5

See Tables A1.12 and A2.14, see also von den Driesch (1976:84-85).

A5.13 Patella

LOC	LAY	BON	ANM	SYM	GB	GL
08JHAT2004	(6)	1532	L.CER	L	50.8	58.1
08JHAT2004	(6)	1533	L.CER	R	46.6	65.3

See Table A1.13, see also von den Driesch (1976:85).

A5.14 Tibia

LOC	LAY	BON	ANM	SYM	Bp	BFp	Bd	Dd
08JHAT2003	(6)	206	L.CER	L	88.0	84.4	—	—
08JHAT2004	(6)	1537	L.CER	L	—	—	—	38.7
08JHAT2505	(6)	209	L.CER	L	—	—	55.4	42.5
08JHDT1105	(5)	494	L.CER	R	—	—	42.8	30.5
08JHAT2004	(4)	1538	L.CER	L	—	—	—	(40.0)
08JHAT2002	(7)	1325	M.CER	R	—	—	27.7	21.8
08JHAT1901	(6)	2013	M.CER	L	44.6	43.4	—	—
08JHAT2004	(6)	1542	M.CER	L	—	—	27.7	22.0
08JHAT2505	(6)	1860	M.CER	R	45.1	42.5	—	—
08JHAT2505	(6)	1861	M.CER	R	—	—	25.3	19.6
08JHDT1005	(5)	780	M.CER	R	—	—	28.8	20.8
08JHDT1204	(5)	849	M.CER	L	—	—	27.4	22.2
08JHDT1305	(5)	1159	M.CER	L	47.0	43.7	—	—
08JHDT1801	(5)	366	M.CER	L	—	—	26.9	20.4
08JHDT1205	(4)	927	M.CER	L	(43.0)	—	—	—
08JHDT1305	(4)	1160	M.CER	L	—	—	26.7	20.0
08JHAT2004	(6)	1543	S.CER	L	—	—	18.4	13.7
Modern Sample		No. 87	C.NIP	L	64.2	61.4	42.6	32.2
Modern Sample		No. 103	M.REE	L	29.7	28.8	18.4	14.3

See Table A1.14, see also von den Driesch (1976:86-87).

A5.15 Carpals

LOC	LAY	BON	ELM	ANM	SYM	GB	Gd
08JHAT1901	(6)	2016	C-2+3	C.UNI	L	26.9	28.5
08JHAT2004	(6)	1555	C-RAD	C.UNI	L	19.4	33.4
08JHAT2004	(6)	1556	C-RAD	C.UNI	R	16.9	32.8
Modern Sample		No. 87	C-2+3	C.NIP	L	19.8	22.3
Modern Sample		No. 87	C-RAD	C.NIP	L	13.9	27.6

See Table A1.15, see also von den Driesch (1976:89).

A5.16 Calcaneus

LOC	LAY	BON	ANM	SYM	GL	GL w/o Ep	Dsustali	SLtuber
08JHAT2004	(6)	1549	L.CER	R	—	102.4	33.4	—
08JHAT2005	(6)	2	L.CER	L	129.0	—	39.6	68.8
08JHAT2106	(6)	1795	L.CER	L	—	—	(35.4)	—
08JHAT2105	(5)	1	L.CER	R	139.1	—	41.5	75.3
08JHDT1003	(5)	690	L.CER	L	—	97.3	33.4	—
08JHDT1305	(5)	1161	L.CER	R	121.9	—	36.7	63.5
08JHAT2004	(4)	1550	L.CER	R	—	104.8	—	—
08JHDT1005	(7)	782	M.CER	L	—	—	19.7	—
08JHAT2003	(6)	1419	M.CER	R	66.0	—	19.1	31.8
08JHDT1305	(5)	1163	M.CER	R	66.6	—	(16.9)	33.1
08JHAT2103	(4)	1747	M.CER	L	66.7	—	19.5	35.9
08JHDT1205	(4)	932	M.CER	L	—	—	19.6	35.2
08JHDT1305	(4)	1162	M.CER	L	65.6	—	22.0	32.9
08JHDT1801	(4)	374	M.CER	R	—	—	20.6	—
08JHDT1801	(4)	375	M.CER	L	64.5	—	18.8	35.9
Modern Sample		No. 87	C.NIP	L	100.2	—	28.1	53.4
Modern Sample		No. 103	M.REE	L	45.0	—	13.6	25.2

See Table A1.16, see also von den Driesch (1976:90-91).

A5.17 Astragalus

LOC	LAY	BON	ANM	SYM	GLl	GLm	Dl	Dm	Bd	BFp
08JHAT2106	(6)	1796	L.CER	L	—	—	—	—	36.4	—
08JHDT1105	(5)	503	L.CER	L	—	54.8	32.4	33.0	—	31.9
08JHDT1105	(5)	504	L.CER	R	49.3	46.5	26.0	26.2	30.0	26.1
08JHAT2505	(6)	1868	M.CER	R	29.5	28.3	16.2	16.6	18.0	15.7
08JHDT1205	(5)	933	M.CER	R	33.0	30.5	17.9	17.0	19.9	17.8
Modern Sample		No. 87	C.NIP	L	47.3	45.1	26.6	26.5	29.1	25.8
Modern Sample		No. 103	M.REE	L	20.5	20.3	11.1	11.4	12.6	11.0

See Table A1.17, see also von den Driesch (1976:88-89).

A5.18 Central and Fourth Tarsal

LOC	LAY	BON	ANM	SYM	GB	Gd
08JHAT2106	(7)	109	C.UNI	L	50.9	42.6
08JHAT2002	(6)	60	C.UNI	R	48.8	38.6
08JHAT2002	(6)	111	C.UNI	L	45.8	39.9
08JHAT2004	(6)	1554	C.UNI	R	48.7	40.8
08JHAT2004	(5)	110	C.UNI	L	46.9	42.8
Modern Sample		No. 87	C.NIP	L	38.0	32.8
Modern Sample		No. 103	M.REE	L	16.0	15.4

See Table A1.18, see also von den Driesch (1976:91).

A5.19 Metacarpal III+IV

a. Proximal End

LOC	LAY	BON	ANM	SYM	GL	Bp	BFp	Dp	DFp	Lfm	Bfm	Lfl	Bfl
08JHAT2005	(7)	1641	C.UNI	L	—	47.3	43.9	33.0	30.6	34.7	27.1	30.1	20.2
08JHAT2003	(6)	192	C.UNI	R	—	48.5	46.1	35.6	30.0	37.5	25.0	29.6	21.4
08JHAT2105	(4)	1773	C.UNI	L	—	44.0	41.9	32.2	27.3	33.5	22.3	23.6	16.5
08JHAT1901	(7)	2019	M.CER	L	—	—	—	—	—	14.6	9.3	—	—
08JHAT1901	(6)	2020	M.CER	R	121.4	21.2	20.3	15.3	14.2	15.9	8.4	12.0	8.9
08JHDT1304	(5)	1047	M.CER	L	—	21.6	21.0	15.5	13.7	16.5	12.0	14.1	8.7
08JHDT1304	(5)	1048	M.CER	R	132.0	24.0	23.2	17.4	16.1	19.4	14.9	14.6	7.6
08JHDT1305	(5)	1166	M.CER	L	—	22.9	22.8	17.4	17.1	18.7	12.7	13.9	10.0
Modern Sample		No. 87	C.NIP	L	206.5	35.8	33.5	24.9	21.6	22.7	18.6	20.4	14.2
Modern Sample		No. 103	M.REE	L	78.8	15.3	15.0	10.5	9.3	11.5	7.0	10.7	5.3

GL = greatest length.
See other abbreviations in Table A1.19, see also von den Driesch (1976:92-93).

A5.19 Metacarpal III+IV (continued)

b. Distal End

LOC	LAY	BON	ANM	SYM	SD	Bd	MS (2)	MS (3)	MS (4)	MS (5)	MS (6)	MS (7)	MS (8)
08JHAT2004	(6)	1563	C.UNI	R	—	46.0	(43.0)	(20.6)	(19.6)	23.3	30.5	(19.5)	30.4
08JHDT1304	(6)	1046	C.UNI	?	22.2	—	—	—	—	—	—	—	—
08JHDT1204	(5)	856	C.UNI	R	—	—	—	—	21.2	—	—	23.5	31.4
08JHAT2002	(4)	1333	C.UNI	R	—	46.2	47.4	21.4	22.4	24.2	30.4	21.9	30.0
08JHDT1801	(4)	377	C.UNI	L	—	42.7	43.2	19.8	19.1	20.2	27.2	20.8	27.3
08JHAT1901	(7)	2019	M.CER	L	12.5	—	—	—	—	—	—	—	—
08JHAT1901	(6)	2020	M.CER	R	13.4	21.4	21.6	10.0	9.6	11.0	13.8	10.1	(12.7)
08JHAT2001	(6)	1216	M.CER	R	13.3	—	—	—	—	—	—	—	—
08JHDT1304	(5)	1047	M.CER	L	13.8	—	—	—	—	—	—	—	—
08JHDT1304	(5)	1048	M.CER	R	13.9	23.4	23.5	10.6	10.9	10.7	13.8	9.5	13.5
08JHDT1305	(5)	1167	M.CER	?	12.8	—	—	—	—	—	—	—	—
Modern Sample		No. 87	C.NIP	L	21.3	34.7	34.9	17.0	16.1	17.5	22.2	16.6	21.8
Modern Sample		No. 103	M.REE	L	9.3	14.9	15.0	6.9	7.2	6.9	9.4	6.7	9.1

See Table A1.20, see also von den Driesch (1976:92-93).

A5.20 Metatarsal III+IV

a. Proximal End

LOC	LAY	BON	ANM	SYM	GL	Bp	BFp	Dp	DFp	Lfm	Bfm	Lfl	Bfl	Bfpl
08JHAT2003	(6)	1421	M.CER	R	158.1	22.3	21.4	21.3	19.5	17.5	8.8	16.6	9.7	6.2
08JHDT1801	(4)	380	M.CER	R	—	21.4	—	22.8	—	16.2	7.5	16.6	9.9	12.0
Modern Sample		No. 87	C.NIP	L	225.8	31.4	30.3	31.3	30.5	20.0	12.0	20.3	14.1	11.9
Modern Sample		No. 103	M.REE	R	108.1	14.2	13.9	14.9	14.0	10.1	5.6	9.8	5.4	7.8

GL = greatest length.
See other abbreviations Table A1.20, see also von den Driesch (1976:92-93).

b. Distal End

LOC	LAY	BON	ANM	SYM	SD	Bd	MS (2)	MS (3)	MS (4)	MS (5)	MS (6)	MS (7)	MS (8)
08JHAT2003	(6)	1421	M.CER	R	14.5	24.2	23.9	10.5	10.7	12.1	15.2	11.3	15.4
08JHDT1305	(5)	1168	M.CER	?	9.9	—	—	—	—	—	—	—	—
Modern Sample		No. 87	C.NIP	L	20.2	36.8	36.2	16.2	16.4	19.2	24.2	16.3	23.5
Modern Sample		No. 103	M.REE	R	8.8	15.8	15.9	7.6	7.4	7.8	9.7	6.5	9.9

See Table A1.20, see also von den Driesch (1976:92-93).

A5.21 Metapodial III+IV

LOC	LAY	BON	ANM	SYM	MS (3)	MS (5)	MS (6)
08JHAT2505	(7)	1873	C.UNI	L	20.6	21.1	28.2
08JHDT1105	(5)	507	C.UNI	?	19.5	24.1	30.4
08JHAT2004	(5)	1564	C.UNI	?	20.9	22.5	28.9
08JHDT1305	(4)	1170	C.UNI	?	20.4	22.5	29.5

See Table A1.20, see also von den Driesch (1976:92-93).

A5.22 First Phalanx

LOC	LAY	BON	ANM	SYM	Glpe	Bp	Dp	SD	Bd	Dd
08JHAT2002	(8)	1335	C.UNI	4	59.4	22.1	26.7	18.9	20.0	17.8
08JHAT2103	(8)	1751	C.UNI	4	—	—	—	17.2	20.7	19.8
08JHDT1005	(8)	785	C.UNI	5	61.9	23.4	29.8	22.0	21.9	19.3
08JHAT2004	(6)	1565	C.UNI	4	—	—	—	21.6	23.3	19.2
08JHAT2004	(5)	1566	C.UNI	5	(63.0)	(24.0)	—	19.9	22.6	19.4
08JHAT2004	(5)	1567	C.UNI	5	—	22.1	29.2	—	—	—
08JHDT1304	(5)	1051	C.UNI	5	62.6	22.0	29.4	18.7	22.4	19.7
08JHAT2103	(4)	1750	C.UNI	4	60.4	22.8	26.7	18.5	21.7	18.9
08JHAT2103	(4)	1752	C.UNI	5	—	—	—	18.7	20.8	19.1
08JHDT1005	(4)	783	C.UNI	4	—	25.0	—	—	—	—
08JHDT1305	(4)	1171	C.UNI	5	65.3	21.7	27.8	—	—	—
Modern Sample		No. 87	C.NIP	4	52.2	19.1	23.2	15.0	16.5	14.5
Modern Sample		No. 103	M.REE	4	19.5	7.7	9.6	6.2	6.7	6.7

See Table A1.22, see also von den Driesch (1976:96-97).

A5.23 Second Phalanx

LOC	LAY	BON	ANM	SYM	GL	Bp	SD	Bd	Dd
08JHDT1005	(8)	786	C.UNI	5	49.5	24.1	18.5	18.4	24.5
08JHDT1304	(8)	1052	C.UNI	4	—	—	18.6	(17.1)	(22.3)
08JHAT1901	(6)	2022	C.UNI	4	49.6	22.7	16.9	20.0	27.6
08JHAT2004	(6)	1570	C.UNI	4	48.6	24.5	18.5	19.2	25.7
08JHAT2004	(6)	1571	C.UNI	4	44.4	21.8	17.2	17.5	23.3
08JHAT2106	(6)	1797	C.UNI	4	49.5	21.0	16.6	17.4	25.2
08JHAT2003	(5)	1425	C.UNI	4	43.9	20.2	18.7	18.2	24.5
08JHAT2102	(4)	1699	C.UNI	4	—	—	12.4	15.2	20.4
08JHAT2103	(4)	1753	C.UNI	4	44.0	(21.6)	4.6	18.2	24.6
08JHAT2002	(6)	1339	M.CER	4	28.5	11.6	9.0	9.9	12.0
Modern Sample		No. 87	C.NIP	4	39.3	16.3	13.5	14.8	20.9
Modern Sample		No. 103	M.REE	4	17.2	7.4	5.2	6.2	7.8

See Table A1.23, see also von den Driesch (1976:98-99).

A5.24 Third Phalanx

LOC	LAY	BON	ANM	SYM	DLS	MBS	Ld	HP
08JHAT2004	(6)	1573	C.UNI	5	57.9	14.6	48.6	26.9
08JHAT2004	(5)	1572	C.UNI	4	(48.5)	14.1	38.3	25.8
08JHDT1205	(5)	939	C.UNI	5	55.7	12.6	47.5	27.4
08JHAT2103	(4)	1754	C.UNI	4	(56.3)	14.6	45.9	25.8
Modern Sample		No. 87	C.NIP	4	41.0	12.6	39.9	24.2

See Table A1.24, see also von den Driesch (1976:101).

Table A6 Measurements of Other Mammals

A6.1 Mandible

a. Mandible (MS 4–10)

LOC	LAY	BON	ANM	SYM	MS 4	MS 5	MS 6	MS 7	MS 8	MS 9	MS 10
08JHAT2005	(6)	1614	V.VUL	L	(87.4)	(87.6)	(87.4)	(58.5)	55.0	49.5	23.5
08JHDT1801	(9)	311	S.CAR	L	—	—	—	—	56.9	48.6	29.6

MS 4–10: See von den Driesch (1976:60-61).

b. Mandible (MS 11, 12, 14, 17–20)

LOC	LAY	BON	ANM	SYM	MS 11	MS 12	MS 14	MS 17	MS 18	MS 19	MS 20
08JHAT2005	(6)	1614	V.VUL	L	32.6	27.2	14.9	6.5	33.0	13.2	11.8
08JHDT1801	(9)	311	S.CAR	L	27.7	24.0	18.0	8.8	—	16.2	13.5
08JHDT1004	(5)	698	S.CAR	L	—	30.3	—	—	—	—	14.0

MS 11, 12, 14, 17–20: See von den Driesch (1976:60-61).

A6.2 Scapula

LOC	LAY	BON	ANM	SYM	DHA	BG
08JHAT2605	(6)	1911	RATTU	L	(42.0)	8.3

See Table A1.7, see also von den Driesch (1976:74-75).

A6.3 Humerus

LOC	LAY	BON	ANM	SYM	GL	GLC	GL w/o PE	Dp	SD	Bd	BT	GLFd	LFd	Dd
08JHDT1305	(6)	1118	MACAC	R	—	—	—	—	12.6	31.9	20.0	—	—	18.2
08JHAT2005	(6)	1627	LEPUS	R	83.5	82.4	—	15.2	4.7	10.3	—	—	—	7.8
08JHAT2004	(5)	1505	LEPUS	R	—	—	—	—	5.8	10.9	—	—	—	8.2
08JHDT1105	(5)	465	LEPUS	L	—	—	—	—	—	10.6	—	—	—	8.0
08JHAT2004	(6)	1506	PAGUM	L	—	—	94.7	—	8.4	24.7	17.6	10.7	7.6	11.8
08JHAT2505	(5)	1847	PAGUM	L	—	—	60.4	—	5.1	21.7	12.8	—	5.6	10.2
08JHAT2505	(5)	1848	PAGUM	R	—	—	61.0	—	5.2	21.5	12.6	—	5.8	10.0
08JHDT1104	(8)	631	S.CAR	L	—	—	—	—	7.8	—	—	—	—	—
08JHDT1104	(8)	632	S.CAR	R	—	—	—	—	7.9	—	—	—	—	—

GL = greatest length.
See other abbreviations in Tables A1.8, A2.9 and A3.3, see also von den Driesch (1976:76-77).

A6.4 Radius

LOC	LAY	BON	ANM	SYM	GL	Bp	Dp	SD	Bd	Dd
08JHDT1004	(5)	709	SUS	L	172.5	34.1	24.9	21.8	41.9	29.2
08JHDT1105	(6)	474	URSUS	R	251.8	37.6	26.3	18.1	46.3	30.1
08JHDT1004	(5)	277	URSUS	R	—	—	—	—	53.8	34.1
08JHAT2102	(4)	1692	S.FEL	L	102.8	11.8	7.9	7.5	17.3	9.1

GL = greatest length.
See other abbreviations in Tables A1.9 and A2.10, see also von den Driesch (1976:79-81).

A6.5 Ulna

LOC	LAY	BON	ANM	SYM	GL	DPA	SDO	BPC
08JHDT1004	(5)	710	SUS	L	238.5	51.4	38.6	26.7
08JHDT1204	(4)	838	RATTU	L	59.2	9.6	—	6.4
08JHDT1305	(6)	1127	HYSTR	L	—	16.4	—	12.2
08JHDT1105	(6)	475	URSUS	R	296.4	43.1	33.5	33.5
08JHDT1802	(8)	574	URSUS	R	—	—	—	38.0

GL = greatest length.
See other abbreviations in Table A1.10, see also von den Driesch (1976:79-81).

A6.6 Pelvis

LOC	LAY	BON	ELM	ANM	SYM	LAR	SH	SB	LFo
08JHAT2002	(5)	1308	PELV	MACAC	R	20.0	18.5	8.6	—
08JHAT2002	(5)	262	PELV	MACAC	L	19.9	18.6	8.9	28.5

See Tables A1.11 and A2.12, see also von den Driesch (1976:82-83).

A6.7 Sacrum

LOC	LAY	BON	ANM	SYM	BFcr	HFcr
08JHAT1901	(6)	2006	SMALL	M	11.1	4.8

See Table A2.13, se also von den Driesch (1976:71).

A6.8 Femur

LOC	LAY	BON	ANM	SYM	GL	GLC	GL w/o BE	Bp	DC	SD	Bd	Btpat	BTr
08JHAT2002	(6)	1321	LEPUS	R	—	—	—	18.7	8.2	—	—	—	17.7
08JHAT2004	(6)	1531	LEPUS	L	—	102.2	—	20.2	7.9	7.8	15.4	6.0	18.8
08JHAT2003	(6)	1415	S.FEL	L	(113.0)	—	—	—	—	7.5	16.8	6.9	—
08JHAT1901	(6)	2012	S.CAR	R	—	—	—	—	—	7.4	17.6	7.8	—
08JHAT2505	(6)	1858	S.CAR	R	—	—	54.1	—	—	6.9	—	—	—

GL = greatest length.
GLC = greatest length from the femoral head.
BTr = (greatest) breadth of the region of the third trochanter.
See other abbreviations in Tables A1.12 and A2.14, see also von den Driesch (1976:84-85).

A6.9 Tibia

LOC	LAY	BON	ANM	SYM	GL	GL w/o PE	Bp	SD	Bd	Dd
08JHAT0303	(7)	2051	LEPUS	L	—	—	15.7	—	—	—
08JHDT1304	(5)	1038	LEPUS	L	—	—	(16.7)	6.6	—	—
08JHAT2102	(4)	1707	RATTU	L	—	29.0	4.5	1.8	3.4	—
08JHAT2102	(4)	1708	RATTU	L	—	30.1	4.6	1.9	3.3	—
08JHAT2102	(4)	1709	RATTU	R	—	29.9	—	1.8	3.2	—
08JHAT2505	(6)	1867	S.FEL	L	—	—	(17.0)	6.0	—	—
08JHAT2103	(4)	1745	S.FEL	R	96.4	—	(18.1)	5.6	13.4	9.7

See Tables A1.14 and A2.15, see also von den Driesch (1976:86-87).

A6.10 Metacarpal

LOC	LAY	BON	ELM	ANM	SYM	GL	Bd
08JHDT1105	(5)	505	MC 4	URSUS	R	59.2	15.9

See Table A3.8, see also von den Driesch (1976:95).

A6.11 First Phalanx

LOC	LAY	BON	ANM	SYM	GL	Bp	SD	Bd
08JHAT2004	(6)	1569	URSUS	4	40.4	16.2	10.7	12.0

GL = greatest length.
See Table A1.22, see also von den Driesch (1976:96-97).

Table A7 Measurements of Birds

A7.1 Coracoids

LOC	LAY	BON	ANM	SYM	GL	Lm	Bb	BF
08JHAT1901	(6)	2023	ANAS	R	(54.4)	49.3	(19.6)	—
08JHAT2004	(6)	1574	ANAS	R	—	—	—	19.5
08JHAT2505	(6)	1883	ANAS	R	55.2	49.8	—	20.3
08JHAT2003	(6)	1429	BIRD	L	(36.7)	34.2	—	11.3
08JHAT2605	(5)	1932	BIRD	L	—	39.7	—	—
08JHAT2605	(5)	1933	BIRD	R	—	41.0	—	—

GL = greatest (diagonal) length.
Lm = medial length.
Bb = (greatest) basal breadth.
BF = breadth of the basal articular surface.
See details in von den Driesch (1976:112-113).

A7.2 Scapula

LOC	LAY	BON	ANM	SYM	GL	Dic
08JHAT2002	(6)	1355	BIRD	?	—	12.0

GL = greatest length.
Dic = (greatest) cranial diagonal.
See details in von den Driesch (1976:115).

A7.3 Humerus

LOC	LAY	BON	ANM	SYM	GL	Bp	SC	Bd
08JHAT2605	(5)	1926	ANATI	L	—	24.2	8.7	—
08JHAT2605	(5)	1927	ANATI	R	—	24.2	8.7	—
08JHAT2002	(8)	1347	ANAS	R	75.5	—	5.1	10.3
08JHAT2005	(6)	1643	ANAS	L	—	20.2	6.7	—
08JHAT2605	(6)	1928	ANAS	L	—	15.9	4.7	10.2
08JHAT2605	(6)	1930	ANAS	L	—	16.8	4.8	—
08JHAT2605	(6)	1931	ANAS	L	—	—	5.3	—
08JHAT2605	(5)	1929	ANAS	L	—	(15.3)	5.2	11.2
08JHAT2003	(6)	1427	BIRD	R	—	15.5	4.8	—
08JHAT2505	(5)	1877	BIRD	L	—	(17.0)	5.2	(10.8)
08JHAT2505	(5)	1878	BIRD	L	—	—	5.0	10.5
08JHAT2505	(5)	1880	BIRD	R	—	15.2	4.5	10.0
08JHAT2102	(4)	1700	BIRD	R	77.7	17.1	5.1	10.7
08JHAT2103	(4)	1755	BIRD	L	(74.9)	14.2	4.7	—

GL = greatest length.
Bp = (greatest) breadth of the proximal end.
SC = smallest breadth of the corpus.
Bd = (greatest) breadth of the distal end.
See details in von den Driesch (1976:116-117).

A7.4 Ulna

LOC	LAY	BON	ANM	SYM	GL	Dip	Bp	SC	Did
08JHAT2102	(4)	1704	ANAS	R	69.2	7.7	6.1	3.7	7.7
08JHAT2605	(5)	1934	CYGNU	R	176.9	15.9	11.7	6.0	11.6
08JHAT2505	(3)	1885	CYGNU	L	—	—	—	6.3	10.8
08JHDT1003	(5)	694	BIRD	R	—	—	—	—	9.1
08JHDT1305	(6)	1185	BIRD	R	—	—	—	—	(8.4)
08JHAT1901	(6)	2025	BIRD	R	66.9	(8.1)	(6.9)	3.8	7.9
08JHAT2005	(6)	1644	BIRD	L	56.8	10.2	6.9	3.5	7.5
08JHDT1305	(5)	1184	BIRD	R	61.8	8.9	6.5	3.7	8.7
08JHDT1801	(4)	398	BIRD	R	—	17.0	12.7	—	—
08JHAT2004	(4)	1576	BIRD	L	—	—	—	3.6	7.7

GL = greatest length.
Dip = (greatest) diagonal of the proximal end.
Bp = (greatest) breadth of the proximal end.
SC = smallest breadth of the corpus.
Did = (greatest) diagonal of the distal end.
See details in von den Driesch (1976:118-119).

A7.5 Femur

LOC	LAY	BON	ANM	SYM	GL	Lm	Bp	Dp	SC	Bd	Dd
08JHAT2103	(4)	1765	GALLUS	R	—	71.2	—	(9.8)	6.4	14.6	11.7
08JHAT2003	(6)	1426	GRUID	L	(83.5)	(80.2)	(12.7)	9.0	6.0	(13.2)	10.7

GL = greatest (diagonal) length.
Lm = medial length.
Bp = (greatest) breadth of the proximal end.
Dp = (greatest) depth of the proximal end.
SC = smallest breadth of the corpus.
Bd = (greatest) breadth of the distal end.
Dd = (greatest) depth of the distal end.
See details in von den Driesch (1976:125).

A7.6 Tibiotarsus

LOC	LAY	BON	ANM	SYM	GL	La	Dip	SC	Bd	Dd
08JHAT2102	(4)	1703	ANAS	L	73.4	69.0	11.3	3.3	7.8	7.7
08JHDT1802	(6)	600	PHASI	?	—	—	—	—	10.5	10.6
08JHAT2102	(4)	1710	BIRD	R	—	—	39.0	14.6	—	—

GL = greatest (diagonal) length.
La = axial length.
Dip = (greatest) diagonal of the proximal end.
SC = smallest breadth of the corpus.
Bd = (greatest) breadth of the distal end.
Dd = (greatest) depth of the distal end.
See details in von den Driesch (1976:126-127).

A7.7 Tarsometatarsus

LOC	LAY	BON	ANM	SYM	GL	Bp	SC	Bd
08JHAT2002	(6)	1352	BIRD	L	—	8.6	4.3	—
08JHAT2003	(6)	1428	BIRD	R	55.3	8.9	4.5	—

GL = greatest (diagonal) length.
Bp = (greatest) breadth of the proximal end.
SC = smallest breadth of the corpus.
Bd = (greatest) breadth of the distal end.
See details in von den Driesch (1976:129)